# Organic Chemistry
# Made Easy

# Organic Chemistry Made Easy

Joachim Nentwig
Manfred Kreuder
Karl Morgenstern

**VCH**

Joachim Nentwig
Ahornweg 11
D-3043 Schneverdingen
Germany

Manfred Kreuder
Karl Morgenstern
Bayer AG, Werk Uerdingen
D-4150 Krefeld-Uerdingen

**Library of Congress Cataloging-in-Publication Data**

Nentwig, J. (Joachim)
    Organic chemistry made easy / by Joachim Nentwig, Manfred Kreuder,
Karl Morgenstern
            p.    cm.    --   (Chemistry made easy)
      Includes index
      ISBN 1-56081-548-5. -- ISBN 3-527-89542-6 (Weinheim). -- ISBN
1-56081-549-3 (set)
         1. Chemistry, Organic.   I. Kreuder, Manfred.   II. Morgenstern,
Karl. III. Title.   IV. Series.
QD253.N38   1991                            91-44866
547'.0077--dc20                              CIP

© 1992 VCH Publishers, Inc.

Printed in the United States of America

ISBN: 1-56081-548-5 VCH Publishers, Inc.
ISBN: 3-527-89542-6 VCH Verlagsgesellschaft

Printing history:

10   9   8   7   6   5   4   3   2   1

VCH Publishers, Inc.
220 East 23rd Street
Suite 909
New York, N.Y. 10010

VCH Verlagsgesellschaft mbH
P.O. Box 10 11 61
D-6940 Weinheim
Federal Republic of Germany

VCH Publishers (UK) Ltd.
8 Wellington Court
Cambridge CB1 1HW
United Kingdom

# PREFACE

In order to make the technique of programmed learning and our subject matter optimally compatible, we used a mixture of "linear programming" (after Skinner) and "branched programming" (after Crowder). Mixed programs of this type have been proven particularly effective. The term *hybrid program* was proposed for them at DIDACTA, an international conference on education and educational methods in Switzerland.

For us, programming the material meant the creation of an eticing stimulus and an educational aid for the reader. A programmed text requires the reader to complete one step before the next can be made, leading to an intensive absorption of the material. The reader is drawn into a question-and-answer game where the solution to every question must be found, thereby enabling the user to take control of the learning performance. The places the reader in a classroom-like situation, with the added advantage of solely determining the rate of progress. A well-programmed text is, to a large extent, capable of replacing a tutor. Therefore, this text is of particular value wherever the opportunity to attend classes or lectures does not exist (or was missed), and in cases where additional self initiative is desired.

We ascribe the huge successes of the original German editions of *General and Inorganic Chemistry Made Easy* and *Organic Chemistry Made Easy* to the fact that all the programs were thoroughly tested in classrooms and in face-to-face tests before they were published. Thus, they are not merely "desk programs," but are instead instructional and learning aids that have proven their value in practice.

This edition was revised, updated, and rearranged. Naturally, we would be grateful for any critical comments or suggestions.

January 1992
J.N.
M.K.
K.M.

# Contents

dissociation constant – Law of Mass Action – ionic product of water –
pH – buffers – hydrolysis – solubility product.

# How to use this book

For those interested in acquiring a thorough grounding in the fundamentals of chemistry, this book provides an effective and simple means. The method of teaching utilized here is based on the proven approach of

*programmed instruction.*

The material we cover is presented in the form of 25 separate Programs. Although at the outset relatively light demands are made on the student, the situation changes as the student begins to advance through the book. Because our method is rather a specialized one, we outline below details of the techniques employed.

The material in each Program has been divided up into numerous small units or *frames*. At the end of every frame you, the student, are asked to answer one or more questions. The intention is to check whether you have really understood the content of the particular frame. Such questions may involve the setting up of a formula or an equation, the calculation of some quantity, the insertion of missing words into sentences and so on. After answering, you are advised to proceed to another frame where the correct answer is to be found or where any mistakes you may have made are analyzed. Only at this stage should you consider moving on to learn something new. By following this procedure you will be conducted step by step through all of the Programs.

To illustrate our approach more fully, we now present two very simple sample frames. These demonstrate the two basic ways in which questions will be posed. We mention in passing that you will continually have need of a pen (or pencil) and paper as you work your way through each Program.

---

**1** Chemistry may be studied by an effective technique known as programmed instruction. The material covered in this volume is presented in the form of 25 separate Programs.

Now write out the completed version of the following sentence: We shall be studying chemistry here by means of .............. instruction.

Continue your reading at  [11]  .

---

You would then be required to write down: "We shall be studying chemistry here by means of *programmed* instruction."

Please do *not* write your answer within the actual frame itself. That would make it impossible for you to revise the material at some later stage! Be sure to write all of your answers and notes on separate sheets of paper and not directly into the book.

After writing down your answer, turn to the next frame indicated; in the above case you would turn to frame number [11] . There you will be able to check your answer. (Do please make sure that in turning up the relevant frame you do not jump over into some other Program!)

The route to be followed through any given Program is indicated not by the page number but rather by the frame number located at the top left hand corner of each frame. Thus, you would check your answer here against the correct solution given in frame [11] . Filled-in words are always *italicized* in our solutions.

---

[11]    We shall be studying chemistry here by means of *programmed* instruction.

It has been demonstrated that programmed instruction is a more effective technique than learning from a conventional textbook.

A Program can facilitate learning by covering the material in small stages, by enabling the student to check each answer instantly and by permitting the student to work at his own individual pace throughout.

What is the most effective way of learning?

● By making use of a conventional textbook        →    [3]

● By making use of a Program                      →    [18]

● There is no difference between the two ways     →    [34]

● I cannot answer this question                   →    [54]

---

The student should now decide in favor of one of the four possible options and then turn to the frame indicated for the option chosen.

If the student had chosen frames [3] or [34] , he would discover that his answer was *not correct*. No one should feel too bad about that: additional help is always provided elsewhere as indicated. The correct answer is, of course, given in frame [18] .

Whenever you feel rather uncertain about how to answer a particular question and would appreciate having some further explanation on it, you are advised to choose a frame such as $\boxed{54}$ . It is always wiser to follow this route than simply to guess at an answer.

During the coming weeks you will begin to realize that learning by means of programmed instruction is not only much easier and more enjoyable than conventional textbook learning, but also that you can progress through the material more rapidly. However, you should not look upon programmed instruction as a miracle method to attain your goal without any real work on your part being necessary! You will certainly *not* be able to learn chemistry "in your sleep" by this means. It is only by conscientiously working your way through each and every Program that success can be achieved.

Below we give some tips which should help to speed you on your way:

1. When commencing work on a Program you should be in a relaxed state and have sufficient time to complete the Program. It is best to work through a Program in *one* sitting if possible and then to re-work it again several times over the next few days. Of course, if you do have to interrupt working your way through a Program, simply jot down the number of the last frame you studied.

2. The individual frames are intended to be *worked* through and not merely *read* through. Do not leave a frame until you are quite certain you have grasped its content. A Program is not designed in such a way that you can work backwards through it again! Check all your answers very carefully against those given in the next frame and correct any errors you may have made. Do not get upset if you make mistakes the first time you work through a Program. However, it would be a great pity if you did not learn from the mistakes you make!

3. Absolute *honesty* is called for at those points where you are asked to select one out of several possible answers to a question. We recognize, of course, that it would be only too easy to cheat here. Our advice, however, is always to make a serious attempt to arrive at the correct answer. Cheating will certainly *not bring* the kind of *success* you could otherwise expect. From time to time you will have the opportunity of choosing as your answer some such phrase as "I do not understand the question" or, simply, "I do not know". Never hesitate to select this answer if you genuinely feel uncertain. You will then be guided to a frame where additional help is available.

4. The continual checking of your progress and the choices you have to make in answering questions serve to introduce a certain excitement into the learning process. The learning process is thereby transformed into something resembling the playing of a game. This effect is an intended one and is highly desirable, even though there is a potential danger that students may thereby be encouraged to work through Programs too fast and perhaps even start guessing at answers. Be on your guard against this possible disadvantage! You should always endeavor to work slowly, to remain calm and to maximize your concentration.

Success in learning requires *active participation* on the part of the student, *immediate checking* of all answers and progression through the material at an *individually paced speed.* Programmed instruction satisfies all of these requirements.

# Organic Chemistry
# Made Easy

# Program 1

## Organic Chemistry (Introduction I)

**1** This and the following program provide a first glimpse in the fascinating world of organic chemistry.

The emphasis of the book lies in the programs 7 to 23. In the 17 chapters we present a much fuller coverage of organic compounds.

Programs 3 to 6 will give you the theoretical background for the better understanding of the subject matter.

In general terms it may be said that:

*Organic chemistry is the chemistry of carbon compounds. The only exceptions to this are a few simple carbon compounds, namely the oxides of carbon and carbonic acid and its salts.*

*The compounds formed by all the remaining elements belong to the realm of inorganic chemistry.*

Examples for such simple carbon compounds are

> Carbon monoxide
> Carbon dioxide
> Carbonic acid
> Sodium carbonate
> Calcium carbonate

Write down now chemical formulas for these compounds.

Continue your reading at [9]

**2** Possibly you got confused here by the shorthand way we used of writing down the phenyl group. To avoid any further confusion, we now present the shorthand notation for both the phenyl group and benzene with the full chemical formula in each case:

Benzene:

or

Note that all vertices in the hexagon are identical and are depicted as

Phenyl group

or

Note that here only five of the vertices forming the hexagon are identical ( ) whereas a hydrogen atom is missing from the remaining vertex.

What is the molecular mass of the phenyl group?

● The same as that of benzene                    →    17

● The same as that of benzene minus 1          →    26

● The same as that of benzene plus 1           →    23

---

**3**        Your answer is wrong. The compounds listed in this particular answer include carbon monoxide and pyrites which belong to the domain of inorganic chemistry.

Read through    9    again, and then attempt this question once more.

---

**4**        1. Organic chemistry is the chemistry of carbon compounds. The only exceptions to this are certain simple carbon compounds, namely the oxides of carbon and carbonic acid and its salts.

2. There are aliphatic compounds (i.e. chain-like compounds) and aromatic compounds (i.e. benzene and similar compounds).

3. Hydrocarbons; these consist of carbon and hydrogen.

4. The methyl, ethyl and phenyl groups.

5.

6. Ethane, acetylene, methanol and diethyl ether.

7. From sugar by means of alcoholic fermentation.

8. Ether and benzene are highly flammable. When working with these and other organic solvents, care must be taken to ensure that there are no naked flames in the vicinity. Methanol is very poisonous.

**End of Program 1.**

---

**5**     We are going to demonstrate here how the structural formulas asked for may be drawn in simple stages.

Start by writing down a row of four carbon atoms thus:

    C   C   C   C

The four carbon atoms are then joined together to form a chain:

    C–C–C–C

Next, draw in the remaining valence bonds for each of the carbon atoms:

$$-\overset{\displaystyle |}{\underset{\displaystyle |}{C}}-\overset{\displaystyle |}{\underset{\displaystyle |}{C}}-\overset{\displaystyle |}{\underset{\displaystyle |}{C}}-\overset{\displaystyle |}{\underset{\displaystyle |}{C}}-$$

From this structural skeleton it should be evident that the end carbon atoms can combine with three hydrogen atoms each, whereas the two middle carbon atoms can combine only with two hydrogen atoms each. In all, therefore ten hydrogen atoms can be attached to this particular skeleton.

Write out now the complete structural formula with the hydrogen atoms. Give also the chemical formula which this structure represents. Check your answers at  |43|  .

---

**6**     Your answer is wrong, probably because of a mistake in your calculation.

Work your way through the problem in  |37|   once again.

---

**7**     Methane has the structural formula

$$H-\overset{\displaystyle H}{\underset{\displaystyle H}{C}}-H$$

If the four atoms of the methane molecule are replaced by chlorine atoms, we obtain the structural formula for carbon tetrachloride thus:

$$Cl-\overset{\displaystyle Cl}{\underset{\displaystyle Cl}{C}}-Cl$$

Accordingly, the chemical formula of carbon tetrachloride is $CCl_4$. The molecular mass of carbon tetrachloride must therefore equal $1 \times 12 + 4 \times 35.5 =$ ...............

Calculate the molecular mass of the $CCl_4$ molecule, and check your answer at   $\boxed{15}$  .

---

| **8** |

Your answer is quite correct.

Now let's examine the properties of these organic compounds.

Methanol or methyl alcohol is an extremely poisonous substance. It is also flammable.

Ethanol or ethyl alcohol is widely used as a beverage and in small amounts it is not harmful. In larger quantities, however, this compound too acts as a poison. Again, it is flammable.

Ethyl alcohol is one of the oldest known organic compounds. It is prepared by so-called "alcoholic fermentation" of sugar or starch.

Another class of organic compound is formed by the *ethers*. These may be regarded as arising from the replacement of both hydrogen atoms in a water molecule by organic groups. One very commonly used ether is diethyl ether. In this particular compound the two hydrogen atoms of water are each replaced by an ethyl group thus:

$$H-\overset{\overset{\displaystyle H}{|}}{\underset{\underset{\displaystyle H}{|}}{C}}-\overset{\overset{\displaystyle H}{|}}{\underset{\underset{\displaystyle H}{|}}{C}}-O-\overset{\overset{\displaystyle H}{|}}{\underset{\underset{\displaystyle H}{|}}{C}}-\overset{\overset{\displaystyle H}{|}}{\underset{\underset{\displaystyle H}{|}}{C}}-H \quad \text{or simply} \quad C_2H_5-O-C_2H_5$$

Diethyl ether is often simply referred to as "ether". **It has a very low boiling point (around 35 °C) and is extremely flammable. When working with ether it is therefore absolutely vital that there are no flames in the vicinity.**

Using keywords list now the most important things you have just learnt about methanol, ethanol and ether.

Then continue your reading at   $\boxed{38}$  .

---

| **9** |

Now see if the formulas you listed check out:

| | |
|---|---|
| Carbon monoxide | CO |
| Carbon dioxide | $CO_2$ |
| Carbonic acid | $H_2CO_3$ |

Organic Chemistry (I)                    **Program 1**

Sodium carbonate       $Na_2CO_3$
Calcium carbonate      $CaCO_3$

Where do the names organic and inorganic chemistry come from?

In the past it was believed that the chemical compounds found in living nature,
e.g. chlorophyll, sugar, vitamins, olive oil, starch, urea, proteins and the
coloring matter in flowers, could be synthesized only in the *organs* of plants and
animals. These and other similar substances were therefore assigned to the
domain of *organic chemistry*.        ⸜

Conversely, those elements and compounds occurring in non living nature, e.g.
rocks, minerals, ores, water, air and metals, were assigned to the domain of
*non-organic* or *inorganic* chemistry.

Nowadays, it is common knowledge that numerous organic compounds can be
synthesized both industrially and in laboratories. Consequently, we do not have to
rely on plant or animal sources for such compounds. In fact, chemists have
succeeded in synthesizing a wide range of organic compounds not found in
nature, though it is fair to point out that industries based on organic chemistry
frequently draw upon raw products originating from plant or animal sources, e.g.
oil and coal.

Interestingly enough *inorganic* substances are also found in *living organisms*. A
number of such compounds, including water, sodium chloride, potassium salts
and hydrochloric acid (gastric juice), are essential for human beings, animals
and plants.

Which of the following substances belong to the domain of organic chemistry?

Olive oil, carbon monoxide, starch, gastric juice, pyrites

● Olive oil, carbon monoxide, starch and pyrites      →    | 3 |

● Olive oil, starch, gastric juice                    →    |22|

● Olive oil, starch, gastric juice
   and carbon monoxide                                →    |29|

● Olive oil and starch                                →    |40|

● Starch, gastric juice and carbon monoxide           →    |18|

| 10 |    Your answer is wrong.

An explanation of this is given in    | 5 | .

11
    Your answer is wrong because you are not working carefully enough.

Read through  1  once again.

12
    Your answer is wrong. Take another careful look at the structural
formula in  35  . Count up the valence bonds emanating from each of the six
carbon atoms present in this formula. You will discover that the number of
such bonds is not four. But, because the valence of carbon is always four, it is
possible to attach hydrogen atoms to the six carbon atoms forming the ring.

The structural formula of benzene is given in  35  .

13
    Your answer is wrong.

An explanation of this is given in  2  .

14
    Your answer is wrong.

An explanation of this is given in  45  .

15
    Correct, the molecular mass of $CCl_4$ is 154. This concludes the present
Program. For general review purposes, you should now see if you can answer
the following eight questions:

1. What is meant by the term organic chemistry?

2. How are organic compounds classified?

3. What are the simplest organic compounds called and what elements do they
   consist of?

4. What are the names of the monovalent groups deriving from methane, ethane
and benzene?

5. Draw out the structural formula for each of the following compounds:
   Methane, benzene, ethanol, phenol, ethylene

6. Give the names of the organic compounds represented below:

$$\begin{array}{c} H\ \ H \\ |\ \ \ | \\ H-C-C-H \\ |\ \ \ | \\ H\ \ H \end{array}$$     $H-C\equiv C-H$     $CH_3OH$     $C_2H_5-O-C_2H_5$

7. From what substance is alcohol produced and by which process?

8. What precautions should be taken when using ether, benzene and methanol?

Check your answers at   4  .

---

**16**    Your answer is wrong.

An explanation of this is given in   5  .

---

**17**    Your answer is wrong. Benzene is quite different from the phenyl group!

Take another careful look at the relevant formulas in   2  .

---

**18**    Your answer is wrong. The compounds listed in this particular answer include carbon monoxide (CO) which we said belongs to the domain of inorganic chemistry.

Read through   9   again, and then attempt this question once more.

---

**19**    Your answer is quite right.

Write down now the complete structural formula for benzene, and then check your answer at   30  .

If you feel somewhat uncertain here, you are advised to turn back to   35  .

---

**20**    Your answer is wrong.

An explanation of this is given in   2  .

---

**21**    You appear to have miscalculated here. The chemical formula of benzene is $C_6H_6$.

Make another attempt at the problem in   30  .

---

| 22 |      No. Although the listed compound gastric juice is indeed found in living organisms, it actually belongs to the domain of inorganic compounds.

Read through  | 9 |  again, and then attempt this question once more.

---

| 23 |      Your answer is wrong. The phenyl group is obtained from the formula for benzene by *omitting* one of its hydrogen atoms!

Read through   | 42 |   once again.

---

| 24 |      Your answer is wrong. Consider here how many valence bonds from carbon atoms will be left over after two (tetravalent!) carbons have joined together to form a double bond:

$$C=C$$

There will clearly be two valence bonds which are unattached on each carbon atom thus:

$$\diagdown C = C \diagup$$

Write down now the complete formula for this hydrocarbon, i.e. with the four hydrogen atoms included. Continue your reading at   | 35 |  .

---

| 25 |      Now check the structural formulas you have written:

| H | H H | H H H |
|---|-----|-------|
| \| | \| \| | \| \| \| |
| H-C-H | H-C-C-H | H-C-C-C-H |
| \| | \| \| | \| \| \| |
| H | H H | H H H |
| Methane | Ethane | Propane |

The generic term used to designate these three compounds is: *hydrocarbons.*

By drawing out the structural formulas of organic compounds in the most symmetrical way possible, the visualization of the structures in question is made much easier. Of course, such formulas could be drawn out in other ways, e.g. the structural formula of methane could be written as:

Similarly, the structural formula of propane could be written as:

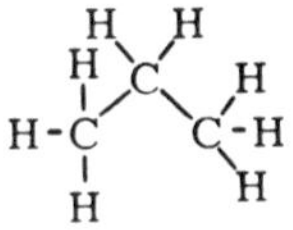

Although the latter two have been drawn correctly – in that all of the carbon atoms are tetravalent – the actual structures are rather difficult to visualize. For this reason all structural formulas are normally drawn with the maximum amount of symmetry possible for the particular structure. They are then not only easier to draw but also easier to visualize. Having said this, however, it should not be forgotten that all such formulas are only *depictions* of the atoms and their bonds. In reality, all atoms and molecules exist in three dimensions and so cannot be properly represented on two dimensional paper.

Draw out now a chain of four carbon atoms, all of which are bonded together.

How many hydrogen atoms would have to be attached to this chain to make a complete structural formula?

- 8 hydrogen atoms      →    10

- 10 hydrogen atoms     →    43

- 9 hydrogen atoms      →    16

- 12 hydrogen atoms     →    33

---

**26**    Your answer is quite right.

We have already mentioned that organic compounds always contain carbon and hydrogen, though they can contain a number of other elements such as oxygen, chlorine, sulfur or nitrogen.

Examples of organic compounds containing oxygen are the *alcohols* and the *phenols*.

Alcohols may be thought of as arising from the water molecule by replacement of one of its hydrogen atoms with an aliphatic group, e.g. with the methyl group. (The actual preparation of alcohols is, of course, carried out in quite a different way!)

      H-O-H       $CH_3$-O-H

                       Methanol or

       Water         Methyl alcohol

Phenols may be thought of as arising from the replacement of one of the hydrogens in water by an *aromatic* group, e.g. by the phenyl group.

H-O-H

Water

OH

Phenol

Copy down now the formulas and names of these two new compounds. Draw also the structural formula of ethanol (ethyl alcohol). It should then be easy for you to calculate the molecular mass of ethanol. The relevant atomic masses are: C = 12, O = 16, H = 1.

What is the molecular mass of ethanol?

- 32        →   14

- 46        →   8

- 47        →   34

---

**27**     Your answer is wrong. Have another good look at the last structural formula in   35   . Count up the valence bonds emanating from each of the carbon atoms in this structural formula. You will quickly see that the number is not four! Now, as carbon always has a valence of four, hydrogen atoms should be attached to the six carbon atoms forming the ring.

The correct structural formula for benzene is given in   35   .

---

**28**     You have either miscalculated here or not taken into account the fact that a *tetra*valent carbon atom can combine with *four* monovalent hydrogen atoms.

Answer the question in   40   once again.

---

**29**     No. Let's check whether the compounds mentioned belong to the domain of organic or inorganic chemistry.

Olive oil has been specifically mentioned as an organic compound. It consists of structurally complex carbon compounds. The same may also be said of starch.

The other compounds mentioned (carbon monoxide and gastric juice) belong to the domain of inorganic chemistry.

Chemically speaking, gastric juice is simply hydrochloric acid, HCl. This clearly belongs to inorganic chemistry. It represents an example of an inorganic compound found in a living organism. Certain inorganic substances are even essential for life in human beings, animals and plants. Examples of such substances include water, sodium chloride and potassium salts.

To which branch chemistry are carbonic acid and potassium carbonate assigned?

● To inorganic chemistry       →    40

● To organic chemistry         →    11

---

30      The complete structural formula for benzene is:

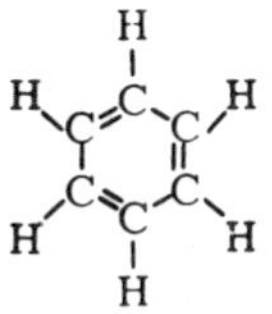

A simplified version of this formula is often used consisting of a hexagon with alternating double bonds thus:

When written in this way, however, it is understood that each vertex of the hexagon symbolizes one carbon atom with an attached hydrogen atom.

Make sure you are familiar with this latter type of representation of benzene.

What is the molecular mass of benzene? The relevant atomic masses are: $C = 12, H = 1$.

● 72   →   49

● 78   →   42

● 82   →   21

● 74   →   54

---

31      No. It shouldn't be too difficult to write down the correct chemical formulas from the structural formulas we have listed. After all, you need only count up the various atoms present!

Read through   ⎡50⎤   once again.

---

⎡**32**⎤     You have either miscalculated here or not taken into account the fact that a *tetra*valent carbon atom can combine with *four* monovalent hydrogen atoms.

Answer the question in   ⎡40⎤   once again.

---

⎡**33**⎤     Your answer is wrong.

An explanation of this is given in   ⎡5⎤  .

---

⎡**34**⎤     Your answer is wrong.

An explanation of this is given in   ⎡45⎤  .

---

⎡**35**⎤     Your answer is quite correct.

The complete structural formula of the simplest hydrocarbon having a double bond between its carbon atoms is as follows:

$$\begin{array}{c} H \diagdown \quad \diagup H \\ C{=}C \\ H \diagup \quad \diagdown H \end{array}$$

This compound is known as *ethylene* or *ethene*.

Finally, it is also possible for carbon atoms to combine with *three bonds* connecting them together. The simplest example is provided by the *acetylene* molecule which has the structural formula:

$$H\text{-}C{\equiv}C\text{-}H$$

Note that the carbon has a valence of four again.

Write down now the structural formulas, chemical formulas and names of the two hydrocarbons mentioned above.

Hydrocarbons having double or triple bonds are referred to as *unsaturated hydrocarbons*.

Within the domain of organic chemistry there are two broad categories of compounds, namely:

1. The *aliphatic* compounds – chain-like hydrocarbon compounds.

2. The *aromatic* compounds – benzene and related compounds.

It is worth committing these two categories to memory.

*Benzene* is a hydrocarbon whose carbon atoms are joined together in the form of a ring. Six carbon atoms are alternately bonded together by three single and three double bonds as illustrated below:

Although this type of representation for benzene is not strictly accurate, we shall be explaining in Program 8 why this is a useful working formula for benzene.

As in all previous instances, the carbon is tetravalent in the benzene molecule. How many hydrogen atoms should therefore be added to complete the structural formula of benzene?

- 12   →   12

- 6   →   19

- 3   →   27

---

36   You appear to have mixed up ethane and propane.

Read through   50   once again.

---

37   All the organic compounds we have discussed so far are fairly flammable. In fact, there are only a few organic compounds which are not flammable. Among the latter are included especially those containing a lot of chlorine, e.g. methylene chloride, chloroform and carbon tetrachloride. Such chlorine-containing organic compounds are frequently used as solvents in laboratories and in industry.

To form carbon tetrachloride, all four of the hydrogen atoms in methane are replaced by chlorine atoms. Now draw out the structural formula of carbon tetrachloride.

What is the molecular mass of carbon tetrachloride? The relevant atomic masses are: $C = 12$, $Cl = 35.5$.

- 117.5                    → $\boxed{44}$

- 154                      → $\boxed{15}$

- 155                      → $\boxed{48}$

- I obtained another result     → $\boxed{6}$

---

$\boxed{38}$
    Methanol:    Very poisonous, flammable

        Ethanol:     Harmless in small amounts, flammable, prepared by alcoholic fermentation

        Ether:       Low boiling point, *extremely flammable,* no flames allowed in the vicinity when working with ether

Did your keywords convey the same meaning as those above? If not, then write out the correct version as above, and be sure to memorize these characteristics of the compounds we have been discussing!

If you still remember the structural formula of diethyl ether, it should be easy for you to draw it out. Do this now and then check your structure at $\boxed{46}$ .

---

$\boxed{39}$
    *Aliphatic* and *aromatic* compounds.

To continue our discussion now, if one of the hydrogen atoms is omitted from the formula of methane, we obtain a *monovalent group* of atoms. This group is known as the *methyl group* and has the formula:

$$
\begin{array}{c}
\text{H} \\
| \\
\text{-C-H} \\
| \\
\text{H}
\end{array}
$$
            (Chemical formula: $-CH_3$)

Analogously, by omitting one hydrogen atom in each case, we may derive the *ethyl group* from ethane and the *phenyl group* from benzene thus:

$$
\begin{array}{c}
\text{H H} \\
| \ | \\
\text{-C-C-H} \\
| \ | \\
\text{H H}
\end{array}
$$
            (Chemical formula: $-C_2H_5$)

            (Chemical formula: $-C_6H_5$)

Copy down these three formulas with their names.

What will be the molecular mass of each of these radicals? The relevant atomic masses are C= 12, H = 1.

| Methyl | Ethyl | Phenyl | | |
|--------|-------|--------|---|----|
| 16 | 30 | 78 | → | 53 |
| 15 | 77 | 29 | → | 47 |
| 15 | 29 | 77 | → | 26 ˇ |
| 15 | 29 | 78 | → | 20 |
| 15 | 29 | 65 | → | 13 |

---

**40**    Your answer is quite right.

The area covered by organic chemistry is extraordinarily large. The number of known organic compounds currently exceeds 5 million. Inorganic chemistry by comparison can claim merely some 200,000 compounds!

*Carbon always has a valence of four in organic compounds.*

The simplest organic compounds contain only carbon and hydrogen. Such compounds are referred to as *hydrocarbons*.

The highly flammable gas *methane* is the simplest of the hydrocarbons. Its molecule consists of *one* carbon atom combined with *four* hydrogen atoms.

Write down now the chemical formula for methane and then calculate the molecular mass of this compound. The relevant atomic masses are C = 12, H = 1.

What is the molecular mass of methane?

- 14   →   55
- 15   →   28
- 16   →   50
- 17   →   32

---

**41**    Your answer is quite right.

We have already discussed three simple organic compounds, namely methane, ethane and propane.

Write down now the structural formula for each of these compounds together with their names. What is the generic term used to designate such compounds?

Check your answers at    25   .

---

**42**     Benzene has a molecular mass of 78.

Let's now briefly review what we have covered so far.

In each of the sentences below jot down the missing words or formulas:

1. The simplest organic compounds consist of the elements .............. and
............

2. The simplest organic compounds are called ..............

3. $CH_4$ is called ..............
   $C_2H_6$ is called ..............

   ⬡ is called ..............

4. Ethylene or ethene has the structural formula ..............
   Acetylene has the structural formula ..............

Check out your answers by going now to    52   .

---

**43**     Your answer is quite correct. The complete structural formula of the hydrocarbon having four carbon atoms is as follows:

$$H-\overset{\displaystyle H}{\underset{\displaystyle H}{C}}-\overset{\displaystyle H}{\underset{\displaystyle H}{C}}-\overset{\displaystyle H}{\underset{\displaystyle H}{C}}-\overset{\displaystyle H}{\underset{\displaystyle H}{C}}-H$$        (Chemical formula: $C_4H_{10}$)

Carbon atoms are not necessarily joined together by *single* bonds to form more or less long chains. It is quite possible for two carbon atoms to be joined together by a *double* bond. We indicate two carbon atoms combined together by means of a double bond as follows:

     $C=C$

How many hydrogen atoms may be attached to such a unit? (Remember that four valence bonds must emanate from each carbon atom.)

- 2 hydrogen atoms  →  24
- 4 hydrogen atoms  →  35
- 6 hydrogen atoms  →  51

---

44   Your answer is wrong.

An explanation of this is given at  7  .

---

45   Just as when a *methyl group* replaces one of the hydrogen atoms in water *methanol* or *methyl alcohol* is formed:

$$
\begin{array}{c}
\quad H \\
\quad | \\
H-C-O-H \\
\quad | \\
\quad H
\end{array}
\quad \text{or} \quad CH_3\text{-}O\text{-}H
$$

replacement by an *ethyl group* will similarly result in *ethanol* or *ethyl alcohol*.

The ethyl radical is derived from the ethane molecule. The latter has the structural formula:

$$
\begin{array}{c}
H \quad H \\
| \quad | \\
H-C-C-H \\
| \quad | \\
H \quad H
\end{array}
$$

By leaving off one of the hydrogen atoms in the ethane molecule, we obtain the formula for the ethyl group:

$$
\begin{array}{c}
H \quad H \\
| \quad | \\
H-C-C- \\
| \quad | \\
H \quad H
\end{array}
$$

It should now be clear that the structural formula of ethyl alcohol will be as follows:

$$
\begin{array}{c}
H \quad H \\
| \quad | \\
H-C-C-OH \\
| \quad | \\
H \quad H
\end{array}
$$

To familiarize yourself with this material, you should now read  26  again carefully.

---

**46**

$$H-\overset{\displaystyle H}{\underset{\displaystyle H}{\overset{|}{\underset{|}{C}}}}-O-\overset{\displaystyle H}{\underset{\displaystyle H}{\overset{|}{\underset{|}{C}}}}-H \quad \text{or simply} \quad CH_3-O-CH_3$$

Was your formula right? If so, continue your reading now at [37].

Was your formula wrong? If so, you should correct it and then continue your reading at [8].

**47**     Your answer is wrong. It is impossible for the ethyl group to have molecular mass greater than that of the phenyl group.

Make another attempt to solve the question in [39].

**48**     Your answer is wrong.

An explanation of this is given in [7].

**49**     Your answer is wrong. It is obvious that you have only taken the atomic masses of the six C atoms into account. Benzene, however, has the chemical formula $C_6H_6$.

You should now attempt the question in [30] once again.

**50**     Your answer is quite right.

Methane has the chemical formula $CH_4$. Its structural formula is as follows:

$$H-\overset{\displaystyle H}{\underset{\displaystyle H}{\overset{|}{\underset{|}{C}}}}-H$$

Carbon has one very special characteristic, namely that:

*Individual carbon atoms can be linked together chemically to form chain-like compounds.*

The hydrocarbon compound containing two bonded carbon atoms has the structural formula:

$$H-\overset{\displaystyle H}{\underset{\displaystyle H}{\overset{|}{\underset{|}{C}}}}-\overset{\displaystyle H}{\underset{\displaystyle H}{\overset{|}{\underset{|}{C}}}}-H$$

This compound is known as *ethane*. Note that in this case too *every carbon atom* has a *valence of four*.

When three carbon atoms combine together to form a chain, we obtain the *propane* molecule. This has the structural formula:

$$\begin{array}{ccc} H & H & H \\ | & | & | \\ H-C-C-C-H \\ | & | & | \\ H & H & H \end{array}$$

What are the chemical formulas for ethane and propane?

| Ethane | Propane | | |
|--------|---------|---|----|
| ● $C_3H_8$ | $C_2H_6$ | → | 36 |
| ● $C_2H_6$ | $C_3H_8$ | → | 41 |
| ● $C_2H_4$ | $C_3H_6$ | → | 31 |

---

**51**    See if you can work out how many valence bonds of (tetravalent!) carbon will be left over when two carbon atoms are joined together by a double bond:

   C = C

Clearly, the answer must be that there are two unattached valences on each carbon atom thus:

   $\diagdown C = C \diagup$

Write out now the full structural formula for this hydrocarbon (including the four hydrogen atoms), and then continue your reading at   35   .

---

**52**    1. The simplest organic compounds consist of the elements *carbon* and *hydrogen*.

2. The simplest organic compounds are called *hydrocarbons*.

3. $CH_4$ is called *methane*.

   $C_2H_6$ is called *ethane*.

   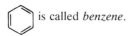 is called *benzene*.

4. Ethylene has the structural formula

$$\begin{array}{cc} H & H \\ \diagdown \diagup \\ C=C \\ \diagup \diagdown \\ H & H \end{array}$$

Acetylene has the structural formula  H-C≡C-H

Organic compounds are classified into two major categories, namely the
.............. compounds (chain-like compounds) and the ...... compounds
(benzene and related compounds).

Jot down the two missing words. → |39|

---

|53|  Your answer is wrong. Instead of calculating the molecular masses of
methane, ethane and benzene, you should have calculated the molecular masses
of the methyl, ethyl and phenyl *groups*. The methyl, ethyl and phenyl groups
may be derived respectively from the formulas for methane, ethane and benzene
by *omitting one hydrogen atom* in each case.

Make another attempt at the question posed in  |39| .

---

|54|  You seem to have miscalculated here. The chemical formula for
benzene is $C_6H_6$.

Now see if you can solve the problem in  |30| .

---

|55|  You have either miscalculated here or have not remembered that a
*tetra*valent carbon atom can form bonds to *four* hydrogen atoms.

Make another attempt at the question posed in  |40| .

# Program 2

## Organic Chemistry (Introduction II)

**1**     Let's start by briefly reviewing the ground we have covered so far.

Organic chemistry is the chemistry of carbon compounds. The only exceptions to this are a few very simple carbon compounds, namely the oxides of carbon, carbonic acid and its salts. All of the remaining elements and their compounds belong to the domain of inorganic chemistry.

Organic chemistry is sub-divided into the chemistry of aliphatic compounds (chain-like compounds) and the chemistry of aromatic compounds (benzene and related compounds). The simplest organic compounds, the hydrocarbons, consist of the elements carbon and hydrogen. Currently, we have discussed *methane* ($CH_4$), *ethane* ($C_2H_6$), *propane* ($C_3H_8$), *ethylene* ($C_2H_4$), *acetylene* ($C_2H_2$) and *benzene* ($C_6H_6$).

Monovalent groups may be derived form the hydrocarbons by omitting one hydrogen atom in each case. Thus, we can derive the methyl group from methane, the ethyl group from ethane and the phenyl group from benzene.

If one of the hydrogen atoms in a water molecule is replaced by an aliphatic group, we obtain the formula for an alcohol. Phenols may also be derived from the formula for water, though in such instances the hydrogen atom is replaced by an aromatic group. As examples, we mention here *methanol* ($CH_3OH$), *ethanol* ($C_2H_5OH$) and phenol ($C_6H_5OH$).

If both of the hydrogen atoms in water are replaced by organic groups we obtain ethers. An important example is *diethyl ether* ($C_2H_5-O-C_2H_5$).

Most organic compounds are readily flammable. In this context diethyl ether is especially dangerous. **Great caution is thus necessary when handling it.**

Phenol and related compounds are highly *corrosive*.

In all organic compounds carbon has a valence of four.

Now write down structural formulas for all the compounds printed in italics in this section. You should not find this too difficult as the chemical formula has been given in each case.

Then continue with your reading at  .

|2|     Yes, the solvents benzene, methanol, ethanol and ether are indeed flammable. Remember that benzene and ether are especially dangerous in this regard.

To conclude the Program, we list below several questions which you should attempt to answer in writing.

1. Write down the name, chemical and structural formula of both an aliphatic and an aromatic amine.

2. What type of reaction do amines display?

3. What is the structural formula for acetic acid and what are its salts called?

4. How is an ester formed?

5. Name one process for the manufacture of alcohol (include the relevant equation).

6. Which two raw materials are of special importance in organic chemistry?

7. What four products are formed when coal is converted into coke?

8. What are the three main types of foodstuff that we need for our nourishment?

Your should check all your answers at   |36|   .

|3|     Your answer is quite correct.

Since all living creatures use up oxygen and produce carbon dioxide when they breathe, the atmosphere should gradually become poorer in oxygen and richer in carbon dioxide. The process of "photosynthesis", however, is able to restore the balance.

In the presence of a light source plants use up the carbon dioxide in the air. They are able to remove the carbon from the carbon dioxide and give off oxygen in the process. The carbon is used by the plants to produce various carbohydrates. The whole process is described as *photosynthesis*.

Now write out the completed version of the sentence below:

In the process of breathing .............. is used up and .............. is produced whereas in the process of photosynthesis .............. is used up and ............... is produced.

Continue your reading at ⬚10 .

---

**4**     Your answer is quite correct.

The structural formula for acetic is as follows:

$$H-\overset{\displaystyle \overset{H}{|}}{\underset{\displaystyle \underset{H}{|}}{C}}-C\overset{\nearrow O}{\underset{\searrow O-H}{}}$$

Acetic acid has in all four hydrogen atoms. In aqueous solution, however, only one of these four atoms dissociates to form a positively charged hydrogen ion. The atom which does this is the one attached to the oxygen atom.

Acetic acid is a monobasic organic acid.

The three hydrogen atoms attached to the carbon atom do not dissociate. If they were to, methane, ethane, benzene and other hydrocarbons would be acids.

We have just discussed the amines. They react as bases.

Write out now the structural and chemical formulas of two amines and of acetic acid, adding the name of each compound alongside.

Check your answer at ⬚34 .

---

**5**     Aliphatic compounds are those organic compounds which have a chain-like structure. Examples include methane, ethane, propane and methanol. Aromatic compounds are those organic compounds which contain a benzene ring.

On this basis how would you classify ethanol and phenol?

● Both are aromatic compounds                    → ⬚17

● Both are aliphatic compounds                    → ⬚12

● Ethanol is an aliphatic compound whereas
  phenol is an aromatic compound                  → ⬚44

---

**6**     $CH_3COOCH_3$, the methyl ester of acetic acid.

We now turn to a completely new topic in organic chemistry, namely to the widespread use of *raw materials*. Such materials are extremely important in the manufacture of organic compounds.

The most important raw materials for the manufacture of organic compounds are *crude oil* and *coal*.

Crude oil consists mainly of hydrocarbons containing more than four carbon atoms.

Crude oil can be separated into different fractions by the process of distillation. For instance, it is possible to produce in this way gasolene (boiling point circa 100 °C) and diesel oil (boiling point circa 200 °C).

Crude oil can also be split up in a so-called *cracking process*. The crude oil is passed over catalysts which are heated to very high temperatures. The products of this cracking process are low boiling hydrocarbons such as methane, ethane and ethylene.

Using keywords write out the information contained in this section. Then continue your reading at   19   .

---

| 7 |

     Your answer is quite right.

Methylamine is an aliphatic compound whereas aniline is an aromatic compound.

Certain amines react as strong bases.

**Accordingly, one must work as carefully with amines as with sodium or potassium hydroxides. Protective goggles should always be worn. Ignoring this simple precaution could have very serious consequences.**

Acids are also known in the domain of organic chemistry; these are the so-called *carboxylic acids*. A very simple and commonly used organic acid is *acetic acid*. Its structural formula is as follows:

$$H-\overset{\overset{\displaystyle H}{|}}{\underset{\underset{\displaystyle H}{|}}{C}}-C\overset{\displaystyle O}{\underset{\displaystyle O-H}{\diagdown}}$$

Acetic is often used in the household as a dilute solution. Pure acetic acid is corrosive.

What type of acid is acetic acid?

● A monobasic acid       →    4

- A tribasic acid     →    14
- A tetrabasic acid    →    18
- I don't know       →    22

---

**8**

$$CH_3COOH + HOC_2H_5 \rightarrow CH_3COOC_2H_5 + H_2O$$

From the reaction of acetic acid and ethanol the products formed are *ethyl acetate* (the ethyl ester of acetic acid) and water.

What is the name of the ester formed when acetic acid and methanol react together?

Jot down your answer and then turn to   24   .

---

**9**      Your answer is wrong. Take another good look at each of the formulas below and then copy them down.

$$\text{H} \atop H\text{-}\overset{\text{H}}{\underset{\text{H}}{C}}\text{-H} \text{ or } CH_4$$

Methane

$$H\text{-}\overset{\text{H}}{\underset{\text{H}}{C}}\text{-}\overset{\text{H}}{\underset{\text{H}}{C}}\text{-H} \text{ or } C_2H_6$$

Ethane

$$H\text{-}\overset{\text{H}}{\underset{\text{H}}{C}}\text{- or } -CH_3$$

Methyl group

$$H\text{-}\overset{\text{H}}{\underset{\text{H}}{C}}\text{-}\overset{\text{H}}{\underset{\text{H}}{C}}\text{- or } -C_2H_5$$

Ethyl group

$$H\text{-}\overset{\text{H}}{\underset{\text{H}}{C}}\text{-OH or } CH_3\text{-OH}$$

Methanol or
Methyl alcohol

$$H\text{-}\overset{\text{H}}{\underset{\text{H}}{C}}\text{-}\overset{\text{H}}{\underset{\text{H}}{C}}\text{-OH or } C_2H_5\text{-OH}$$

Ethanol or
Ethyl alcohol

The addition of water to ethylene produces ethanol.

Continue your reading at   15   .

**10** In the process of breathing *oxygen* is used up and *carbon dioxide* is produced whereas in the process of photosynthesis *carbon dioxide* is used up and *oxygen* is produced.

We repeat here a few of the statements discussed earlier on. The relevant missing words should be inserted at the appropriate points.

Organic chemistry is the chemistry of .............. compounds with the exception of the oxides of carbon, carbonic acid and its salts.

Inorganic chemistry is the chemistry of all the remaining elements and compounds.

Carbon always has a valence of .............. Carbon atoms are able to combine together to form chains and rings. The carbon atoms may be bonded together by single, double or triple bonds.

The simplest organic compounds are the .............. From these we can derive the alcohols, phenols, ethers, amines and carboxylic acids.

Continue your reading at 31 .

---

**11** Your answer is wrong.

An explanation of this is given in 5 .

---

**12** Your answer is wrong again.

Ethane, $CH_3-CH_3$, and propane, $CH_3-CH_2-CH_3$, are examples of aliphatic compounds.

Aromatic compounds contain the benzene ring:

Benzene

The formula for ethanol is $CH_3-CH_2-OH$.

The formula for phenol is —OH

Read through 5 once again.

---

**13**     The salts of acetic acid are called acetates. Sodium acetate is the sodium salt of acetic acid. It has the chemical formula $CH_3COONa$. In water it dissociates into the negatively charged acetate ion and the positively charged sodium ion:

$$CH_3COO^\ominus + Na^\oplus$$

The acetate group is thus a singly charged negative acid group. Now write down the chemical formulas for potassium and calcium acetates. → 9

---

**14**     Your answer is wrong. Perhaps this question was a little too difficult for you. An explanation of the relevant points is to be found in 27 .

---

**15**     Your answer is quite correct.

The addition of water to ethylene is used on a large scale industrially for the synthesis of ethanol.

Another important synthetic process for alcohol is based on the alcoholic fermentation procedure.

You can no doubt still recall the starting material used for the production of alcohol by the alcoholic fermentation procedure. What is the material in question?

● Protein    →    32

● Sugar     →    26

● Fat       →    28

---

**16**     You have not been working carefully enough.

Read through 26 once again.

---

**17**     Your answer is wrong again.

Ethane, $CH_3-CH_3$, and propane $CH_3-CH_2-CH_3$, are examples of aliphatic compounds.

All aromatic compounds contain the benzene ring:

Benzene

The formula for ethanol is $CH_3-CH_2-OH$.

The formula for phenol is ⟨◯⟩-OH·

Read through ⑤ once again.

---

**18**    Your answer is wrong. Perhaps this question was a little too difficult for you. An explanation of the relevant points is to be found in ⟨27⟩ .

---

**19**    Now check your answers:

*The raw materials for making organic compounds are crude oil and coal. Crude oil consists of hydrocarbons. Both gasolene and diesel oil may be obtained by means of distillation. Crude oil can be split up in a cracking process which involves special catalysts and high temperatures. The products of such cracking include methane, ethane and ethylene.*

If what you wrote was incomplete, copy down the above text in italics.

The cracking products may be converted by subsequent chemical reactions into a variety of other valuable chemical products. Let's look at just one example.

If ethylene is reacted with water using catalysts and high temperatures, the following reaction takes place:

$$\begin{array}{ccc} H\ H & & H\ H \\ |\ \ | & & |\ \ | \\ C=C\ +\ H\text{-}O\text{-}H & \rightarrow & H\text{-}C\text{-}C\text{-}OH \\ |\ \ | & & |\ \ | \\ H\ H & & H\ H \end{array}$$

You should now copy down this equation.

What is the reaction product according to this equation?

- Methanol      →   ⟨9⟩

- Ethanol      →   ⟨15⟩

- Phenol      →   ⟨25⟩

- Acetic acid      →   ⟨30⟩

**20**    Now check out your answers and correct any mistakes:

Methane

$$H-\overset{\displaystyle H}{\underset{\displaystyle H}{C}}-H$$

Ethane

$$H-\overset{\displaystyle H}{\underset{\displaystyle H}{C}}-\overset{\displaystyle H}{\underset{\displaystyle H}{C}}-H$$

Propane

$$H-\overset{\displaystyle H}{\underset{\displaystyle H}{C}}-\overset{\displaystyle H}{\underset{\displaystyle H}{C}}-\overset{\displaystyle H}{\underset{\displaystyle H}{C}}-H$$    or    $CH_3-CH_2-CH_3$

Ethylene

$$\overset{H}{\underset{H}{}}C=C\overset{H}{\underset{H}{}}$$

Acetylene    $H-C\equiv C-H$

Benzene

   or   

Methanol

$$H-\overset{\displaystyle H}{\underset{\displaystyle H}{C}}-O-H$$

Ethanol

$$H-\overset{\displaystyle H}{\underset{\displaystyle H}{C}}-\overset{\displaystyle H}{\underset{\displaystyle H}{C}}-O-H$$    or    $CH_3-CH_2OH$

Phenol

Diethyl ether

$$H-\overset{\displaystyle H}{\underset{\displaystyle H}{C}}-\overset{\displaystyle H}{\underset{\displaystyle H}{C}}-O-\overset{\displaystyle H}{\underset{\displaystyle H}{C}}-\overset{\displaystyle H}{\underset{\displaystyle H}{C}}-H$$    or    $CH_3-CH_2-O-CH_2-CH_3$

Another important class of substances in organic chemistry is known as the amines. We can imagine them to be formed by the *replacement of one or more of the hydrogen atoms* in the simplest compound of nitrogen and hydrogen, ammonia ($NH_3$), by *organic groups*.

Examples of simple amines include:

1. *Methylamine:*

$$H-\overset{\displaystyle H}{\underset{\displaystyle H}{\overset{|}{\underset{|}{C}}}}-NH_2 \quad \text{or} \quad CH_3-NH_2$$

2. *Phenylamine,* which is usually called *aniline:*

or   $C_6H_5-NH_2$

To which class of organic substances do each of the above amines belong?

● The first is an aliphatic compound whereas the second is an
aromatic compound                                  →    7

● Both are aromatic compounds                          →    11

● Both are aliphatic compounds                         →    29

---

**21**    Now check your answers:

Potassium acetate: $CH_3COOK$

Calcium acetate: $(CH_3COO)_2Ca$

Carboxylic acids react with alcohols to form esters. In the process, one water molecule is split off from one molecule of acid and one molecule of alcohol. The reaction of acetic acid and ethanol is typical of this type of reaction:

Write out the equation for this reaction using ordinary chemical formulas thus:

$$CH_3COOH + \text{.............} \quad \rightarrow \quad 8$$

---

---

**22**    It is just as well that you didn't guess here but admitted you did not know the answer. The question was actually quite difficult.

An explanation of the relevant points is given in    27   .

---

**23**    No. Careful inspection of the diagram in    33    should lead you to the correct answer.

Now read through    33    once again.

---

**24**    The ester formed from acetic acid and methanol is the

*methyl ester of acetic acid or methyl acetate.*

Write down the formula for methyl acetate.    →    6

---

**25**    Your answer is wrong.

Phenol is derived from benzene by replacing one of its hydrogen atoms by an OH group thus:

Benzene                        Phenol

At this point we repeat the equation for the reaction of ethylene with water:

$$
\begin{array}{ccc}
\text{H H} & & \text{H H} \\
| \ | & & | \ | \\
\text{C=C} + \text{H}_2\text{O} & \rightarrow & \text{H-C-C-OH} \\
| \ | & & | \ | \\
\text{H H} & & \text{H H}
\end{array}
$$

What is the reaction product according to this equation?

● Methanol        →    9

● Ethanol         →    15

● Acetic acid     →    30

---

**26**    Yes, ethanol is produced by the alcoholic fermentation of sugar.

In addition to crude oil, coal is also an important source of raw materials for the manufacture of organic compounds. There are four principal types of coal:

*Peat, Brown coal, Bituminous coal* and *Anthracite.*

These various types of coal have been formed from plants in a gradual transformation process which has lasted tens of thousands of years. Thus, the extensive forests covering the earth's surface in primordial times, which were laid waste by violent earthquakes, are found today in the form of coal deposits.

As plants begin to decay the first product which forms is *peat.* Peat still contains a great deal of water and can be used as a fuel only after it has been dried out.

*Brown coal* is considerably richer in carbon. It can be used as a fuel after the removal of some of its water. Briquettes can be made from brown coal which has been dried out.

*Bituminous coal* contains very little water and consists mainly of carbon – up to 90% is carbon. It is an extremely valuable fuel.

Copy down the three types of coal we have described here and write out the basic characteristics of each using keywords. Include the carbon content, water content and suitability as a fuel.

What is the change in the carbon content as we go from *bituminous coal* through *brown coal* to *peat?*

- The carbon content increases          →     43

- The carbon content remains the same     →     16

- The carbon content decreases           →     33

---

27     The structural formula for acetic acid is:

$$H-\overset{\overset{\textstyle H}{|}}{\underset{\underset{\textstyle H}{|}}{C}}-C\overset{\nearrow O}{\underset{\searrow OH}{}}$$

The acetic acid molecule has four hydrogen atoms. However, only one of these four hydrogen atoms dissociates in aqueous solution to form a positively charged hydrogen ion. This is the one attached to the oxygen atom.

Acetic acid is a *mono*basic organic acid.

The three hydrogen atoms attached to the carbon atom do not dissociate. If they did, methane, ethane and benzene as well as other hydrocarbons would also be acids.

We have just discussed the amines and, as we said, they react as bases.

Write down now both the chemical and structural formulas for two amines and for acetic acid. Give the names of the compounds you choose.

See if your formulas are correct by turning now to ⬛34 .

---

**28**     No, *ethanol is obtained from the alcoholic fermentation of sugar.*

Copy down this statement and then continue your reading at ⬛26 .

---

**29**     Your answer is wrong.

An explanation of this is given in ⬛5 .

---

**30**     Your answer is wrong.

Continue your reading at ⬛27 .

---

**31**     *carbon*

       *four*

       *hydrocarbons*

Among the important solvents are benzene, methanol, ethanol and ether.

Are these solvents we have mentioned flammable?

● No       →    ⬛38

● Yes      →    ⬛2

---

**32**     No, *ethanol is obtained from the alcoholic fermentation of sugar.*

Copy down this statement and then continue your reading at ⬛26 .

---

**33**     Your answer is quite right.

Coal is of great importance as a fuel. However, when it is burnt, many of the valuable substances it contains are lost. They could, of course, be put to much better use.

In a process known as the *degasification of coal* such valuable substances are not lost but are retained for further use. The process is carried out in gas works and coking plants.

How many different products are formed during the degasification of coal?

You should be able to answer this question after taking a look at the diagram below.

- Two      →    |23|

- Three    →    |37|

- Four     →    |41|

The following figure illustrates a miniature version of a gas works.

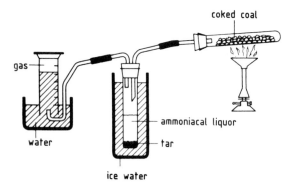

Copy down this diagram – it need not be a work of art, but at least get the details correct!

---

|34|

H
|
H–C–NH₂   or   CH₃–NH₂
|
H

Methylamine

NH₂ or C₆H₅–NH₂

Aniline

H
|   ⫶O
H–C–C⟨    or   CH₃–COOH
|   OH
H

Acetic acid

If your answer was wrong, you should in your own interest now return to 20 . If your answer was right, you can go on to 13 .

| 35 | | | | |
|----|----|----|----|----|
| $CH_4$ | $H_2$ | CO | $NH_3$ | ⬡ |
| methane | hydrogen | carbon monoxide | ammonia | benzene |

The manufacture of water gas: $C + H_2O \rightarrow CO + H_2$

Our lives have been profoundly changed over the past century by products of the chemical industry based on crude oil and coal. By means of manifold different reactions, it is now possible to transform these raw materials into a very wide range of new products.

Plastics and synthetic fibers, dyes and drugs, detergents and paints are but some examples of products of organically based chemical industry. It is inconceivable that we could now manage without them.

We should also mention here that organic substances play a vital role in all living processes.

For our nourishment *three main kinds of foodstuff* are absolutely indispensable, namely:

> *Fats* (esters of "fatty acids" and glycerol)
> *Protein* (nitrogen-containing compounds)
> *Carbohydrates* (sugar and starch)

Be sure to memorize this important list.

Fats and proteins are provided for us by animals and plants. There are both animal and plant fats and animal and plant proteins. Carbohydrates on the other hand are obtained mainly from plants. Sugar is extracted from sugar cane or sugar beet, for instance, and starch may be extracted from potatoes or grain.

The food that we eat is partly "burnt" in our bodies. In this way the energy that we need is provided. To burn up our food we must breathe in oxygen. This oxygen oxidizes the foodstuffs and produces carbon dioxide in the process. In fact, all living creatures breathe in oxygen and breathe out carbon dioxide.

What three kinds of foodstuff are indispensable for human nourishment?

● Sugar, starch and fat                    → 42

● Plant oils, animal fats and carbohydrates        → 39

- Proteins, carbohydrates and fats      →   3

- Plant proteins, animal proteins and carbohydrates    →   40

---

**36**

1. Methylamine    $CH_3-NH_2$       
$$H-\underset{\underset{H}{|}}{\overset{\overset{H}{|}}{C}}-NH_2$$

   Aniline      $C_6H_5-NH_2$

2. Amines react as bases; with acids they form salts. Caution is necessary when working with amines!

3. $H-\underset{\underset{H}{|}}{\overset{\overset{H}{|}}{C}}-\overset{\overset{O}{\diagup\!\!\!\diagup}}{\underset{\diagdown}{C}}_{OH}$

   The salts of acetic acid are called acetates.

4. By reacting together an alcohol and an acid and splitting off a water molecule.

5. The reaction of ethylene with water in the presence of catalysts:

$$\underset{\underset{H}{|}}{\overset{\overset{H}{|}}{C}}=\underset{\underset{H}{|}}{\overset{\overset{H}{|}}{C} } + H-O-H \longrightarrow H-\underset{\underset{H}{|}}{\overset{\overset{H}{|}}{C}}-\underset{\underset{H}{|}}{\overset{\overset{H}{|}}{C}}-OH$$

6. Crude oil and coal

7. Ammoniacal liquor, tar, coal gas and coke.

8. Protein, fats and carbohydrates.

**End of Program 2.**

---

**37**      Your answer is wrong. Take another careful look at the diagram in   33   and you should be able to come up with the right answer.

Read through   33   once again.

**38**    Your answer is not correct.

*The solvents benzene, methanol, ethanol and ether are flammable. Benzene and ether are especially dangerous in this regard.*

Copy down these two statements in italics and then continue reading at [2].

---

**39**        Your answer is wrong. One foodstuff which is essential for human nourishment is missing here.

Read through [35] once again.

---

**40**        Your answer is wrong. One foodstuff which is essential for human nourishment is missing here.

Read through [35] once again.

---

**41**        Your answer is quite right. Let's consider each of the four products in turn:

1. The gas, *coal gas*, contains hydrocarbons (e.g. methane), hydrogen and carbon monoxide (which of course makes it poisonous!).

2. The *ammoniacal liquor* is essentially a solution of ammonia ($NH_3$) in water. We shall be discussing ammonia in Program 20.

3. The *tar* contains many valuable organic compounds, e.g. benzene.

4. The *residue* consists of coke. From the chemical viewpoint coke is mainly carbon. Coke plays a vital role in our economy. It is used not only as a fuel but also in numerous chemical processes. An example is the manufacture of water gas. In this particular process, steam is passed over red-hot coke and the products hydrogen and carbon monoxide are produced.

In this section we have mentioned a number of compounds present in coal gas and tar. Write down now the names and formulas of these compounds. Also write out the equation for the manufacture of water gas.

Check your answers at [35] .

---

**42**        Your answer is wrong. One foodstuff which is indispensable for human nourishment is missing here.

Read through [35] once again.

---

**43**     You have not been working carefully enough.

Read through   26   once again.

**44**     Your answer is quite correct.

Ethanol is an aliphatic compound whereas phenol is an aromatic compound.
Take another good look at the formulas which are given at the end of
section  9 , and then see if you can solve the problem there.

# Program 3

## The Structure of the Atom

[1]     This program will commence with a somewhat more detailed account of the structure of the atom, since without it a good understanding of the properties of elements and chemical compounds is not possible.

We had always to keep in mind that all atomic models are just pictures to explain the experimental results of natural science. With the increasing knowledge in chemistry and physics the atomic model had to be developed accordingly.

More than 2000 years ago Greek philosophers developed the idea that matter is composed of finite "atoms" (atomos, greek = indivisible).

The modern atomic model dates from Dalton who in 1903 suggested that each element consists of different, indivisible small particles, the atoms. This model was sufficient to explain, for example, the laws of stoichiometry. To explain the chemical bond satisfactorily, the Dalton model had to be modified.

The atoms are, in fact, divisible into even smaller particles. Following the model introduced by Niels Bohr, a Danish scientist, every atom consists of a nucleus and an outer electron cloud:

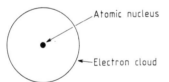

The nucleus consists of protons and neutrons.

As the name suggests, electrons make up the electron cloud.

Complete the following scheme and check your answer at  [11]  .

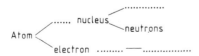

| 2 |   This atom has five electrons. The first shell has two electrons and is fully occupied. The second shell has three electrons.

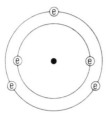

Whereas the K-shell is already full with two electrons, the L-shell can take 8.

The figure shows an atom with a total of eight electrons in the electron shell:

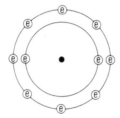

In this atom the first shell (K-shell) has 2 electrons, and the second or L-shell 8 electrons.

It is not possible to have more than 10 electrons in these two inner shells. However, as one moves to the outer shells, more and more electrons can be accepted. Thus, the third shell (M-shell) is fully occupied when it contains 18 electrons.

How many electrons are to be found in the electron cloud of an atom whose K-, L-, and M-shells are full?

● The atom has 10 electrons → | 13 |

● The atom has 18 electrons → | 16 |

● The atom has 28 electrons → | 5 |

| 3 |    The N-Shell can accommodate a maximum of 32 electrons. To repeat:

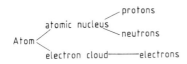

First the inner shells of the electron cloud are filled with electrons. The $n^{th}$-shell can accept $2 \cdot n^2$ electrons.

After this short revision of what we already know of the atom, we can proceed with more details. Protons and electrons possess equal but opposite charges:

> Protons are positively (+) charged
> Electrons are negatively (–) charged

Neutrons are electrically neutral. Thus, all atoms have an overall electric charge of zero, i.e. they are electrically neutral.

It follows, therefore, that the number of protons in any atom is the same as the number of electrons. The positive charges of the protons cancel the negative charges of the electrons.

The charge of a proton or electron is the smallest charge possible. This is why it is called an *elementary charge.*

A particular nucleus contains 4 neutrons and 4 protons. How many elementary charges does it have?

Can you predict the atomic number of an atom which has 4 protons and 5 neutrons in the nucleus?

● Its atomic number is 4 → | 19 |

● Its atomic number is 5 → | 6 |

● Its atomic number is 9 → | 14 |

| 4 |    An element consisting of atoms with 6 protons, 6 neutrons and 6 electrons has a relative atomic mass of 12.

The electrons are not taken into consideration since the mass of an electron is only 1/1850 that of a proton.

**3**

When both the relative atomic mass and the number of protons of an element
are known, it is then possible to calculate the number of neutrons. You only
need to subtract the number of protons from the atomic mass.

We will now take a look at the atomic structures of the different elements.

We will use the following symbols in our discussion:
Protons = p
Neutrons = n
Electrons = e

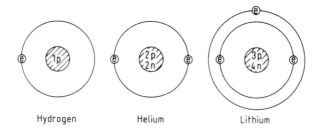

Hydrogen          Helium          Lithium

It should be noted here that the atoms are not drawn to scale. The nuclei are
shown far too large in relation to the electron shells.

Draw the atomic structure of the carbon atom. Carbon has 6 protons and 6
neutrons in its nucleus.

Check your answer at  17

---

**5**     Correct; an atom whose K-, L- and M-shells are filled possesses 28
electrons.

It is not necessary to memorize the number of electrons any one shell can accept
since there is a rule which allows one to calculate it. Here it is!

The 1$^{st}$ shell (K-shell) can accept a maximum of 2 electrons: $2 = 2 \cdot 1 \cdot 1$

The 2$^{nd}$ shell (L-shell) can accept a maximum of 8 electrons: $8 = 2 \cdot 2 \cdot 2$

The 3$^{rd}$ shell (M-shell) can accept a maximum of 18 electrons: $18 = 2 \cdot 3 \cdot 3$

The n$^{th}$ shell can accept a maximum of x electrons:   $x = 2 \cdot n \cdot n$

Thus, it is only necessary to remember the formula $2 \cdot n^2$.

Now calculate how many electrons the N-shell can accept.

Write down the figure and compare at $\boxed{3}$

---

$\boxed{6}$     You did not pay attention to the fact that the atomic number is concerned *only with the number of protons*. Read $\boxed{3}$

---

$\boxed{7}$     The atomic number tells us *nothing* about the number of neutrons in the nucleus. The atomic number is equal to the number of protons in the nucleus and to the number of electrons as well. All the atoms are electrically neutral. Due to the fact, that any atom has the same number of protons and electrons, they have neither a positive nor a negative charge. The atomic number has nothing to do with the number of neutrons in the nucleus.

Continue reading at $\boxed{15}$

---

$\boxed{8}$     The following words are missing.

    in 1: *neutrons*
    in 2 and 3: *electrons*
    in 4: *protons*

If you made any mistakes, read $\boxed{17}$ again.

The hydrogen atom has *one* proton in its nucleus and *one* electron in its electron cloud. It has a relative atomic mass of 1.

There are other hydrogen atoms, those of heavy hydrogen or deuterium.

A deuterium atom has *one* proton and *one* neutron in the nucleus, as well as *one* electron in the electron cloud.

Deuterium has a relative atomic mass of 2.

Deuterium has the chemical symbol D.

An even heavier hydrogen isotope exists, namely tritium (chemical symbol T). A tritium nucleus contains 1 proton and 2 neutrons. There is one electron in its electron cloud.

**3**

The relative atomic mass of tritium is 3.

Below are shown the structures of the three different kinds of hydrogen.

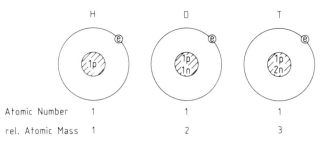

| | H | D | T |
|---|---|---|---|
| Atomic Number | 1 | 1 | 1 |
| rel. Atomic Mass | 1 | 2 | 3 |

Chemically they cannot be distinguished from each other because they all have the same electron cloud.

> Atoms with the same atomic number but different relative masses are known as *isotopes*

Can isotopes be separated by chemical methods?

- Yes      →    9
- No      →    18
- I don't know   →    12

---

**9**     The different isotopes of an element differ only in the number of neutrons in the nuclei. The electron cloud, responsible for the chemical properties, is the same for all the isotopes of a given element. Because of this, isotopes can be separated by .............. methods only. → 8

---

**10**     The 4th, 5th, and 6th shells are also known as the N-, O- and P-shells.

The following figure shows a simplified picture of the nucleus and electron shells of an atom.

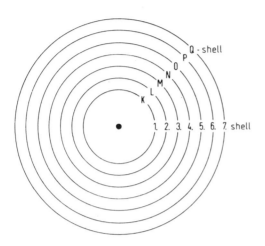

Correct any mistakes you made.

Protons, neutrons and electrons are not randomly distributed within the atom. The arrangement of these elementary particles is subject to certain fixed rules. The following rules control the behavior of the electrons:

Let us look at an example, an atom which has two electrons moving around its nucleus:

The electrons are represented by e. Both are occupying the innermost shell, the K-shell.

This shell is already fully occupied when it contains 2 electrons. If the atom possesses a third electron, this has to occupy a place in the next shell, the L-shell. Show an atom which has 5 electrons.

Check your result at ☐2

|11|

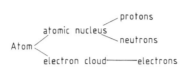

Protons, neutrons, and electrons are the building blocks which go to make up the atoms of every element.

The number and arrangement of these building blocks or elementary particles varies from one element to another.

Before we go on to discuss individual elements, we need to take a more detailed look at the electron cloud.

Even with our present knowledge it is quite complicated and difficult to visualize the structure of the atoms. For the sake of simplicity, we can assume, however, that the electrons move in well-defined orbits around the nucleus. This simple model is able to explain many observations on the nature of matter. However, a more accurate description of these observations needs the application of more sophisticated theories.

Please fill in the missing words in the following sentence:

...................move in well-defined ..........around the atomic nucleus, which consists of protons and ....................

Check your answer at |20|

|12|      Only the number of *neutrons* in the nuclei distinguishes one isotope from another. The electron cloud, which is responsible for the chemical properties, is the same for all the isotopes of a particular element.

Therefore, it is possible to separate the isotopes of any element from each other by ................... methods only. → |18|

|13|      You considered the electrons in the K- and L-shells, but not those in the M-shell.

Read |2| again carefully.

|14|      You did not remember that *only the protons* contribute to the atomic number. → |3|

|15|      The atomic number tells us *nothing* about the number of neutrons in the nucleus. The atomic number is equal to the number of protons in the nucleus and to the number of electrons as well.

Now that we have looked at the charges of the elementary particles, let us consider their masses. Both a proton and a neutron have a relative mass of 1.

The mass of an electron is only 1/1850 that of a proton. Therefore the mass of the electrons can be neglected because it is so small.

It is possible to calculate the relative mass of any one element. One only needs to know the number of protons and neutrons and to add them together.

What is the relative mass of an atom which has 6 protons, 6 neutrons and 6 electrons?

● 12    →      |4|

● 18    →      |21|

|16|      You only took the electrons of the 3$^{rd}$ or M-shell into consideration. You forgot the electrons in the other shells.

Please read |2| again carefully.

|17|      Compare with the following drawings:

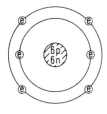

Carbon

It is possible to draw the atomic structure for every element in this way. The

**3**

number of protons and neutrons in the nucleus and the number of electrons in the electron cloud are of particular importance.

The number of electrons in the outermost shell of the electron cloud is responsible for the chemical properties of that particular element.

Carbon has 4 electrons in the outer shell. This is the reason why carbon atoms play such an important part in building up the big variety of organic compounds. The outer shell electrons determine the valence of the particular elements. *Carbon is tetravalent.*

For a short repetition, please fill in the missing words:

1. The relative atomic mass is the sum of the relative mass of the protons and that of the .................

2. The number of ..............in the outermost shell of an atom determines the chemical properties of any one element.

3. The number of protons in the nucleus and the total number of ...................
   in the electron cloud are always equal.

4. The atomic number is equal to the number of ............. in the nucleus.

Check your answers at  8  .

---

**18**    Your answer is quite correct. Isotopes have nearly identical chemical properties. So it is impossible to separate them by chemical methods.

Most of the elements exist in form of two or more isotopes. The atomic weights of the two natural existing carbon atoms are 12 and 13. The natural abundance is 98.9%, for the $^{12}$C-isotope and 1.1% or the $^{13}$C-isotope.

From this it is possible to calculate the relative mass for the naturally occurring isotope mixture:

$$12 \cdot \frac{98.9}{100} + 13 \cdot \frac{1.1}{100} = 12.01$$

For all practical calculations of organic molecular weights we use the figure 12 for the atomic mass of the carbon atom.

This completes the 3rd program.

---

**19**     An atom possessing 4 protons and 5 neutrons has the *atomic number* 4. The number of neutrons does not affect the magnitude of the atomic number.

Is it possible to predict the number of neutrons in the nucleus of an atom, knowing only the atomic number?

● Yes → 7

● No → 15

---

**20**     *Electrons* move around the nucleus, which contains the *protons* and *neutrons*, in well-defined *orbits*.

These orbits may also be described as "shells" because the electrons are free to move in them in all directions with respect to the nucleus. As an approximation one may visualize them as spherical shells. Altogether there are seven such shells.

The shell nearest to the nucleus is designated the first shell. Then comes the 2nd shell. The 7th shell is therefore the one farthest from the nucleus.

The letters of the alphabet are also used to name the shells. The letter K and not A, however, is the first one used. Therefore, the 1st shell is also the K-shell, the 2nd is the L-shell, the 3rd the M-shell, and the last shell, the 7th, is the Q-shell.

Which letters are used to name the 4th, 5th, and 6th shells? → 10

---

**21**     Your answer is wrong.

The mass of an electron is only 1/1850 that of a proton. Therefore the mass of the electrons can be neglected because it is so small.

Both a proton and a neutron have a relative mass of 1.

It is possible to calculate the relative mass of any one element. One only needs to know the number of protons and neutrons and to add them together.

What is the relative mass of an atom which has 6 protons, 6 neutrons and 6 electrons?

Note the figure and check at 4 .

# Program 4

## The Chemical Bond (I)

---

**1**     The structure of the atoms was described in program 3.

We have read there that the number of electrons in the different shells is limited. The first shell is fully occupied with 2 electrons, the second shell with 8 electrons.

The corresponding elements are

   *Helium* (He) and *Neon* (Ne)

Both of these elements are noble gases. They exist free in nature, and normally do not form chemical compounds.

Other elements with 8 electrons in the outer electron shell show the same behavior.

What do you conclude from this?

Atoms possessing 8 electrons in the outer electron shell are particularly reactive/unreactive (delete whichever does not apply). → 23

---

**2**     You gave the electron pairs which are *not* common to the two atoms. → 11

---

**3**     If you forgot the details repeat 15   .

---

**4**     The following designations are correct:

$$·\ddot{O}·   \quad \text{or} \quad   ·\ddot{O}:$$

Oxygen has six electrons in the outermost shell. Four of these form two electron pairs. The remaining two electrons are single or un-paired.

A further example:

$$:\ddot{F}·   \quad \text{and} \quad   :\ddot{Ne}:$$

**4**

Fluorine has three electron pairs and one unpaired electron, neon has four electron pairs in the valence shell.

The constellation of eight electrons in the form of four pairs is referred to as ...................................

Fill in the missing words and go on at $\boxed{8}$ .

---

$\boxed{5}$

$$H \overset{\displaystyle \overset{..}{\underset{|}{N}}}{\phantom{|}} {}_{H} \, H$$

The $CH_4$ molecule forms a tetrahedron with a hydrogen atom at each corner. The carbon atom is in the center of the tetrahedron.

If you put the tetrahedron on the table in front of you, it will be standing on its base with three hydrogen atoms in the plane of the table, the forth projecting vertically above it.

$$\overset{\displaystyle \overset{H}{\underset{|}{C}}}{\phantom{|}} H \underset{H}{\phantom{x}} H$$

Complete the following sentence:

The water molecule is angular, ammonia forms a ................., and methane $(CH_4)$ a ........................... $\rightarrow$ $\boxed{25}$

---

$\boxed{6}$     The chloride ion is *negatively* charged.

When sodium and chlorine react with each other, the chlorine atom accepts one electron which is given up by the sodium.

$$Na^{\cdot} + {\cdot}\overset{..}{\underset{..}{Cl}}: \longrightarrow Na^{\oplus}\overset{..}{\underset{..}{Cl}}:^{\ominus}, \quad \text{sodium chloride}$$

**4**

Both the ions have obtained noble gas configuration with 8 electrons in the outer electron shell.

Ions are charged particles which, depending on the charge, repel or attract each other. The *ionic bond* is due to the electrostatic attraction between oppositely charged species.

This law is known from physics.

The positive sodium ion, $Na^\oplus$, attracts the negative chloride ion, $Cl^\ominus$, (and, of course, each chloride ion attracts sodium ions).

The *ionic bond* arises through this mutual attraction.

All positively charged ions are called *cations,* the negatively charged ones *anions*. These names come from electrolysis, in which the positive ions go to the *cathode* (= the negative electrode), the negative ions to the *anode* (= the positive electrode).

Which anions and cations did you get to know in this program? Note them down; there have been two so far. → $\boxed{10}$ .

---

$\boxed{7}$        You are mistaken. Remember that each chlorine atom only needs one more electron in order to obtain noble gas configuration.

Read $\boxed{35}$ once more.

---

$\boxed{8}$        The correct answer is *noble gas configuration*. All atoms which do not have this configuration attempt to obtain it by forming compounds.

Elements can be joined in different ways in chemical compounds. The most important bonding types are:

    a) the ionic bond
    b) the covalent bond.

Write these names down. We will describe them one after the other.

To understand the ionic bond let's have a look on the construction of the sodium atom:

**4**

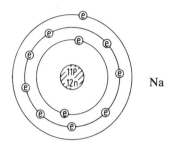

Na

The simplest way in which sodium can attain noble gas configuration is to release one electron:

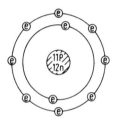

How is the sodium atom charged after the release of an electron?

● Positively (+) → 27

● Negatively (–) → 31

● No charge    → 19

● I don't know  → 17

---

**9**     Although there are sodium ions in the diagram which are surrounded by only four chloride ions, these sodium ions do not have chloride ion neighbors on all sides. The diagram shown is only a small part of the ion lattice. Consider a sodium ion in the center, and you will find the correct answer. → 28

---

**10**    Sodium cation ($Na^{\oplus}$) and chloride anion ($Cl^{\ominus}$).

Sodium and chloride ions combine together to form a sodium chloride salt crystal.

The arrangement of ions in a crystal is called a *lattice*. In order to understand the steric arrangement of ions in the lattice, let us first consider the following plane in which the sodium ions are symbolized by full circles (●), the chloride ions by open ones (○):

**4**

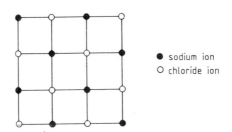

● sodium ion
○ chloride ion

What is the number of chloride ions having the same distance to one single sodium ion in this crystal plane?

●    Two      → 37

●    Three →    41

●    Four     → 28

---

| 11 | :N:::N: |

We have now studied the covalent bonds in the molecules $Cl_2$, $O_2$, and $N_2$.

How many shared electron pairs are there in $O_2$, in $N_2$, and in $Cl_2$?

| | $O_2$ | $N_2$ | $Cl_2$ | | |
|---|---|---|---|---|---|
| ● | 2 | 1 | 3 | → | 2 |
| ● | 4 | 2 | 6 | → | 24 |
| ● | 2 | 3 | 1 | → | 21 |
| ● | 2 | 2 | 1 | → | 34 |

---

| 12 | Correct. There are no single NaCl molecules in a sodium chloride crystal. Each sodium ion is surrounded by six chloride ions, and each chloride by six sodium ions.

**4**

Let us now look at some further examples of the ionic bond.

·Mg·

Magnesium has two electrons in the third shell (the M-shell).

In order to attain noble gas configuration, magnesium must release both of these electrons, thereby becoming a doubly positive magnesium ion.

The reaction between magnesium and chlorine is an example. The two electrons which are given up by magnesium are accepted by the two chlorine atoms.

Write down the reaction between magnesium and chlorine using the simple notation. → 29

---

**13**     Helium has two electrons.

Let us now consider covalent compounds formed from two different elements.

Insert diagram D1 from old page 40

$H_2 + Cl_2 \longrightarrow 2\ HCl$, or

H–H + :C̈l–C̈l: ⟶ H–C̈l: + H–C̈l:

The water molecule can be designated

:Ö:     or     ·Ö:     or     Ö
H  H        H   H      H   H

Similarly, ammonia:

H:N̈:H    or    H–N̈–H    or    H–N̄–H
  H               H           H

Thus, also in compounds consisting of two different elements, each atom acquires noble gas configuration.

Write the formula for $CH_4$ in the same manner. → 20

---

**14**     The characteristic feature of the covalent bond is the *shared electron pair*.

Now a few words about the notations used: Instead of two points (dots) one frequently uses a line (bond) to designate the electron pair common to two atoms, e.g.:

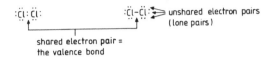

Also the unshared electron pairs (lone pairs) can be designated by a horizontal or vertical bar. This is used particularly frequently in German literature:

$$\mathrm{I\bar{C}l\text{-}\bar{C}lI}$$

For many purposes only the bonding electron pair(s) is (are) of interest. Thus, the chlorine molecule is written:

Cl–Cl.

Now write the oxygen atom, depicting the valence electrons by dots → 16 .

---

15    Each sodium atom in the sodium chloride lattice is surrounded by six equidistant chloride ions.

The crystal lattice of sodium chloride is shown again in the figure below:

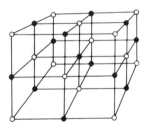

From this it is seen that sodium chloride does not exist in the form of single molecules of NaCl in the solid state, but rather as an extended crystal lattice in which each sodium ion is surrounded by six chloride ions. Likewise, each chloride ion is surrounded by six sodium ions.

To summarize, complete the following sentences:

**4**

The elements sodium and chlorine react to form the compound ...................
To do so, the sodium atom releases an electron, which is accepted by the
chlorine atom. A positively charged ...............ion and a negatively
charged ..............are formed. Each ion has..............electrons in
the valence shell. → 33

---

16

·Ö·          Oxygen has six outer electrons.

How can oxygen achieve noble gas configuration?

Oxygen occurs in nature as the molecule $O_2$.

When two oxygen atoms approach each other so as to form a molecule, the two
outer electron shells will overlap to some extent. Thus, the two "free" electrons
of each atom will also belong to the other atom. In this way each atom attains
noble gas configuration, and the oxygen molecule is formed:

·Ö· + ·Ö· → :Ö::Ö:

Now write the oxygen molecule using lines in place of points for two-electron
bonds and electron pairs.     →     22

---

17          A sodium atom is overall electrically neutral. In the nucleus it
contains twelve neutrons and eleven protons, the latter being positively charged.
There are eleven electrons in the electron cloud, which carry the negative
charges.

If this sodium atom releases an electron, thereby obtaining noble gas
configuration, the remainder must be electrically charged. → 8

---

18          The chlorine atom is electrically neutral. If it accepts one electron, thus
attaining noble gas electron configuration, it also acquires a charge.

After the addition of one electron, the chlorine atom becomes
positively/negatively charged.

Delete as appropriate and check your answer at 6 .

---

19          Protons are positive; electrons are negative.

When the outwardly neutral sodium atom releases an electron, it must become charged → $\boxed{8}$ .

---

$\boxed{20}$

$$\overset{\displaystyle H}{\underset{\displaystyle \ddot{H}}{H\!:\!\ddot{C}\!:\!H}} \qquad \overset{\displaystyle H}{\underset{\displaystyle H}{H\!-\!\underset{|}{\overset{|}{C}}\!-\!H}}$$

This way of writing formulas for compounds like HCl, $H_2O$, $NH_3$, and $CH_4$ does not always correspond to the real arrangement of the atoms. Therefore let us look at it more closely:

HCl      is a linear ("stretched") molecule: H–Cl

$H_2O$      is bent:  (the three atoms,

H, O, and H, form the corners of a triangle).

$NH_3$      has the shape of a pyramid in which the base is an equilateral triangle. The hydrogen atoms are at the corners of the base, and the nitrogen atom at the top of the pyramid:

base of pyramid

Draw the ammonia molecule in this sterical manner. → $\boxed{5}$

---

$\boxed{21}$

In $O_2$      there are two shared electron pairs between the two atoms of the molecule.

In $N_2$      there are three shared electron pairs between the two nitrogen atoms.

In $Cl_2$      there is one shared electron pair between the two atoms of the molecule.

Note:      Each shared electron pair corresponds to a chemical bond.

The bonds are named as follows:

      one shared electron pair: single bond
      two shared electron pairs: double bond

**4**

three shared electron pairs: triple bond

A hydrogen atom has one single electron. In the hydrogen *molecule* the *two* electrons belong to both nuclei at the same time:

= H : H = H–H or $H_2$

The hydrogen molecule thus has noble gas electron configuration, since the K-shell is fully occupied with two electrons.

We have already discussed a noble gas possessing two electrons. Do you remember it?   |13|

---

|22|

Did you make any mistakes?

The horizontal lines (bonds or valence bonds) in this molecule correspond to the two shared electron pairs.

Nitrogen has five electrons in the valence shell (one electron pair and three single (unpaired) electrons). Thus, the nitrogen molecule looks like this:

    |N≡N|

Now draw the $N_2$ molecule depicting all the electrons by dots.→ |11|

---

|23|    Atoms with eight electrons in the outer shell are particularly *unreactive.*

Thus, eight electrons in the outer shell result in a particularly stable state.

All noble gases with the exception of helium have eight electrons in the outer shell. It is for this reason that such an electron configuration is described as a *noble gas configuration.*

To understand the chemical bond it is very important to know that

> All atoms which do not have noble gas
> configuration attempt to obtain it.

**4**

For this reason they form compounds and therefore many elements are not found free in nature.

The following notation is used to indicate the number of electrons in the outer shell (the valence electrons). Each such electron is indicated by a dot beside the chemical symbol of the element.

Examples:     Li·    ·Be·    ·Ḃ·    ·Ċ·

When the number of such electrons exceeds four, two electrons will get together in the form of electron pairs. This is the case with nitrogen:

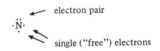

electron pair

·Ṇ·

single ("free") electrons

Write down the oxygen atom with its outer electrons symbolized by dots. → ☐4

---

**24**     Contrary to the question, you gave the number of *non-bonding* electron pairs, i.e. those which are *not* common to the two atoms → ☐11 .

---

**25**     The water molecule is bent; ammonia forms a *pyramid*, and $CH_4$ (methane) has the shape of a regular *tetrahedron* with C in the center.

Compounds which are bound together by covalent bonds exist in the form of single molecules. Many of them are quite volatile. Substances like $Cl_2$, $N_2$, $O_2$, $NH_3$, $CH_4$ and many of the organic compounds with lower molecular weights are gases at ordinary temperature and pressure. Others are liquids, which often show high volatility.

In the solid state the single molecules of organic compounds are held together in crystal lattices. In contrast to ionic inorganic substances they have a much lower melting point and most of them are distillable without decomposition.

Therefore the technique of distillation is very often used for purification and separation of organic compounds.

Non-volatile organic products are substances with higher molecular weights and of course polymeric products.

Write down the missing words:

**4**

At ordinary temperature and pressure many organic compounds
are ............. Even solid products are rather ...............They can be
purified by............... This doesn't apply for products with
high molecular masses. → |38|

---

|26|    Each chlorine atom has seven electrons in the outer shell (three
electron pairs and one free electron).

It would be false to say that only one electron is common to the two atoms in
the chlorine molecule. Where is the other "free" electron?

Study |35| again.

---

|27|    When a sodium atom has given up an electron, it carries a positive
charge and has then become an *ion* (pronounced like iron, dropping the r):

$$Na\cdot \xrightarrow[\text{one electron}]{\text{loss of}} Na^{\oplus}$$

$Na^{\oplus}$ is the sodium ion.

Chlorine has seven outer electrons (valence electrons). By accepting one more
it obtains noble gas configuration:

$$:\overset{..}{\underset{..}{Cl}}\cdot \xrightarrow[\text{of one electron}]{\text{acceptance}} :\overset{..}{\underset{..}{Cl}}:^{\ominus}$$

A chloride ion is formed.

How is the chloride ion charged?

- Negatively (−)     →   |6|

- Positively (+)     →   |18|

- Neutral           →   |36|

---

|28|    Correct, the sodium ion in the plane has four equidistant chloride ion
neighbors. In the crystal there are several such planes disposed above each other
and arranged in such a way that there is a chloride ion above each sodium ion,
and a sodium ion above each chloride ion:

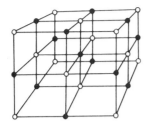

This is called a *lattice*. It shows the steric arrangement of ions in a crystal. The above figure shows the crystal lattice of sodium chloride.

What is the number of chloride ions having the same distance to one and the same sodium ion in this crystal?

● Four     →          9

● Six      →          15

● Eight    →          39

---

29

$$:\overset{..}{\underset{..}{Cl}} \cdot + \cdot Mg \cdot + \cdot \overset{..}{\underset{..}{Cl}}: \longrightarrow :\overset{..}{\underset{..}{Cl}}:^{\ominus} \ Mg^{\oplus\oplus} \ :\overset{..}{\underset{..}{Cl}}:^{\ominus}$$

In a simplified notation one can omit the outer electrons, only specifying the charges:

$$\longrightarrow \ Cl^{\ominus} \ Mg^{\oplus\oplus} \ Cl^{\ominus}$$

or, simpler still, just write the global formula:

$$\longrightarrow \ MgCl_2$$

Instead of $Mg^{\oplus\oplus}$ one can also write $Mg^{2\oplus}$.

It is easy to see the valence of the ions from such equations.

For positively charged ions the valence is equal to the number of electrons released; for negatively charged ions the number of electrons accepted.

For positively charged ions the valence is equal to the number of electrons released; for negatively charged ions the number of electrons accepted.

**4**

the sodium ion has the valence 1+
the magnesium ion has the valence 2+
the chloride ion has the valence 1–

Having discussed the ionic bond, we now come to the *covalent bond*.

The covalent bond is completely different from the ionic bond. However, the driving force is again the achievement of ..............gas configuration for the atoms involved. → 35

---

30      Correct, two electrons are shared by the two atoms. In this way, each chlorine atom acquires eight electrons in the outer (valence) shell:

six from its three electron pairs, and another two from the common electron pair.

This can also be shown as follows:

Two electrons belong to both chlorine atoms simultaneously. The two electrons thus form a *shared electron pair,* or *bonding electron pair.*

Complete the following sentence:

The covalent bond is characterized by a ..............electron
pair. → 14

---

31      A sodium atom is overall electrically neutral. If it releases an electron in order to attain noble gas configuration, it will become charged.

The sodium atom will become positively/negatively charged.

Which is correct? → 27

---

32      Wrong. Remember that each chlorine atom only needs one more electron in order to get noble gas configuration.

Repeat 35 .

**4**

---

**33**

The missing words were:
sodium chloride (NaCl)
sodium ion
chloride ion
eight

The sodium and chloride ions attract each other and form an ion lattice, the sodium chloride lattice.

Do molecules of NaCl exist as such in this lattice?

•     Yes         →   40

•     No          →   12

•     I don't know   →   3

---

**34**     Your answer is correct for oxygen and chlorine but not for nitrogen. Take a closer look at the formula. 11

---

**35**

*Noble gas configuration.*

Here is the chlorine atom again:

:Cl·

Chlorine does not occur as free atoms but as molecules, $Cl_2$, in which two chlorine atoms are bound together. How does this take place?

The chlorine atom has seven electrons in the outer shell; three electron pairs and a single electron. When two chlorine atoms approach each other so that the electron shells overlap, the two single electrons originating from the two atoms will be *shared* in the molecule. Thus, each atom in the molecule has noble gas configuration. This can be visualized as follows:

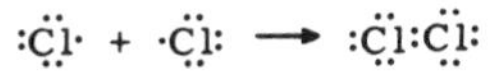

:Cl· + ·Cl: ⟶ :Cl:Cl:

In the chlorine molecule each atom has eight electrons in the outer electron shell. How many electrons are *shared* by the two atoms?

- One electron → 26
- Two electrons → 30
- Seven electrons → 7
- Eight electrons → 32

---

**36**
Protons are positively charged, electrons negatively charged. When the electrically neutral chlorine atom accepts a further electron, it becomes an electrically charged chloride ion. → 27

---

**37**  It is true that some of the sodium ions in the picture have only two neighboring chloride ions. However, you should realize that this picture is only a small fragment of the crystal lattice.

In order to answer the question correctly you must consider a sodium ion in the interior of the lattice. → 10

---

**38**  *Gases, volatile, distillation*

To complete the picture about the chemical bond, we should mention that besides the ionic bond and the covalent bond there are two other possibilities to hold chemical compounds together: the complex bond and the metallic bond.

The achievement of noble gas configuration is the driving force in the more complicated, complex substances too.

The metallic bond has much in common with the ionic bond and allows an understanding of the physical properties of metals.

The atomic model of Niels Bohr which we used in this program was sufficient for an explanation of the different forms of the chemical bond dealt with.

However, we need a more sophisticated atomic model to explain some phenomena in the organic chemistry, for example the double bond or the character of aromatic organic compounds.

You will find out more about these interesting topics in the next program (number 5).

This is the end of program 4.

**4**

|39|    Count again !

Look at the sodium ion in the center of the diagram and consider only those chloride ions which are the same distance removed. → |28|

---

|40|    Wrong. Read |15| again.

---

|41|    Admittedly, there are some sodium ions on the surface which have only three neighboring chloride ions. However you should realize that our picture is only a small fragment of the crystal surface.

In order to answer correctly you should consider a sodium ion in the interior of the surface |10|

# Program 5

## The Chemical Bond (II)

---

**1**    You already know about different atomic models. We
had always to keep in mind that all models of the atoms are
just pictures to explain the experimental results of natural
science. Dalton's indivisible atoms suffice, for example, to explain
the laws of stoichiometry.

To explain the chemical bond, this model had to be modified.
In Program 3 the rigid shell model was introduced.
The atoms are, in fact, divisible into even smaller particles.

What are the three particles of which an atom is composed?    →    7

---

**2**    Your answer is wrong.

The Heisenberg uncertainty principle stipulates that the exact and simultaneous
determination of the position and momentum of an electron is impossible.

According to the electron shell model described earlier, the electrons move in
rigid shells with sharrply defined radii. These would exactly determine the
distance of the electrons from the nucleus. This model would also allow the
precise calculation of the momentum of the electron.

Therefore, the electron shell model does not satisfy the uncertainty
principle.    →    8

---

**3**    Your answer is wrong.

The K-shell can have a *maximum* of only *two* electrons.    →    20

---

**5**

| 4 | No, the probability of finding an electron is not the same everywhere. The figure below illustrates the

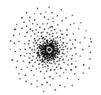

probability of an electron residing within a given region of space. The higher the density of the points, the higher the probability.

Seen from the area of highest density, how does the probability vary?

- Decreases slowly toward the outside,
  rapidly toward the center        →    ☐ 32

- Decreases slowly toward the center,
  rapidly toward the outside        →    ☐ 14

- The electron moves only on the surface
  of a sphere                  →    ☐ 50

---

| 5 | No, an atom can form four bonds only when four singly occupied orbitals are available.    →    ☐ 26

---

| 6 | The ground state of the carbon atom is described by the following electron configuration:

$$2s^2, 2p_x^1, 2p_y^1$$

If, as you suggest, a 2s electron were promoted into the $p_x$ orbital, this orbital would become doubly occupied. However, four *singly* occupied orbitals have to be created.

Repeat    ☐ 26    .

**5**

| 7 |

Atoms are composed of three elementary particles,

protons,

neutrons, and

electrons.

The atomic nucleus contains protons and neutrons only.

According to the rigid shell model, electrons move on
sharply defined spherical surfaces in the space surrounding
the nucleus. This model was sufficient to understand
some chemical properties.

What determines the chemical properties of an element?

- The number of protons                    →   | 15 |

- The number of electrons                   →   | 22 |

- The number of outer electrons
  (i.e. in the outer electron "shell")      →   | 29 |

- I don't know                              →   | 36 |

| 8 |

The simultaneous and exact determination of the position and momentum of an electron is not possible. Therefore, the "shell" model cannot be
correct.

This impossibility of precisely locating the electron and determining its momentum
is not just a question of inadequate measuring equipment. The experimental
setup itself will influence the momentum of the electron in an unpredictable
manner. The more accurate the method the greater the influence. Hence, an
exact measurement becomes impossible.

The impossibility described above is a *fundamental natural law*. It applies to all
sub-atomic particles, not just electrons.

Give the name normally used for this principle together with the name of the
discoverer.   →   | 16 |

| 9 |

No, the expression $2(sp^3)^4$ has the following meaning:

2: the number off the shell to which the orbitals belong (here the L-shell).

3: the number of p orbitals used to form the four $sp^3$ orbitals.

4: the number of electrons occupying the four hybrid orbitals.

How many $sp^3$ orbitals are there?    →    37

---

| 10 |

You confused the ground state of carbon with its excited state. Continue at    27    .

---

| 11 |

Your answer is wrong.

An $sp^2$ hybridized carbon atom possesses three $sp^2$ orbitals and one p orbital. A σ bond is formed by the overlapping of two $sp^2$ orbitals belonging to two different carbon atoms. How many unused orbitals remain on the two carbon atoms? Two p orbitals and ............... $sp^2$ hybrid orbitals.    →    46

---

| 12 |

The K-shell can accept a total of two electrons.    →    20

---

| 13 |

Your answer is not entirely correct. It does not suffice that there are four outer shell electrons. In order to form four bonds, the electrons must be associated with four singly occupied orbitals.    →    26

---

| 14 |

Your answer shows that you do not understand the matter. Repeat from    16    .

---

| 15 |

No, the number of electrons in the outer shell is responsible for the chemical properties of an element. The number of protons in the nucleus is only of indirect consequence inasmuch as this number determiens the filling of the shells and hence the number of outer shell electrons.

Continue at    29    .

**16**    *The Heisenberg uncertainty principle.*

**5**

In keeping with the uncertainty principle, the *orbital* model does not confine the electron to moving on a rigid sphere, but instead defines the probability of finding the electron within a given region of space. This probability can be calculated mathematically. We shall limit ourselves to a simplified pictorial description, however.

Figure a shows the orbital of the electron associated with a hydrogen atom. The orbital defines the space (in this case spherical) within which there is a high probability of finding the electron (e.g. 90%). That is, in 90 out of 100 hydrogen atoms, the electron would actually be residing within this sphere.

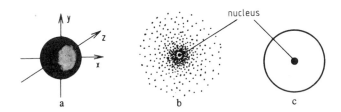

Figure b shows a cross-section of the probability sphere. The densitiy of the points corresponds to hypothetical snapshots of the electron. The probability of finding the electron is not the same everywhere within the sphere. It is highest in a region close to the nucleus and slowly decreases toward the exterior of the sphere. The probability is zero in the center, the position of the nucleus.

Figure c is a very rough representation of an orbital, defining only the boundary of the probability sphere within which the electron is likely to be found (e.g. with 90% probability).

How does the probability of finding the electron vary within the sphere?

● Decreases slowly toward the outside,
rapidly toward the center       →    $\boxed{32}$

● Decreases slowly toward the center,
rapidly toward the outside       →    $\boxed{23}$

● The probability is the same everywhere    →    $\boxed{4}$

● The electron moves on the surface of
a sphere       →    $\boxed{41}$

**5**

---

|17|   Correct.

The new state of the carbon atom formed is called an *excited state*. Give the electron configuration in the excited state!

Ground state of                          Excited state of
the carbon atom                          the carbon atom

$2s^2$                                   $2s^1$
$2p_x^1$                                 ......
$2p_y^1$                                 ......  →   |27|

---

|18|     No, the expression $2(sp^3)^4$ has the following meaning:

2:   the number of the shell to which the orbitals belong (here the L-shell).

3:   the number of p orbitals used to form the four $sp^3$ orbitals.

4:   the number of electrons occupying the four hybrid orbitals.

How many $sp^3$ orbitals are there?     →     |37|

---

|19|     In order to answer the question, you must know that only singly occupied orbitals are used in the formation of atomic bonds.     →     |71|

---

|20|   Correct.

The K-shell can accept a total of two electrons. They are s electrons with opposite spin.

Electron configurations are described by short-hand notations, e.g. for the hydrogen atom:

       $1s^1$

The first "1" means that the electron is in the first shell (the K-shell); the "s", that it is in an s orbital. The exponent "1" indicates that there is one electron in the s orbital.

What is the meaning of the following configuration, and which element does it describe?

$1s^2$                                         →   28

**5**

---

21    You confused the ground state with the hybrid state!   →   27

---

22    Your answer is incomplete: the number of *outer shell electrons* determines the chemical properties.   →   29

---

23    No, the probability of finding the electron varies as shown in the figure:

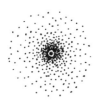

The higher the density of the points, the higher the probability.

Seen from the area of highest density, how does the probability vary?

● Decreases slowly toward the outside,
   rapidly toward the center          →   32

● The probability is the same everywhere     →   50

● The electron moves on the surface of a sphere   →   14

---

24    Your answer is wrong.

When the six carbon atoms in the benzene ring are sp$^2$ hybridized, how many p oribitals and how many p electrons are left?   →   73

|25|

    Remember that the helium atom has a *doubly* occupied orbital. You should also know that helium – being a noble gas – possesses a *filled* shell.

**5**   Repeat   |35| .

|26|     Correct.

In order to form four atomic bonds, the atom must posses *four* singly occupied orbitals. The figure shows the ground state

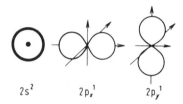

$2s^2$        $2p_x^{\,1}$        $2p_y^{\,1}$

configuration of the carbon atom in which there are only *two* singly occupied orbitals ($2p_x^1$, $2p_y^1$).

How is it possible for carbon to form four bonds? In order to obtain the four singly occupied orbitals, one of the 2s electrons is promoted into the empty 2p orbital.

Which is the empty p orbital?

- $2p_x$   $\rightarrow$   |6|

- $2p_y$   $\rightarrow$   |34|

- $2p_z$   $\rightarrow$   |17|

|27|

| Ground state of the carbon atom | Excited state of the carbon atom | |
|---|---|---|
| $2s^2$ | $2s^1$ | |
| $2p_x^1$ | $2p_x^1$ | $2p_y^1$ |
| $2p_y^1$ | $2p_z^1$ | |

The excited state configuration will allow the formation of four atomic bonds. The model is not yet complete, however. We know that the four bonds in methane, for example, are *identical*. The above configuration would lead to two different types of bonds since two different types of orbitals are present, s and p.

In order to solve this dilemma, the four orbitals are *hybridized,* that is, the one 2s and the three 2p orbitals are mixed in such a way that *four new identical* orbitals arise. These are called *sp³ hybrid orbitals.*

Pictorially, the hybridization corresponds to the following transformation:

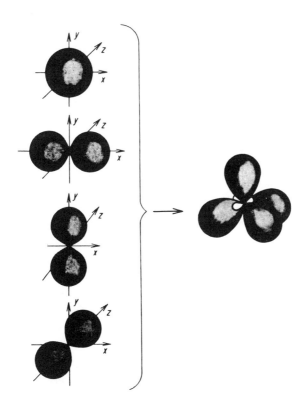

The short-hand notation is:

Excited state configuration of the carbon atom:

$$2s^1$$
$$2p_x^1$$
$$2p_y^1$$
$$2p_z^1$$

$$\xrightarrow{\text{hybridization}}\quad 2(sp^3)^4$$

How many orbitals are associated with an $sp^3$-hybridized carbon atom?

**5**

- One → ⟨9⟩
- Two → ⟨18⟩
- Three → ⟨79⟩
- Four → ⟨37⟩

---

⟨28⟩ $1s^2$.

This notation means that the $s$ orbital in the *first* shell (K-shell) contains *2* electrons. The notation therefore describes the element *helium*.

The filling of the second shell (the L-shell) starts with the element lithium. Lithium has two electrons in the (filled) K-shell and one further electron in the L-shell. The third electron is also in an s orbital. Hence the configuration:

$1s^2, 2s^1$

The fourth element is *beryllium*. Here, there are two electrons in the L-shell. Give the electron configuration for Be. → ⟨39⟩

---

⟨29⟩ Correct; the number of outer shell electrons determines the chemical properties.

The "shell model" is of limited use. For example, it does not explain the existence of double and triple bonds or aromatic compounds.

*The orbital model* allows a deeper understanding of the chemical bond. In order to understand what an orbital is, we must first consider a law of atomic physics known as the *Heisenberg uncertainty principle*. This principle stipulates that the exact and simultaneous determination of the position and momentum (= mass times velocity) of an electron is impossible. On the other hand, the shell model assumes that the electron is to be found on a well-defined spherical surface at a particular distance from the nucleus. Furthermore, the electron is assumed to have a well-defined mass, and the calculation of its momentum would be possible by the laws of classical physics. It is important to appreciate the difference between the predictions of the shell model and the Heisenberg uncertainty principle.

Do these two principles contradict each other?

- No               →   2

- Yes             →   8

- I need more information    →   38

---

**30**    Two p orbitals are still unused:

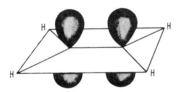

The steric arrangement of the p orbitals is shown below:

The p orbitals are oriented perpendicular to the plane described by the C–C σ bond and the four C–H σ bonds (the σ orbitals are omitted for clarity).

How many electrons are contained in the two p orbitals?

- One      →   81

- Two      →   44

- Four     →   74

---

**31**    Correct.

The two electrons in helium occupy the same orbital but differ in one important respect: they have opposite spins. The electron *spin* is a quantum mechanical number which we can liken to the rotation of a spinning top. There are two possible directions of the rotation, and the spin is, therefore, denoted by arrows: ↑ or ↓. When two electrons have opposite spins, the notation is: ↑↓.

Since the two electrons in helium have opposite spins, the *Pauli principle* is satisfied.

**5**

Formulate the *Pauli principle!*     →     43

---

**32**     Correct.

The probability of finding an electron decreases slowly toward the exterior, rapidly toward the center of the sphere. The slow outward decrease is clearly seen in Figure b. The inward decrease is more difficult to show pictorially as it occurs at very small nuclear distances. The probability becomes zero in the nucleus itself, i.e. in the center of the sphere.

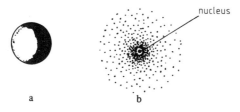

a                         b

The spherical probability distribution of an electron is called an *s orbital.*

Since the orbital does not describe any exact position in space but only a probability, it is in agreement with the Heisenberg uncertainty principle.

Formulate this principle!     →     40

---

**33**     An $sp^2$ hybridized carbon atom possesses three $sp^2$ hybrid orbitals and one p orbital. The formation of a single bond between two carbon atoms is the result of an overlap of two such $sp^2$ orbitals. Two p orbitals and .................... $sp^2$ hybrid orbitals remain unused in this operation. Give the number!
     →     46

---

**34**     No; the ground state of carbon is decribed by the electron configuration

$$2s^2, 2p_x^1, 2p_y^1.$$

If you promote a 2s electron into the $2p_y$ level, this level will become doubly occupied. Therefore, the desired creation of four singly occupied orbitals is not achieved.

Repeat     26     .

---

**5**

**35**   According to the *Pauli principle* two electrons residing in *one* orbital must have different *spins*.

The doubly occupied s orbital in the helium atom corresponds to the K-shell described in Program 1.

What is the maximum number of electrons in the K-shell?

● One            →   25

● Two            →   20

● Eight          →   3

● I don't know   →   12

---

**36**   It was stressed in Program 1 that the number of outer shell elctrons is responsible for the chemical properties of an element.

Continue at   29  .

---

**37**   Correct; the four sp$^3$ orbitals are shown below:

This orbital model agrees with our knowledge of the stereochemistry of carbon compounds. The four orbitals are directed toward the corners of a tetrahedron. When they overlap with the four s orbitals associated with four hydrogen atoms, the methane molecule is formed.

How are mixed orbitals such as the ones in the figure called?   →   70

**5**

**38**     The Heisenberg uncertainty principle stipulates that it is *impossible* at a particular time to determine both the exact location and the momentum of an electron.

The electron shells introduced in Program 1 have exact radii and therefore define an exact distance between the electron and the nucleus. The shell model would also, in principle, allow the exact determination of the momentum of the electron.

It is clear therefore that the shell model contradicts the Heisenberg uncertainty principle.

Continue at    8   .

**39**     $1s^2$, $2s^2$.

The s orbitals cannot contain more than two electrons. This is in agreement with the Pauli principle which allows two electrons to occupy an s orbital provided they have opposite spins. The next element in the periodic table is boron. Now there is no vacancy in the s orbital, so the additional electron must go into a p orbital.

The electron configuration of boron is, therefore:

$1s^2$, $2s^2$, $2p^1$

What is the shape of a p orbital?    →    48

**40**     The *Heisenberg uncertainty principle* stipulates that at any one time it is impossible to determine both the exact location and the momentum of an electron.

According to the Pauli principle, which is another law of atomic physics, two electrons in the same atom cannot be identical in all their characteristics.

The helium atom has two electrons. Can these two electrons occupy the same orbital (i.e. the same probability sphere)?

● Yes, provided that the two electrons differ
  in another respect                              →    31

● No                                              →    49

● I don't understand the problem                  →    58

|41|

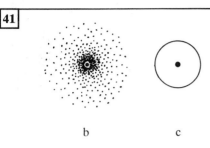

b                    c

No, Figure c is only a simplified version of Figure b and does *not* represent a
"shell" in the sense of Program 1. Since the probability distribution shown in
Figure b is cumbersome to draw – and at any rate incomplete – one uses the
simpler form c.

By careful inspection of Figure b you will see the answer to the question. The
density of the points illustrates the probability of finding the electron. There is
no sharp boundary: Figure b could be extended further into space with a slow
decrease in the density of the points.

How does the probability of finding the electron vary?

● Slowly toward the outside, rapidly
  toward the center                    →    |32|

● Slowly toward the center, rapidly
  toward the outside                   →    |14|

● It does not vary                     →    |50|

---

|42|      Your answer is wrong.

Two of the L-shell electrons are in a doubly occupied s orbital. The other two
are in two singly occupied p orbitals:

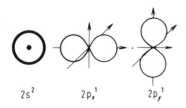

2s$^2$          2p$_x^1$          2p$_y^1$

According to this picture we would expect carbon always to be *divalent*.

What is the prerequisite for the formation of four atomic bonds?

- Four electrons → 5
- Four outer shell electrons → 13
- Four singly occupied orbitals → 26

**5**

---

**43** The *Pauli principle* stipulates that no two electrons within an atom can be entirely identical in all their properties.

Since helium has two electrons in one orbital, they must differ in some respect: they are of opposite spin.

Let us describe the helium atom in the simplified manner shown below

nucleus

where the heavy circle designates a *doubly* occupied orbital.

What is the *shape* of the orbitals containing the electrons in the hydrogen and helium atoms? Check your answer at 52 or, if in doubt, go instead to 16 .

---

**44** There are *two* p electrons.

A second C–C bond is formed by overlapping the two p orbitals as shown in the figure:

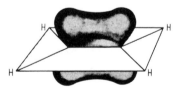

This bond, formed by the parallel orientation of two p orbitals, is called a π bond.

What is a σ bond and what is a π bond?

Check your answer at 67 or, if in doubt, go instead to 69  .

---

**5**

**45**     No, the $2p_x^2$ orbital is doubly occupied and is not, therefore, engaged in bond formation. Singly occupied orbitals are required. Repeat 71 .

---

**46**     *Four.*

The four hybrid orbitals form the four C–H bonds in the ethene molecule. The overlap pattern is indicated below:

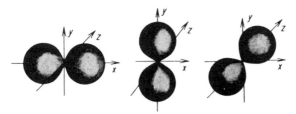

Which carbon orbitals are available for the formation of the C=C double bond? Check your answer at 30 or, if in doubt, go instead to 69 .

---

**47**     The drawing represents the *helium atom which has two electrons* in an *s orbital*.

Being in the same orbital, the electrons must, according to the Pauli principle, differ in some respect.

How do they differ?    →    35

---

**48**     The p orbital has a dumb-bell shape as shown below:

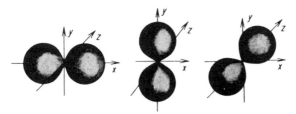

There are three such orbitals at right angles to each other.

From this, how many electrons can the L-shell accept?

- Three     →    63

- Six     →    56

- Eight     →    72

| 49 | Your answer is wrong. |

The Pauli principle stipulates that no two electrons in an atom can have all their characteristics identical. The two eletrons in the helium atom can, however, occupy one and the same orbital provided that they differ in some other respect.

Continue at  31  .

| 50 | Your answer shows that you do not understand the subject. Repeat |
from  16  .

| 51 | Correct. |

What is the prerequisite for the formation of four atomic bonds?

● The availability of four electrons   →   5

● Four electrons in the outer shell   →   13

● Four singly occupied orbitals   →   26

| 52 |   The orbitals associated with the hydrogen and helium atoms are
spherical. They are called *s orbitals*.

Orbitals may also have other shapes. Two possibilities are shown below:

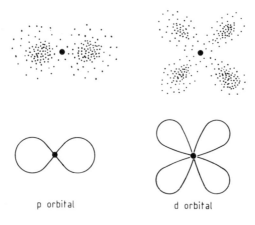

p orbital          d orbital

The upper figures illustrate the probability of finding the electrons within a region of space. The lower figures are simplified repesentations of these orbitals.

The orbital to the left is a *p orbital*, the one to the right a *d orbital*. Examine the upper representation of the p orbital and describe the region(s) of space with the highest probability of finding the electron.   →    60

---

53      Correct. – There are *six* $sp^2$ orbitals in all. Six C–H bonds are formed by overlapping with the six s orbitals of the hydrogen atoms.

In benzene there are six C–C σ bonds and six C–H σ bonds, all in the same plane. The carbon atoms also possess a total of six p orbitals. What is the orientation of these p orbitals with respect to the plane of the σ bonds?   →    73

---

54      Correct. Only the singly occupied orbitals are used in the formation of atomic bonds.

In the case of the oxygen atom the two mutually perpendicular $2p_y^1$ and $2p_z^1$ orbitals are singly occupied. A water molecule is formed by overlapping with two hydrogen $1s^1$ orbitals as shown:

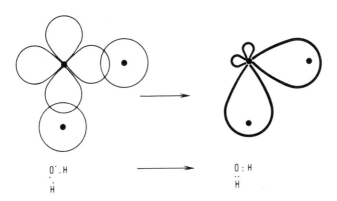

$$O^\cdot.H$$
$$\overset{..}{\underset{H}{}}$$

$$O : H$$
$$\overset{..}{..}$$
$$H$$

The orbitals to the left of this figure are *atomic orbitals* (AO's), those to the right *molecular orbitals* (MO's).

How large is the H–O–H angle in the resulting water molecule?   →    62

**55**        No; your answer describes the excited state configuration of carbon, not the hybridized state.

**5**

Repeat  69 .

**56**        Correct. – The L-shell can accept *six p* electrons plus the *two s* electrons, giving the familiar total of eight. The four orbitals making up the "L-shells" are shown below:

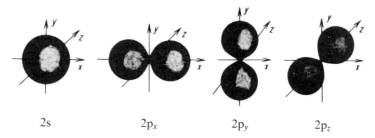

2s              2p$_x$              2p$_y$              2p$_z$

These four orbitals have a common center, the nucleus. The three p orbitals combine to give the following picture:

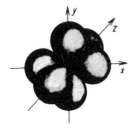

As you see, the so-called "shells" are not shells at all, but rather collections of orbitals.

The electron configurations of the other elements in the periodic table are built up step by step in a similar manner: new orbitals are added for each "shell".

Let us now describe the formation of chemical bonds with the aid of the orbital model.

When two singly occupied orbitals belonging to different atoms *overlap*, a bond consisting of a new orbital common to the two atoms is formed. The two electrons in the new bonding orbital must have different spins. Why?    →    64

**57**
*Hydrogen atom;*
*one electron;*
*s orbital.*

What does the following scheme imply?

Which atom does it represent?

How many electrons are there?

What is the orbital called?

Check your written answer at 47 . (If you do not know the meaning of the heavy circle, see the explanation at 65 .)

---

**58**      The Pauli principle stipulates that no two electrons in an atom can have all their characteristics identical. The two electrons in the helium atom can, however, occupy one and the same orbital provided that they differ in some other respect.

Continue at 31 .

---

**59**      Your answer is wrong. After $sp^2$ hybridization of the six carbon atoms in benzene, how many p orbitals are left?    →    73

---

**60**      The probability space of a p electron consists of two almost spherical lobes separated from each other by the nucleus:

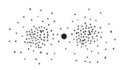

The probability of finding the electron is highest in the *center* of each lobe.

Note that a p orbital is *directional:* it forms bonds only in certain directions.

Draw a p and an s orbital in simplified form.    →    68

---

| 61 |
|---|

Correct, $2s^2$, $2p_x^1$, $2p_y^1$.

Now specify the excited state configuration.    →     69

**5**

| 62 |
|---|

Since the p orbitals are perpendicular, an H–O–H angle of 90 ° is predicted. However, the two occupied molecular orbitals between oxygen and hydrogen repel each other, thereby leading to an expansion of the angle to 107 °:

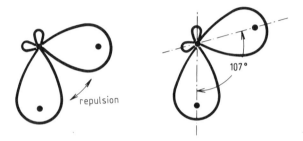

Consider now the carbon atom. The L-shell configuration is:

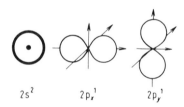

$2s^2$        $2p_x^1$        $2p_y^1$

We know that carbon is tetravalent in its compounds. Is this fact in agreement with the orbital picture?

● Yes        →    42

● No         →    51

| 63 |
|---|

No, there are three p orbitals, but each of these can contain two electrons (of opposite spin). How many p electrons, therefore, can the L-shell contain?    →    56

| 64 |
|---|

According to the Pauli principle two electrons occupying the same orbital cannot be entirely identical. They must, therefore, have opposite spins.

Since no orbital can accept more than two electrons, bonding orbitals are – on average – formed by the union of two *singly occupied* atomic orbitals. This is illustrated in the formation of the hydrogen molecule:

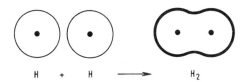

Note that the new *bond* between two atoms corresponds to a *bonding orbital*. This is also called a *molecular orbital* (MO) in contrast to the two *atomic orbitals* (AO's) from which it is derived.

Describe, in words, the formation of the hydrogen molecule: the orbitals containing the valence electrons of the hydrogen atoms ..................　→　$\boxed{71}$

---

$\boxed{65}$　　The solid circles depict *doubly* occupied orbitals.　→　$\boxed{57}$

---

$\boxed{66}$

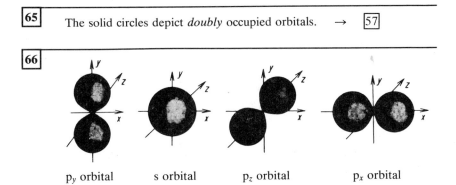

$p_y$ orbital　　　s orbital　　　$p_z$ orbital　　　　$p_x$ orbital

Let us now examine the building up of some atoms in the periodic table using the orbital model. Which atom is represented by the following drawing?

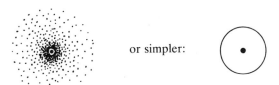

or simpler:

Give the symbol of the element. How many electrons does it possess? What is the orbital called?

Check your answer at　$\boxed{57}$　or, if in doubt, go instead to　$\boxed{16}$　.

|67|

σ bond: overlap of two sp² orbitals.
π bond: overlap of two parallel p orbitals.

The orbital model shows that the two bonds forming the C=C double bond are *non-identical*, a fact which is not seen from the valence bond formula:

Due to a less complete overlap of the p orbitals, the π bond is weaker and more *reactive* than the σ bond. For example, the π bond is involved in addition reactions (Program 8).

The bonding in benzene was explained in an approximate manner in Program 9. The orbital model allows a more satisfactory description. We start with an sp² hybridized carbon atom. Which orbitals are available in the L-shell of such an atom?    →    |82|

|68|

s orbital          p orbital

Remember that these drawings are approximations! An electron cloud has no precise boundary.

The spherical s orbital extends equally in all directions of space. In contrast, the p orbitals are directional, extending along the three principal axes of a coordinate system. For this reason, the three p orbitals are labeled $p_x$, $p_y$, and $p_z$.

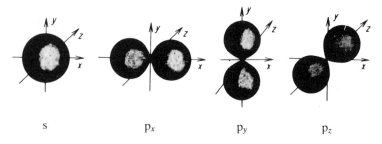

s         $p_x$         $p_y$         $p_z$

All four orbitals have, of course, a common center, the nucleus.

Describe in words the shapes of the orbitals!

s:

$p_x$:

$p_y$:

$p_z$:                                                    →   76

---

**69**     The L-shell of the carbon atom has the following electron configuration:

$2s^1, 2p_x^1, 2p_y^1, 2p_z^1$

The formation of sp$^3$ hybrid orbitals is only one of the possible ways of mixing. The four orbitals above can also be mixed so as to obtain sp$^2$ hybrids. An sp$^2$ hybrid orbital is formed from one s and two p orbitals.

Indicate the electron configuration resulting from the creation of an sp$^2$ orbital!

- $2s^1, 2p_x^1, 2p_y^1, 2p_z^1$     →     55

- $2(sp^2)^3, 2p^1$          →     77

- $2(sp^3)^4$             →     83

---

**70**     The mixed orbitals are called *hybrid* orbitals.

The bonding in ethane is similar to that in methane. Each carbon is bonded to three hydrogens by overlapping between the hydrogen 1s orbitals and three sp$^3$ orbitals per carbon atom.

Each carbon atom possesses a fourth sp$^3$ orbital. Overlapping of these two results in the C–C single bond, termed a σ *orbital* or σ *bond*.

In order to describe the C=C double bond we start again with the L-shell of the carbon atom in the ground state. Indicate the electron configuration:

- $2s^2, 2p_x^1, 2p_y^1$     →     61

- $2s^1, 2p_x^1, 2p_y^1, 2p_z^1$     →     10

- $2(sp^3)^4$          →     21

|71|    The atomic orbitals (AO's) of the two hydrogen atoms *overlap* to form a new orbital common to the two atoms (bonding or molecular orbital, MO).

**5**

The formation of the water molecule can be understood in similar terms. Let us start with the oxygen atom. The electron configuration in the L-shell is as follows:

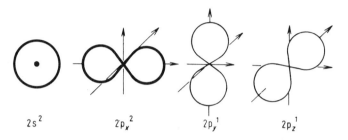

$$2s^2 \qquad 2p_x^{\,2} \qquad 2p_y^{\,1} \qquad 2p_z^{\,1}$$

There are six electrons and four orbitals. Two of these are doubly occupied. (A reason for the particular electron distribution shown can be found in Hund's rule. Lack of space prevents us from going into further details, but the interested reader should consult a textbook on atomic physics.)

Which of the four orbitals shown above can engage in the formation of chemical bonds?

● $2s^2$                          →     |78|

● $2p_x^2$                         →     |84|

● $2p_y^1$, $2p_z^1$                  →     |54|

● $2p_x^2$, $2p_y^1$, $2p_z^1$            →     |45|

● I don't know                →     |19|

---

|72|    You are mistaken. The L-shell can accept a *total* of eight electrons, but two of these are s electrons.

How many *p* electrons can it, therefore, accept?     →     |48|

---

| 73 | The p orbitals are *perpendicular* to the plane of the benzene ring: |

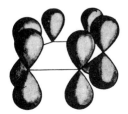

The six C–C σ bonds are drawn as lines for the sake of clarity. For the same reason, the hydrogen atoms and associated bonds are omitted.

The parallel p orbitals overlap sidewise, giving rise to an extended π bond:

How many electrons are engaged in π bond formation?

- Three      →    |24|
- Six        →    |80|
- Twelve    →    |59|

| 74 | No, both of the two p orbitals are *singly* occupied. Remember that we started with four singly occupied orbitals for each carbon atom. After hybridization, three were used in the formation of σ bonds. |

Repeat   |30| .

| 75 | Your answer is wrong. An $sp^2$ hybridized carbon atom has *three* $sp^2$ orbitals. Six carbon atoms are joined in a ring by six σ bonds. To form six σ bonds 12 atomic orbitals are required, i.e. 12 electrons. How many $sp^2$ orbitals remain unused?   →   |53| |

| 76 | *Orbital* | *shape* |
|----|-----------|---------|
|    | s         | sphere  |
|    | $p_x$     | "dumb-bell" |
|    | $p_y$     | "dumb-bell" |
|    | $p_z$     | "dumb-bell" |

Indicate the names of the four orbitals shown above.  →   66

---

77     Correct; an $sp^2$ hybridized carbon atom is described by the orbital configuration $2(sp^2)^3$, $2p^1$. The steric arrangement of the orbitals is shown below:

The p orbital is perpendicular to the plane described by the three $sp^2$ orbitals.

A p orbital extends into the two halves of space surrounding the nucleus. In contrast, hybrid orbitals are largely unidirectional, protruding only into one half of space. The smaller parts of the hybrid orbitals extending into the opposite direction – the so-called back lobes – are clearly seen in the $sp^3$ orbitals shown at  27  . For the sake of clarity the back lobes of the $sp^2$ orbitals were omitted in the drawing above.

In the molecule of ethene two carbon atoms are joined by a σ bond formed by overlapping two $sp^2$ orbitals. How many $sp^2$ orbitals remain on the two carbon atoms?

- Two       →   $\boxed{33}$

- Four      →   $\boxed{46}$

- Six       →   $\boxed{11}$

---

$\boxed{78}$      No, the notation $2s^2$ implies a *doubly* occupied 2s orbital. *Singly* occupied orbitals are used in the formation of chemical bonds. Repeat $\boxed{71}$ .

---

$\boxed{79}$      No, the expression $2(sp^3)^4$ has the following meaning:

2: number of the electron shell (the L-shell).

3: the hybrid orbitals are formed from *three* p orbitals (and one s orbital).

4: the number of (valence) electrons occupying the four hybrid orbitals.

How many $sp^3$ orbitals are there?     →     $\boxed{37}$

---

$\boxed{80}$      Yes, there are six $\pi$ electrons in benzene. The cyclic arrangement of the 6 $\pi$ electron system in benzene is particularly favorable energetically and leads to the particular stability of aromatic compounds (cf. Program 9). The total stabilization of the benzene $\pi$ system is higher than that of three isolated olefinic $\pi$ bonds.

We have illustrated only a few uses of the orbital model. The model is equally important in other areas of chemistry, e.g. for the description of the complexes formed by the transition elements.

Important though models are, it should be realized that they are only approximations of reality. Each model is useful in order to describe certain observations and for making new predictions. There are cases, however, where several different models or postulates must be employed in order to adequately explain all the facts. If a model contradicts experimental observations, it must be abandoned or modified. We can only hope to create models which will be reasonably close to reality. What Nature is really like, we may never know.

This is the end of Program 5.

---

$\boxed{81}$      Your answer is wrong.

The two p orbitals contain one electron each, i.e. a total of two.     →     $\boxed{44}$

---

**5**

---

**82**     The sp$^2$ hybridized carbon atom has the following L-shell configuration:

$2(\text{sp}^2)^3, 2\text{p}^1.$

The six sp$^2$ hybridized carbon atoms in benzene form six σ bonds with each other.

How many unused sp$^2$ orbitals are left among the six carbon atoms after this operation?

- Six          →    53

- Twelve          →    75

- I don't know          →    69

---

**83**     No; you indicated the sp$^3$ hybridized carbon atom. The question was concerned with sp$^2$ hybridization!

Repeat    69  .

---

**84**     No, the 2p$_x^2$ orbital is *doubly* occupied. However, *singly* occupied orbitals are used in the formation of atomic bonds.

Repeat    71  .

---

# Program 6

## Chemical Equilibria

---

**1** The knowledge about chemical equilibria is very important for reactions both in inorganic and organic chemistry. We will start this program with the electrolytic dissociation.

Consider the following experiment:

In three vessels are placed:

a) distilled water,
b) a solution of sodium chloride (NaCl) in water,
c) a solution of sugar in water.

There is no visible difference between the three liquids. They differ, however, in their ability to conduct electric current. The distilled water and the sugar solution have practically no conductivity. The sodium chloride solution, in contrast, is a good conductor.

Complete the following table:

|  | conducts electricity | |
|---|---|---|
|  | yes | no |
| Distilled water | | |
| Salt solution | | |
| Sugar solution | | |

$\rightarrow$ 13

---

**2** Yes, [CH₃COOH] is very large compared with [CH₃COO⁻] and [H⁺].

Let us now continue with the explanation and use of the law of mass action.

Like acetic acid, water dissociates to a very small extent.

Therefore, in the equation

$$\frac{[H^{\oplus}]\cdot[OH^{\ominus}]}{[H_2O]} = K$$

$[H^\oplus]$, $[OH^\ominus]$, and also the dissociation constant K are very small.

Since the concentration of undissociated water is very large compared to $[H^\oplus]$ and $[OH^\ominus]$, $[H_2O]$ can be taken as a constant as well. Multiplying this by K, a new constant k is obtained:

$$[H^\oplus] \cdot [OH^\ominus] = K \cdot [H_2O] = k$$

k has the value $10^{-14}$

Thus

$$[H^\oplus] \cdot [OH^\ominus] = 10^{-14}$$

This relation is known as the ionic product of water, or

*ion product of water.*

Copy down the mathematical equation for the ion product and then complete the following sentence:

The concentration of the protons multiplied by the concentration of ................. always equals $10^{-14}$.   →   [20]

---

**3**

$$Na_2SO_4 \rightarrow 2\ Na^\oplus + SO_4^{2\ominus}$$

Not only salts, but also acids and bases dissociate in aqueous solution.

$$HCl \rightarrow H^\oplus + Cl^\ominus$$
$$NaOH \rightarrow Na^\oplus + OH^\ominus$$

The process of ionization upon dissolution is called *electrolytic dissociation.*

All compounds which dissociate into ions in aqueous solution are called electrolytes. This term is also used for the resulting aqueous solutions. The water soluble acids, bases, and salts are electrolytes.

We have already become familliar with the behavior of the following compounds in aqueous solution:

sodium chloride, sugar, potassium chloride, sodium nitrate, alcohol, sodium sulfate, hydrochloric acid.

How many of these compounds are electrolytes?

- Seven     →     $\boxed{15}$
- Six       →     $\boxed{21}$
- Five      →     $\boxed{17}$
- Four      →     $\boxed{33}$
- Three     →     $\boxed{37}$
- Two       →     $\boxed{38}$

---

$\boxed{4}$
     Right, sugar is non-dissociating. The sugar solution scarcely conducts any electricity.

Thus, not all compounds dissociate upon dissolution in water.

Most of the water soluble organic compounds behave in the same manner as sugar. Alcohol is an example.

In contrast, many salts dissociate, just as NaCl.

Complete the equations of dissociation of potassium chloride and sodium nitrate:

       KCl   →
       $NaNO_3$ →                                        →  $\boxed{35}$

---

$\boxed{5}$    $\dfrac{[NH_3]^2}{[N_2]\cdot[H_2]^3} = K$

A very important example of the use of the law of mass action in organic chemistry is the esterification of carboxylic acids with alcohols.

You will find more about this topic in Program 17.

This is the end of Program 6.

---

| 6 | $CH_3COOH \rightleftarrows CH_3COO^\ominus + H^\oplus$ |
|---|---|

Did you write the equation with the *double arrow*?

The arrows indicate the direction of the reactions:

→   Dissociation of a molecule into ions.
←   Recombination of ions to form molecules.

Both reactions take place simultaneously, and at equilibrium with equal rates.

This is illustrated by an example:

When acetic acid is dissolved in water, an electrolytic dissociation immediately starts:

$$CH_3COOH \quad \rightarrow \quad CH_3COO^\ominus \quad + \quad H^\oplus$$
acetic acid          acetate ion        proton

The extent of the back reaction,

$$CH_3COOH \quad \leftarrow \quad CH_3COO^\ominus \quad + \quad H^\oplus$$

is very small at the onset, but increases as more and more acetic acid molecules dissociate. After a certain time the number of recombinations of ions is exactly equal to the number of dissociations of acetic acid molecules.

When this condition is reached, one speaks of dissociation .............. .
Complete the sentence.   → |22|

---

| 7 |
|---|

You need to exercise the calculation with powers.

$$10^{-7} \cdot [OH^\ominus] = 10^{-14}$$

$$[OH^\ominus] = \frac{10^{-14}}{10^{-7}} = ?$$

Try the following calculations:

$$\frac{10^{-8}}{10^{-4}} = 10^{-8-(4)} = 10^{-8+4} = 10^{-4}$$

$$\frac{10^{-10}}{10^{-2}} = 10^{-10-(2)} = 10^{-10+2} = 10^{-8}$$

Now try and solve

$$[OH^{\ominus}] = \frac{10^{-14}}{10^{-7}} = ? \quad \rightarrow \boxed{16}$$

**6**

---

**8**  Your answer is wrong.

Given $[OH^{\ominus}] = 10^{-9}$ mol/L

we calculate from the ion product of water

$$[H^{\oplus}] = \frac{10^{-14}}{10^{-9}} = 10^{-5} \, \text{mol/L}$$

thus, pH = 5.

Now, is the solution with pH = 5

● alkaline      →  $\boxed{41}$

● or acidic     →  $\boxed{39}$

---

**9**      There are two weak electrolytes (HCN and NH$_3$ (aq.)) and one of medium strength (Ba(OH)$_2$) in this row

Go back to  $\boxed{17}$  and try again

---

**10**    $$\frac{[CH_3COO^{\ominus}] \cdot [H^{\oplus}]}{[CH_3COOH]} = K$$

The undissociated acetic acid which is written to the left in the reaction equation appears in the denominator in the law of mass action equation above. The corresponding ions (proton and acetate ion) are to the right in the reaction equation, and in the numerator in the law of mass action equation.

K is a constant, known as the dissociation constant.

When acetic acid is dissolved in water, one has

$$[CH_3COO^\ominus] = [H^\oplus]$$

**6**

since one acetate ion and one proton are formed from each molecule of dissociating acetic acid.

Work out the magnitude of $[CH_3COOH]$ relative to $[H^\oplus]$ and $[CH_3COO^\ominus]$. Remember that acetic acid is a weak electrolyte.

Relative to $[H^\oplus]$ and $[CH_3COO^\ominus]$ the concentration $[CH_3COOH]$ will be

- very large                       $\rightarrow$    2
- very small                       $\rightarrow$    25
- of the same order of magnitude   $\rightarrow$    18

---

11    The three compounds $H_2S$, HCN, and $NH_3$ (aq.) are all weak electrolytes. The question was to find the row containing two strong and one weak electrolyte, so your answer is wrong.   $\rightarrow$   17

---

12

Your calculation is erroneous. To get more confidence in the calculation with powers, work through the following detailed examples:

$$\frac{10^{-8}}{10^{-4}} = 10^{-8-(-4)} = 10^{-8+4} = 10^{-4}$$

$$\frac{10^{-10}}{10^{-2}} = 10^{-10-(-2)} = 10^{-10+2} = 10^{-8}$$

Now try to solve the problem once more:

given $10^{-7} \cdot [OH^\ominus] = 10^{-14}$

$$[OH^\ominus] = \frac{10^{-14}}{10^{-7}} = ? \quad \rightarrow \quad 16$$

---

13

Compare your answer with the correct table:

| | conducts electricity | |
|---|---|---|
| | yes | no |
| Distilled water | | x |
| Salt solution | x | |
| Sugar solution | | x |

Only the sodium chloride solution conducts electricity. Why?

Sodium chloride in aqueous solution is dissociated into ions:

$$NaCl \rightarrow Na^{\oplus} + Cl^{\ominus}$$

Solutions containing ions conduct electricity.

From this, do you think that sugar molecules are dissociated into ions in aqueous solution?

● Yes        →    24

● No         →    4

---

14    The compounds NaCl, $H_2SO_4$, and NaOH all strong electrolytes. Therefore, your answer is wrong.  →  17

---

15    Wrong; read the explanation at    40

---

16    $[OH^{\ominus}] = 10^{-7}$ mol/L

In pure water both the proton and the hydroxide ion concentrations equal $10^{-7}$ mol/L.

$$10^{-7} \cdot 10^{-7} = 10^{-14}$$

The ion product of water ($10^{-14}$) is a constant. If, for example, the proton concentration $[H^{\oplus}]$ is increased by the addition of a little hydrochloric acid, the concentration $[OH^{\ominus}]$ will decrease, so that

$$[H^{\oplus}] \cdot [OH^{\ominus}] = 10^{-14} \text{ remains true.}$$

Suppose that the $H^{\oplus}$ ion concentration on water has been increased to $10^{-3}$ mol/L by the addtion of an acid. What will be the value of $[OH^{\ominus}]$?  →    32

| 17 |   Correct, five of the compounds are electrolytes and dissociate in aqueous
solution. Two of the compounds, sugar and alcohol, are non-electrolytes.

A continuous spectrum of possibilities exists between the extremes:

> nondissociating in solution (sugar, alcohol) and fully dissociated
> in solution (NaCl, HCl).

The corresponding compounds are described as

> Fully dissociated = strong electrolyte.
> Partly dissociated = medium strength electrolyte.
> Only slightly dissociated = weak electrolyte.

*Strong electrolytes:*
> almost all salts, e.g. $NaCl$, $KCl$, $Na_2SO_4$;
> the strong acids, e.g. $HCl$, $H_2SO_4$, $HNO_3$, $HClO_3$;
> the strong bases, e.g. $NaOH$, $KOH$.

*Medium strength electrolytes:*
> acids like $H_3PO_4$ and $H_2SO_3$;
> bases like $Ca(OH)_2$ and $Ba(OH)_2$

*Weak electrolytes:*
> organic acids, e.g. $CH_3COOH$;
> weak inorganic acids, e.g. $H_2CO_3$, $HCN$, $H_2S$;
> the weak base $NH_3$ (aq.).

Select from the six rows below the one row of compounds which contains two
strong and one weak electrolyte:

- $NaCl$, $H_2SO_4$, $NaOH$       →   | 14 |

- $HCN$, $Ba(OH)_2$, $NH_3$       →   | 9 |

- $H_2SO_3$, $NaOH$, $Na_2SO_4$       →   | 34 |

- $H_2SO_4$, $NH_3$, $Na_2SO_4$       →   | 19 |

- $CH_3COOH$, $Ca(OH)_2$, $H_3PO_4$       →   | 31 |

- $H_2S$, $HCN$, $NH_3$       →   | 11 |

|18|     Your answer is wrong.

Acetic acid is a weak electrolyte. Less than 1% of the molecules dissociate into acetate ions and protons in aqueous solution.

Therefore, $[CH_3COOH]$ is very (large/small) in comparison with $[CH_3COO^\ominus]$ or $[H^\oplus]$. Delete as appropriate.   $\rightarrow$  |2| .

**6**

|19|
    Correct; there were two strong ($H_2SO_4$ and $Na_2SO_4$) and one weak electrolyte ($NH_3$(aq.)) in this row.

The *degree of dissociation* of a particular compound is defined as the fraction of molecules dissociated:

$$\text{Degree of dissociation} = \frac{\text{number of dissociated molecules}}{\text{total number of dissolved molecules}}$$

or

$$\text{Degree of dissociation} = \frac{\text{number of dissociated molecules in \%}}{\text{all dissolved molecules } (= 100\,\%)}$$

Write down these two formulas.

If a compound dissociates to an extent of 20% upon dissolution in water, then the

$$\text{degree of dissociation} = \frac{20}{100} = \underline{0.2}$$

Now calculate the degree of dissociation in the case where 35% of the molecules dissociate upon dissolution.   $\rightarrow$  |28|

|20|     The product of the proton and hydroxide ion concentrations always equals $10^{-14}$,

$$[H^\oplus] \cdot [OH^\ominus] = 10^{-14}$$

Knowing the ion product of water and the value of $[H^\oplus]$, one can always calculate $[OH^\ominus]$ and vice versa.

In pure water we have $[H^{\oplus}] = 10^{-14}$ mol/L (moles per liter). Inserting this into the ion product, we obtain

$$10^{-7} \cdot [OH^{\ominus}] = 10^{-14}$$

**6**

Now calculate $[OH^{\ominus}]$. The value will be

- $10^2$ mol/L     →    $\boxed{12}$

- $10^{-2}$ mol/L     →    $\boxed{7}$

- $10^{-7}$ mol/L     →    $\boxed{16}$

---

$\boxed{21}$     Wrong. Read the explanation at $\boxed{40}$

---

$\boxed{22}$

    Dissociation *equilibrium*

When equilibrium has been reached, there will be well defined concentrations of acetic acid and the two ions formed from it, i.e. acetate ions and protons.

$$CH_3COOH \quad \rightleftharpoons \quad CH_3COO^{\ominus} + H^{\oplus}$$

    acetic acid              acetate ion     proton

Square brackets are used to denote concentrations of ions and molecules:

    Compare:

$[CH_3COOH]$ is the concentration of acetic acid.

$[CH_3COO^{\ominus}]$   is the concentration of acetate ion.

$[H^{\oplus}]$          is the proton concentration.

The *law of mass action* can be applied to any equilibrium reaction.

This law describes the mathematical relationship between the concentrations of dissociated and undissociated molecules.

> The product of the ion concentrations divided by the concentration of undissociated molecules equals a constant, K.

**6**

Applied to the electrolytic dissociation of acetic acid,

$$CH_3COOH \rightleftarrows CH_3COO^\ominus + H^\oplus$$

the law of mass action expression is

$$\frac{[\quad] \cdot [\quad]}{[\quad]} = K \quad \rightarrow \quad \boxed{10}$$

---

$\boxed{23}$

$[H^\oplus] = 10^{-11}$ mol/L, i.e. pH = 11.

Complete the following table.

| $[H^\oplus]$ | | pH |
|---|---|---|
| $10^{-5}$ | mol/L | .................... |
| $10^{-13}$ | mol/L | .................... |
| ......... | mol/L | 7 |
| ......... | mol/L | 2 |

$\rightarrow \boxed{26}$

---

$\boxed{24}$      Since a sugar solution does not conduct electricity, it does not contain any ions either.

Hence, sugar does not dissociate.    $\rightarrow$    $\boxed{4}$

---

$\boxed{25}$      Wrong. Acetic acid is a weak electrolyte. Less than 1% of the molecules dissolved in water dissociate into acetate ions and protons.

[CH₃COOH] is therefore very (large/small) compared with $[CH_3COO^\ominus]$ or $[H^\oplus]$.    $\rightarrow$    $\boxed{2}$

---

$\boxed{26}$     Compare:

| $[H^\oplus]$ | pH |
|---|---|
| $10^{-5}$ mol/L | 5 |
| $10^{-13}$ mol/L | 13 |
| $10^{-7}$ mol/L | 7 |
| $10^{-2}$ mol/L | 2 |

Distilled water is neutral and has $[H^{\oplus}] = [OH^{\ominus}]$. Since $[H^{\oplus}] \cdot [OH^{\ominus}] = 10^{-14}$, it follows that $[H^{\oplus}] = [OH^{\ominus}] = 10^{-7}$ mol/L in distilled water.

Since $[H^{\oplus}] = 10^{-7}$, the pH of distilled water equals 7.

**6**

> In neutral solutions pH equals 7

Aqueous solutions of acids have pH smaller than 7.

Aqueous solutions of bases have pH above 7.

$$
\begin{array}{c}
\text{neutral} \\
\downarrow \\
pH = 1\ 2\ 3\ 4\ 5\ 6\ 7\ 8\ 9\ 10\ 11\ 12\ 13\ 14 \\
\longleftarrow \quad | \quad \longrightarrow \\
\text{increasing acidity} \mid \text{increasing basicity}
\end{array}
$$

When $[OH^{\ominus}] = 10^{-9}$ mol/L, is the solution

- neutral?    →   $\boxed{30}$

- acidic?    →   $\boxed{39}$

- or alkaline? →   $\boxed{8}$

---

$\boxed{27}$    When, for a compound dissolved in water, only 1% is dissociated into ions, then

$$
\alpha = \frac{1}{100} = \underline{\underline{0.01}}
$$

There is a continuous transition from strong to weak electrolytes. the limits can be difined as follows:
    strong electrolytes: $\alpha = 1 - 0.6$
    medium strength electrolytes: $\alpha = 0.6 - 0.01$
    weak electrolytes: $\alpha$ smaller than 0.01

Acetic acid, $CH_3COOH$, is an example of a weak electrolyte. Less than 1% of the molecules dissolved in water are dissociated. The resulting solution therefore contains both $CH_3COO^{\ominus}$ and $H^{\oplus}$ ions, as well as undissociated molecules of $CH_3COOH$.

---

Write down the equation for the electrolytic dissociation of acetic acid.   →   $\boxed{6}$

|28|

$$\text{Degree of dissociation} = \frac{35}{100} = \underline{\underline{0.35}}$$

The Greek letter $\alpha$ (alpha) is often used to denote the degree of dissociation.

*Examples:*

100% dissociation; $\alpha = \dfrac{100}{100} = 1$

80% dissociation; $\alpha = \dfrac{80}{100} = 0.8$

50% dissociation; $\alpha = \dfrac{50}{100} = 0.5$

*Problem:*

1% dissociation; $\alpha = ?$

Check your calculation at |27|

---

|29|

$$\frac{[H^{\oplus}]^2 \cdot [S^{2\ominus}]}{[H_2S]} = K$$

For the tribasic acid phosphoric acid we have at complete dissociation:

$$H_3PO_4 \rightleftarrows 3\ H^{\oplus} + PO_4^{3\ominus} \text{ or}$$

$$\frac{[H^{\oplus}]^3 \cdot [PO_4^{3\ominus}]}{[H_3PO_4]} = K$$

The law of mass action can be used not only for dissociation equilibria but also for chemical reactions. The necessary condition is always that equilibrium is maintained. The synthesis of ammonia can be given as an example:

$$N_2 + 3\ H_2 \rightleftarrows 2\ NH_3$$

$$\frac{[NH_3]^2}{[N_2] \cdot [\ldots\ldots]} = K$$

Write down and complete the law of mass action expression above.   → |5|

**6**

### 30

Your answer is wrong.

If $[OH^{\ominus}] = 10^{-9}$ mol/L

then $[H^{\oplus}] = \dfrac{10^{-14}}{10^{-9}} = 10^{-5}$ mol/L

and pH = 5.

A solution is neutral when pH = 7.

Is the solution with pH = 5

● alkaline?   →   41

● or acidic?   →   39

### 31

One weak electrolyte, $CH_3COOH$, is given in this row, but the other two compounds, $Ca(OH)_2$ and $H_3PO_4$, are medium strength electrolytes. Go back to 17 and try again.

### 32

$[OH^{\ominus}] = 10^{-11}$ mol/L.

This follows from

$[H^{\oplus}] \cdot [OH^{\ominus}] = 10^{-14}$, and

$[H^{\oplus}] = 10^{-3}$ mol/L.

Therefore

$10^{-3} \cdot [OH^{\ominus}] = 10^{-14}$, or

$$[OH^{\ominus}] = \frac{10^{-14}}{10^{-3}} = 10^{-11}$$

If, on the other hand, a base is added, $[OH^{\ominus}]$ will increase.

Problem:
A solution has $[OH^{\ominus}] = 10^{-4}$ mol/L. Compute $[H^{\oplus}]$   →   36

|33| Wrong. Read the explanation at |40|

---

|34|      Two strong electrolytes are given in this row, namely NaOH and $Na_2SO_4$, but $H_2SO_3$ is only a medium strength electrolyte.

**6**

Go back to |17| and try again.

---

|35|

KCl      $\rightarrow K^{\oplus} + Cl^{\ominus}$
$NaNO_3 \rightarrow Na^{\oplus} + NO_3^{\ominus}$

The dissociation of a salt in water is also known as *electrolytic dissociation* because the solution is an electrolyte, i.e. it conducts an electric current.

Sodium sulfate ($Na_2SO_4$) dissociates to two sodium ions and one sulfate ion. The latter carries two negative charges and is written

$SO_4^{2\ominus}$ or $SO_4^{\ominus\ominus}$

Use the symbol $SO_4^{2\ominus}$ in the following.

Write down the equation for the electrolytic dissociation of $Na_2SO_4$.  →|3|

---

|36|

$[H^{\oplus}] = 10^{-10}$ mol/L.

This follows from

$[H^{\oplus}] \cdot [OH^{\ominus}] = 10^{-14}$ mol/L
and $[OH^{\ominus}] = 10^{-4}$ mol/L.

Therefore,

$[H^{\oplus}] \cdot 10^{-4} = 10^{-14}$, or

$[H^{\oplus}] = \dfrac{10^{-14}}{10^{-4}} = 10^{-10}$

Until now, all concentrations have been given as powers of 10. Although this is much simpler than using decimals, a further simplification is the use of the pH value to denote the hydrogen ion concentration. pH is the exponent of $[H^{\oplus}]$ with the sign reversed.

**6**

Example:
$[H^{\oplus}] = 10^{-9}$ mol/L in a given solution. The exponent of this concentration is $-9$. Reversing the sign of the exponent gives the pH value. Thus, pH = 9.

Further examples:

let $[H^{\oplus}] = 10^{-2}$ mol/L; then, pH = 2
let $[H^{\oplus}] = 10^{-7}$ mol/L; then, pH = 7
let $[H^{\oplus}] = 10^{-10}$ mol/L; then, pH = 10.

Problem:
$[H^{\oplus}] = 10^{-11}$ mol/L. Compute pH.     →   23

---

37     Your answer is wrong. Read the explanation at 40 .

---

38     No. There are two non-electrolytes, sugar and alcohol, among the compounds given. However, the number of electrolytes was asked for.

Read the explanation at 40 .

---

39     Correct.
When $[OH^{\ominus}] = 10^{-9}$ mol/L then $[H^{\oplus}] = 10^{-5}$ mol/L and pH = 5, i.e. acidic solution.

Generally:
Acidic solutions have pH below 7
Alkaline solutions have pH above 7.

The law of mass action can equally well be applied to compounds which dissociate into more than two ions.

$H_2SO_4$ is an example.

$$H_2SO_4 \rightleftarrows 2\ H^{\oplus} + SO_4^{2\ominus}$$

$$\frac{[H^{\oplus}]^2 \cdot [SO_4^{2\ominus}]}{[H_2SO_4]} = K$$

The $2\ H^{\oplus}$ in the chemical equation appear as the second power of the proton concentration in the mathematical expression.

Generally:

> The *number* of ions of as particular kind appears as the exponent of the concentration in the law of mass action expression.

Another example is

$$H_2S \rightleftarrows 2\,H^{\oplus} + S^{2\ominus}$$

Formulate the law of mass action equation for this equilibrium.    → 29

---

**40**    Electrolytes are compounds which dissociate into ions in aqueous solution. Both salts, acids, and bases can be electrolytes. The presence of ions causes the solutions to be conductors of electricity. Compounds which do *not* dissociate are *non-electrolytes.*    → 3

---

**41**    No. Read the explanation at 26 .

# Program 7

## Hydrocarbons (I)

**1**    As indicated by their name, *hydrocarbons* are compounds of *carbon* and *hydrogen*. Although only *two* elements are present in the hydrocarbons, there are many ways in which these two elements can combine to form molecules. Our first task is to bring order and clarity to this multitude of compounds.

Let us first repeat a few facts, however. If these facts are not yet known to you, go to  6  for a more detailed presentation.

Carbon is always *tetravalent*. A carbon atom can form bonds to other carbon atoms. In this way chains (straight or branched) and rings of carbon atoms are formed. The remaining valences are used to form bonds to hydrogen, thus leading to the hydrocarbons.

The simplest hydrocarbon is methane, $CH_4$.

The next one is ethane, $C_2H_6$.

A chain of three carbon atoms is present in propane, $C_3H_8$.

Butane, $C_4H_{10}$, has a four-carbon chain, and in pentane, $C_5H_{12}$, there are five carbon atoms in a row.

Write down the structural formulas for the five hydrocarbons mentioned above.   →   7

If more details are required, go instead to  6 .

---

**2**    There is a branching in this structure:

$$\begin{array}{c} \quad \ \ |\ \ \ \ | \\ \quad -\!C\!-\!-\!C\!- \\ \ \ |\quad |\quad | \\ -C\!-\!C\!-\!C\!- \\ \ \ |\quad |\quad | \end{array}$$

and the compound is therefore different from the corresponding linear one:

$$\begin{array}{c} \ \ |\quad |\quad |\quad |\quad | \\ -C\!-\!C\!-\!C\!-\!C\!-\!C\!- \\ \ \ |\quad |\quad |\quad |\quad | \end{array}$$

Repeat  17  *in detail.*

---

**3**      Correct; butene is *hydrogenated* to butane. Propane is *dehydrogenated* to propene.

Following hydrogenation, we shall now examine another very important type of reaction of the alkenes:

The *polymerization* of unsaturated compounds. In this reaction, *very* large molecules (macromolecules) are formed from small ones (e.g. ethylene). The macromolecules or polymers can have molecular masses of 10 000, 100 000, or more.

The following scheme shows the formation of polymers from many small molecules:

Before polymerization:
many small molecules

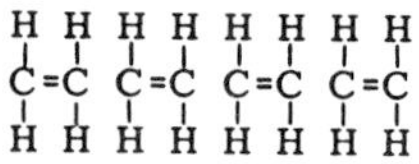

$$\begin{matrix} H & H & H & H & H & H & H & H \\ | & | & | & | & | & | & | & | \\ C=C & C=C & C=C & C=C \\ | & | & | & | & | & | & | & | \\ H & H & H & H & H & H & H & H \end{matrix}$$

After polymerization:
macromolecule

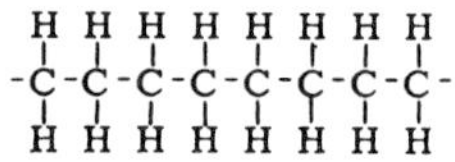

$$\begin{matrix} H & H & H & H & H & H & H & H \\ | & | & | & | & | & | & | & | \\ -C-C-C-C-C-C-C-C- & \ldots etc. \\ | & | & | & | & | & | & | & | \\ H & H & H & H & H & H & H & H \end{matrix}$$

The polymer (macromolecule) shown should be visualized as extending further in both directions.

Which of the compounds below can polymerize?

- Ethene and butane                    → [18]

- Ethane, propane, and butane          → [21]

- Ethene and propene                   → [10]

- Ethane and propene                   → [28]

---

**4**      Two hydrogen atoms are absorbed in the hydrogenation of ethene:

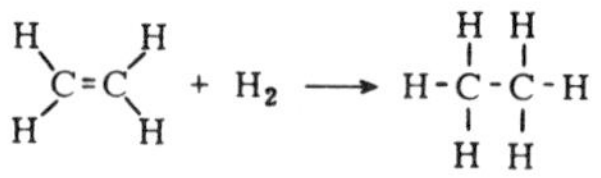

$$\begin{matrix} H & & H \\ \backslash & & / \\ & C=C & \\ / & & \backslash \\ H & & H \end{matrix} \quad + H_2 \longrightarrow \begin{matrix} H & H \\ | & | \\ H-C-C-H \\ | & | \\ H & H \end{matrix}$$

molecular mass 28        molecular mass 30

What is formed by the hydrogenation of propene and what is the molecular mass of the reaction product?    → [40]

---

**5**        Reconsider your answer. The following reaction equations will be of
help:

a)

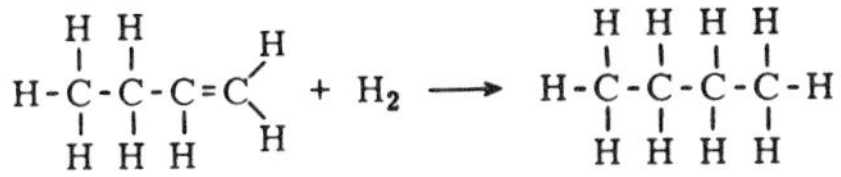

$$H-\underset{\underset{H}{|}}{\overset{\overset{H}{|}}{C}}-\underset{\underset{H}{|}}{\overset{\overset{H}{|}}{C}}-C=C\overset{H}{\underset{H}{<}} \;+\; H_2 \;\longrightarrow\; H-\underset{\underset{H}{|}}{\overset{\overset{H}{|}}{C}}-\underset{\underset{H}{|}}{\overset{\overset{H}{|}}{C}}-\underset{\underset{H}{|}}{\overset{\overset{H}{|}}{C}}-\underset{\underset{H}{|}}{\overset{\overset{H}{|}}{C}}-H$$

b)

$$H-\underset{\underset{H}{|}}{\overset{\overset{H}{|}}{C}}-\underset{\underset{H}{|}}{\overset{\overset{H}{|}}{C}}-\underset{\underset{H}{|}}{\overset{\overset{H}{|}}{C}}-H \;\longrightarrow\; \overset{H}{\underset{H}{>}}C=C\underset{\underset{H}{|}}{\overset{\overset{H}{|}}{-C}}-H \;+\; H_2$$

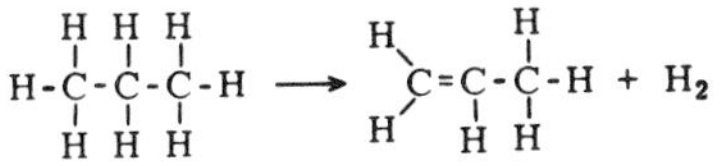

Copy down these two equations and return to [23] .

---

**6**        Carbon is always *tetra*valent, hydrogen *uni*valent.

How many hydrogen atoms can be bound to one carbon atom?    →   [13]

---

**7**

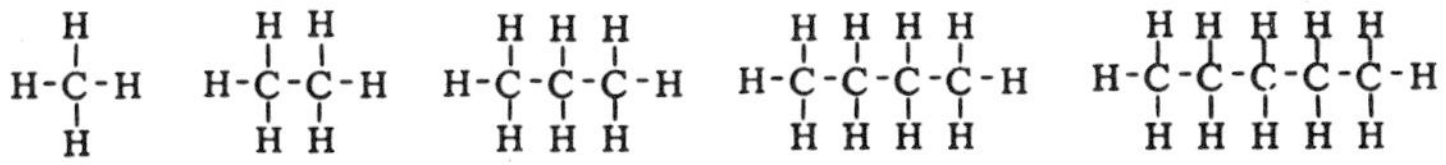

$$H-\underset{\underset{H}{|}}{\overset{\overset{H}{|}}{C}}-H \qquad H-\underset{\underset{H}{|}}{\overset{\overset{H}{|}}{C}}-\underset{\underset{H}{|}}{\overset{\overset{H}{|}}{C}}-H \qquad H-\underset{\underset{H}{|}}{\overset{\overset{H}{|}}{C}}-\underset{\underset{H}{|}}{\overset{\overset{H}{|}}{C}}-\underset{\underset{H}{|}}{\overset{\overset{H}{|}}{C}}-H \qquad H-\underset{\underset{H}{|}}{\overset{\overset{H}{|}}{C}}-\underset{\underset{H}{|}}{\overset{\overset{H}{|}}{C}}-\underset{\underset{H}{|}}{\overset{\overset{H}{|}}{C}}-\underset{\underset{H}{|}}{\overset{\overset{H}{|}}{C}}-H \qquad H-\underset{\underset{H}{|}}{\overset{\overset{H}{|}}{C}}-\underset{\underset{H}{|}}{\overset{\overset{H}{|}}{C}}-\underset{\underset{H}{|}}{\overset{\overset{H}{|}}{C}}-\underset{\underset{H}{|}}{\overset{\overset{H}{|}}{C}}-\underset{\underset{H}{|}}{\overset{\overset{H}{|}}{C}}-H$$

  Methane       ethane        propane          butane              pentane

The series can be continued. Hydrocarbon chains with as many as eighty carbon
atoms are known.

For the present, the first five members of the series will suffice. Please write
down their names and *empirical formulas*.    →   [14]

---

**8**        Wrong. Go back and study [17]   *thoroughly.*

---

**9**        You added only *one* hydrogen atom to ethene and propene.

*Two* hydrogen atoms are added to the double bond in a hydrogenation reaction.

Go back to   [47]  .

---

**10**     Correct; only *alkenes* such as *ethylene* and *propylene* (ethene and propene) can polymerize.

On polymerization,
ethylene gives polyethylene;
propylene gives polypropylene.

**7**

Transparent packing material, toys, buckets, and other utensils are made from polyethylene. You will learn more about this and other synthetics in Program 23.

Let us now continue the presentation of unsaturated hydrocarbons.

Until now, we have been concerned only with alkenes possessing *one* double bond, e.g. ethene and propene. Compounds with two or more double bonds are also known.

Butadiene is an example (pronounced: buta-die-ene). It is derived from butane by the introduction of two double bonds.

Write down the empirical formula for butadiene and compute its molecular mass (using atomic mass of carbon = 12; of hydrogen = 1).

The molecular mass of butadiene is

- 58                 →   77

- 56                 →   56

- 54                 →   35

- I need further information   →   75

---

**11**     You gave the correct empirical formula for butane – $C_4H_{10}$ – only.

Ethane has the empirical formula $C_2H_6$
   (not $CH_4$ which is methane!)

Pentane is $C_5H_{12}$
   (not $C_6H_{11}$ which is hexane!).

Repeat from   7   .

**12**

a) 
$$
\underset{\substack{H\ H\ H}}{\overset{\substack{H\ H}}{H-C-C-C}}\!=\!\underset{H}{\overset{H}{C}} + H_2 \longrightarrow \underset{\substack{H\ H\ H\ H}}{\overset{\substack{H\ H\ H\ H}}{H-C-C-C-C-H}}
$$

b) 
$$
\underset{\substack{H\ H\ H}}{\overset{\substack{H\ H\ H}}{H-C-C-C-H}} \longrightarrow \underset{H}{\overset{H}{C}}\!=\!\underset{\substack{H\ H}}{\overset{H}{C-C-H}} + H_2
$$

Copy down these two equations and go back to   |23|   .

---

**13**  *Four* hydrogen atoms. The empirical formula for the compound is $CH_4$.

Write down the *structural* formula!   →   |20|

---

**14**

  Methane  $CH_4$
  ethane   $C_2H_6$
  propane  $C_3H_8$
  butane   $C_4H_{10}$
  pentane  $C_5H_{12}$

Examining this series, you will see that each compound has *one* carbon atom and *two* hydrogen atoms more than its predecessor.

Butane has *one* $CH_2$-group *more* than propane and one *less* than pentane.

What, therefore, is the empirical formula for the hydrocarbon which follows pentane?

● $C_6H_{12}$        →   |32|

● $C_6H_{14}$        →   |22|

● $C_4H_{10}$        →   |33|

● I don't understand the question  →   |34|

---

**15**  Reconsider your answer. The following equations will be of help:

a) 
$$
\underset{\substack{H\ H\ H}}{\overset{\substack{H\ H}}{H-C-C-C}}\!=\!\underset{H}{\overset{H}{C}} + H_2 \longrightarrow \underset{\substack{H\ H\ H\ H}}{\overset{\substack{H\ H\ H\ H}}{H-C-C-C-C-H}}
$$

b)

$$H\text{-}\overset{\displaystyle H}{\underset{\displaystyle H}{C}}\text{-}\overset{\displaystyle H}{\underset{\displaystyle H}{C}}\text{-}\overset{\displaystyle H}{\underset{\displaystyle H}{C}}\text{-}H \longrightarrow \underset{H}{\overset{H}{>}}C\text{=}C\text{-}\overset{\displaystyle H}{\underset{\displaystyle H}{C}}\text{-}H + H_2$$

Write down these equations and return to $\boxed{23}$ .

---

**7**

$\boxed{16}$      You gave the correct empirical formulas for ethane ($C_2H_6$) and pentane ($C_5H_{12}$) only.

Butane is $C_4H_{10}$
(not $C_3H_8$ which is propane).

Repeat from $\boxed{7}$ .

---

$\boxed{17}$      Your structural formulas are not yet quite correct.

The formulas for the three pentane isomers are:

         (a)                      (b)                      (c)

There are no other pentane isomers, but the formulas (a), (b), and (c) can be written in other ways, e.g.:

    ($a_1$)          ($a_2$)          ($a_3$)          ($a_4$)          ($a_5$)

All these formulas are equivalent to formula (a)! The formulas ($a_1$) − ($a_5$) are only bent chains, *not branched ones*. Stretching the chains gives (a). Check this for each structural formula!

Similarly, the following are all equivalent to (b):

    ($b_1$)          ($b_2$)          ($b_3$)          ($b_4$)          ($b_5$)

Stretching or turning will reproduce formula (b). Verify this for each formula.

Is the formula below a variant of (a), (b), or (c)?

$$\begin{array}{c}\quad\mbox{-C--C-}\\ \text{-C—C—C-}\end{array}$$

● (a)        →    $\boxed{2}$

● (b)        →    $\boxed{24}$

● (c)        →    $\boxed{8}$

7

---

$\boxed{18}$    You forgot an important requirement for polymerization: the molecule must possess a double bond.

Thus, which compounds can polymerize?

● Alkenes      →    $\boxed{10}$

● Alkanes      →    $\boxed{37}$

---

$\boxed{19}$

| | | |
|---|---|---|
| $\begin{array}{c}\text{H}\\ |\\ \text{H-C-H}\\ |\\ \text{H}\end{array}$ | methane; | empirical formula $CH_4$ |
| $\begin{array}{c}\text{H H}\\ |\ |\\ \text{H-C-C-H}\\ |\ |\\ \text{H H}\end{array}$ | ethane | $C_2H_6$ |

continuing the series:    $\begin{array}{c}\text{H H H}\\ |\ |\ |\\ \text{H-C-C-C-H}\\ |\ |\ |\\ \text{H H H}\end{array}$    propane    $C_3H_8$

and then:             butane       $C_4H_{10}$

                  pentane     $C_5H_{12}$

Memorize these names!

Now write down the structural formulas for butane and pentane.    →    $\boxed{7}$

---

**20**

$$\begin{array}{c} H \\ | \\ H-C-H \\ | \\ H \end{array}$$

Carbon has the important property of forming bonds between two or several atoms of the element. In the case of two carbon atoms:

$$-\overset{|}{\underset{|}{C}}-\overset{|}{\underset{|}{C}}-$$

Also three, four, five, six, seven, etc. carbon atoms can be joined in a chain, e.g.

$$-\overset{|}{\underset{|}{C}}-\overset{|}{\underset{|}{C}}-\overset{|}{\underset{|}{C}}-$$

Now write down carbon chains with four and five carbon atoms.    →    27

---

**21**        You forgot an important requirement for polymerization: the molecule must possess a double bond.

Thus, which compounds can polymerize?

● Alkenes       →    10

● Alkanes       →    37

---

**22**        Yes, $C_6H_{14}$ follows after $C_5H_{12}$.

$C_6H_{14}$ is hexane.

It is important to memorize the names and formulas of the six hydrocarbons discussed so far.

Which of the three lines below gives the correct formulas for ethane, butane, and pentane?

| Ethane | butane | pentane | | |
|--------|--------|---------|---|---|
| ● $CH_4$ | $C_4H_{10}$ | $C_6H_{14}$ | → | 11 |
| ● $C_2H_6$ | $C_3H_8$ | $C_5H_{12}$ | → | 16 |
| ● $C_2H_6$ | $C_4H_{10}$ | $C_5H_{12}$ | → | 29 |

**23**    *Hydrogenation*

The opposite process, the elimination of hydrogen with formation of a double bond, is also known.
This reaction is called *dehydrogenation:*

Are the following two reactions hydrogenations or dehydrogenations, or one of each?

a) butene → butane
b) propane → propene

● Both are dehydrogenations                                    →   15

● Both are hydrogenations                                      →   39

● a) is a dehydrogenation;
   b) is a hydrogenation                                       →   5

● a) is a hydrogenation;
   b) is a dehydrogenation                                     →   3

● I need the reaction equations in order to answer            →   12

**24**    You are right. Stretching and turning the formula leads to the structure (b):

Now try to write down the structural formulas for the three pentane isomers once more.

Compare your result at    38  .

**25**    You calcuted the molecular masses of ethene and propene.

However, the masses of the compounds formed by *hydrogenation* of ethene and propene were required.

Go back to    47  .

26

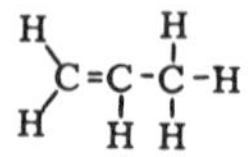

H          H
 \        |
  C=C-C-H
 /    |  |
H     H  H

The empirical formula is $C_3H_6$, and the name is propene (*not* propane, $C_3H_8$!).

There is also another name for propene. What is it?    →    47

7

27

    |  |  |  |
  -C-C-C-C-
    |  |  |  |

    |  |  |  |  |
  -C-C-C-C-C-
    |  |  |  |  |

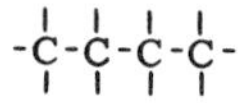

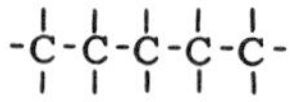

A chain of six carbon atoms is

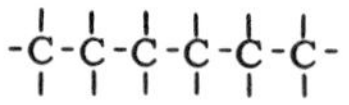

    |  |  |  |  |  |
  -C-C-C-C-C-C-
    |  |  |  |  |  |

When the free valences in these *carbon* chains are saturated by bonding to hydrogen, *hydrocarbons* result:

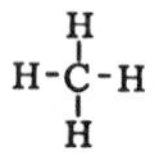

         H
         |
     H-C-H        is methane
         |
         H

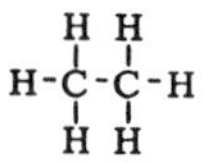

      H  H
      |  |
    H-C-C-H        is ethane
      |  |
      H  H

Write down the empirical formulas for these two compounds.    →    19

28    You forgot an important requirement for polymerization: the molecule must possess a double bond.

Thus, which compounds can polymerize?

● Alkenes      →    10

● Alkanes      →    37

**29**    Correct!

The series of hydrocarbons

| methane | $CH_4$ |
|---------|--------|
| ethane | $C_2H_6$ |
| propane | $C_3H_8$ |
| butane | $C_4H_{10}$ |
| pentane | $C_5H_{12}$ |
| hexane | $C_6H_{14}$ |
| etc. | |

is called a *homologous* series because two consecutive members of the series always differ by a constant amount – in the present case a $CH_2$ group.

The general formula for this homologous series is

$C_nH_{2n+2}$

This general formula allows you to write down the empirical formula of any compound belonging to the homologous series of open chain hydrocarbons. An example:

What is the empirical formula for a hydrocarbon composed of six carbon atoms?

The general formula $C_nH_{2n+2}$ with n = 6 gives $C_6H_{2\cdot6+2}$

So, what is the number of hydrogen atoms?    →    36

**30**    Alkanes are *saturated* hydrocarbons.
Alkenes are *unsaturated* hydrocarbons with at least one *double bond*.

What is the addition of hydrogen to alkenes called?    →    23

**31**

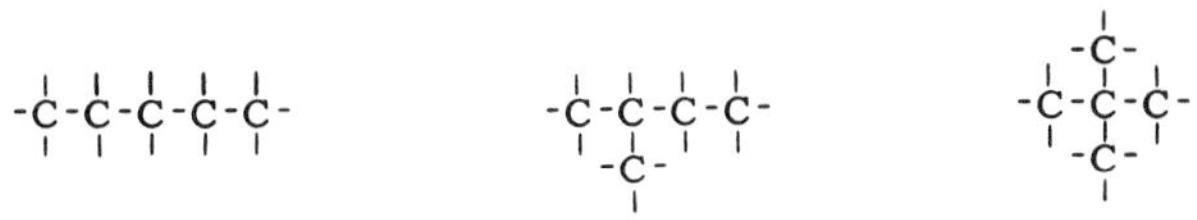

Compare with your own answer for the pentane isomers.

Are yours identical with the above?

If yes, go on to  [38]  .

If no, go now to  [17]  .

---

**7**

**32**      It is true that the hydrocarbon following pentane has six carbon atoms. However, your number of hydrogen atoms is wrong.

Repeat  [14]  in detail.

---

**33**      You didn't read the question properly!

The empirical formula for the hydrocarbon *following* pentane was asked for. Butane ($C_4H_{10}$) is, however, just *before* pentane in the series of open-chain hydrocarbons.

Repeat  [14]  in detail.

---

**34**      The series

| | |
|---|---|
| methane | $CH_4$ |
| ethane | $C_2H_6$ |
| propane | $C_3H_8$ |
| butane | $C_4H_{10}$ |
| pentane | $C_5H_{12}$ |

can be continued. The next compound has one $CH_2$ group more than pentane. What is its empirical formula?     $\rightarrow$    [22]

---

**35**      Correct, butadiene has the empircal formula $C_4H_6$ and a molecular mass of 54.

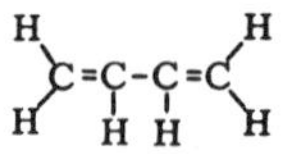

Butadiene

The names of unsaturated compounds with *one* double bond always have the ending -ene, e.g. ethene, propene.

For compounds with *two* double bonds, the syllable "di" (= two) is inserted before the "ene", i.e. buta-di-ene.

Write down your answers to these questions:

a) How many hydrogen atoms will be added to butadiene on complete hydrogenation?

b) What is the name of the compound formed by complete hydrogenation of butadiene?   →   42

**7** ·

---

**36**     The answer is $2 \cdot 6 + 2 = 14$ hydrogen atoms.

Hence, the empirical formula for the open-chain hydrocarbon with 6 carbon atoms (hexane) is $C_6H_{14}$.

You can memorize the rule as follows: the number of hydrogen atoms is *twice* the number of carbons *plus two*.

You will immediately see why by inspection of the formula for hexane, for example.

$$\mathbf{H}-\overset{\displaystyle H}{\underset{\displaystyle H}{C}}-\overset{\displaystyle H}{\underset{\displaystyle H}{C}}-\overset{\displaystyle H}{\underset{\displaystyle H}{C}}-\overset{\displaystyle H}{\underset{\displaystyle H}{C}}-\overset{\displaystyle H}{\underset{\displaystyle H}{C}}-\overset{\displaystyle H}{\underset{\displaystyle H}{C}}-\mathbf{H}$$

Each carbon atom is bonded to *two* hydrogen atoms, except for the *end groups* which each have one hydrogen atom more. (The end-H's are printed in boldface above).

Hence, the total is twice the number of carbon atoms *plus two*.

What we have elaborated here is also expressed in the general formula which you already know.

What is the general formula for the homologous series of open-chain hydrocarbons?

$C_n$ ..................   →   43

---

**37**     You do not yet understand the difference between alkanes and alkenes.

Repeat from   40   .

---

**38**

Your answer is correct if you wrote the following formulas:

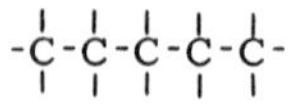

$$
\begin{array}{ccc}
-\overset{|}{\underset{|}{C}}-\overset{|}{\underset{|}{C}}-\overset{|}{\underset{|}{C}}-\overset{|}{\underset{|}{C}}-\overset{|}{\underset{|}{C}}- &
\begin{array}{c}-\overset{|}{\underset{|}{C}}-\overset{|}{\underset{|}{C}}-\overset{|}{\underset{|}{C}}-\overset{|}{\underset{|}{C}}-\\ -\overset{|}{\underset{|}{C}}-\end{array} &
\begin{array}{c}-\overset{|}{C}-\\ -\overset{|}{\underset{|}{C}}-\overset{|}{\underset{|}{C}}-\overset{|}{\underset{|}{C}}-\\ -\overset{|}{C}-\end{array}\\
(a) & (b) & (c)
\end{array}
$$

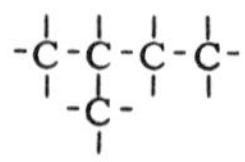

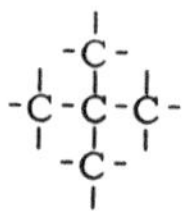

**7**

The longer the carbon chain, the larger the number of isomers. This is exemplified below:

        butane:     4 carbon atoms,  2 isomers
        pentane:   5 carbon atoms,  3 isomers
        hexane:    6 carbon atoms,  5 isomers
        decane:   10 carbon atoms, 75 isomers.

(If you like, try to construct the hexane isomers. They are shown at  **44**  .)

Since there are *two* isopentanes (see formulas (b) and (c) above) different names are required for them.

The isopentane (b) is called methylbutane.
The isopentane (c) is called dimethylbutane.

In order to understand these names we must define the methyl group.

The methyl group may also be called a methyl radical or residue.

Do you know what a methyl group is?

If yes, write down the formula, including the hydrogen atoms.  →  **45**

If not, look it up at  **46**  .

---

**39**

Reconsider your answer. The following equations will be of help:

a) 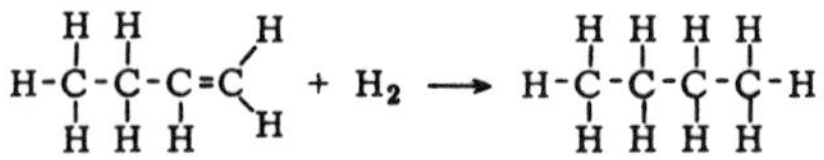     $H-\overset{H}{\underset{H}{C}}-\overset{H}{\underset{H}{C}}-\overset{H}{\underset{H}{C}}=C\overset{H}{\diagdown_{H}}$ + H$_2$  →  $H-\overset{H}{\underset{H}{C}}-\overset{H}{\underset{H}{C}}-\overset{H}{\underset{H}{C}}-\overset{H}{\underset{H}{C}}-H$

b) 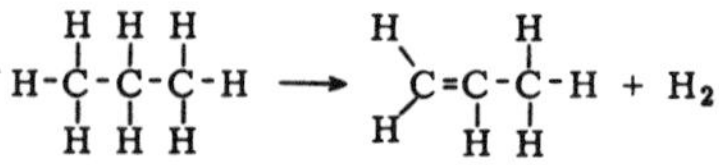     $H-\overset{H}{\underset{H}{C}}-\overset{H}{\underset{H}{C}}-\overset{H}{\underset{H}{C}}-H$  →  $\overset{H}{\diagup}C=\overset{H}{\underset{H}{C}}-\overset{H}{\underset{H}{C}}-H$ + H$_2$

Write down these equations and return to  **23**  .

**40**  The hydrogenation of
ethene gives ethane (molecular mass 30);
propene gives propane (molecular mass 44).

Since ethane and propane have no double bonds, no further hydrogen can be added. They are said to be *saturated* with hydrogen, or to be *saturated* hydrocarbons.

**7**

Complete the statements:

Alkanes are ................... (satured/unsaturated) hydrocarbons.

Alkenes are ................... (saturated/unsaturated) hydrocarbons.   →   30

---

**41**

```
    H H H H                    H H H
    | | | |                    | | |
 H-C-C-C-C-H              H-C-C-C-H
    | | | |                    |   |
    H H H H                    H   H
                            H-C-H
                               |
                               H
```

     n-butane            isobutane

Since, in order to distinguish the isomers only the carbon atoms are necessary, we can for many purposes omit the hydrogen atoms in the structures. This gives much clearer formulas:

```
   | | | |              | | |
  -C-C-C-C-            -C-C-C-
   | | | |              | | |
                         -C-
                          |
```

     n-butane            isobutane

n-Butane (normal butane) may simply be called butane. There is only one isomeric butane, namely isobutane. Attempts to write down the branched chain in other ways will only lead to the same compound. You will see this by studying the formulas given below – in principle there is no difference between these various ways of designating the isobutane molecule. Rotation of the formulas makes them identical:

```
    |                |                  |              | | |
   -C-              -C-                -C-            -C-C-C-
    |                |  |               |  |          | | |
  -C-C-            -C-C-              -C-C-C-            -C-
    |                |  |             | | |              |
   -C-              -C-
    |                |
```

In the case of *pentane* there are *three* ways of arranging the five carbon atoms. This gives the normal n-pentane and *two* isopentanes. Write down the structures of these three pentane isomers, and make sure they are really different. (The hydrogen atoms may be omitted.)   →   31

| 42 |

a) *Four* hydrogen atoms
b) *Butane.*

*Acetylene is another important unsaturated compound. It has two carbon atoms joined through a triple bond.*

| **7** |

Write down the structural formula for acetylene. How many hydrogen atoms are there?

- 8     →     | 82 |
- 6     →     | 63 |
- 4     →     | 81 |
- 2     →     | 51 |

| 43 |

The general formula for this homologous series is

$$C_nH_{2n+2}$$

Another example:

Give the formula for the ninth member of this series. With n = 9 we have

$$C_9H_{2 \cdot 9+2} = C_9H_{20}$$

*Problem:*

Give the empirical formula for the 15th member of this series of hydrocarbons.
    →     | 50 |

| 44 |

The structures of the five hexane isomers are given below (omitting hydrogen atoms).

Now return to   | 38 | .

**45**

```
      H
      |
   H-C-
      |
      H
```

The name "methylbutane" should now be self-explanatory: it is derived from butane by replacement of a hydrogen atom by a methyl group.

```
   H H H H                      H H H H
   | | | |                      | | | |
 H-C-C-C-C-H                  H-C-C-C-C-H
   | | | |                      |   | |
   H H H H                      H   H H
                                |
                              H-C-H
                                |
                                H
```

       Butane                  methylbutane

As you know, the formula for methylbutane can be drawn in various ways which are, however, all equivalent. The following simplified formulas all represent the *same* molecule:

```
    | | | |           -C-              -C-              -C-
  -C-C-C-C-         | | |            | | |            | |
    | | | |         -C-C-C-        -C-C-C-          -C—C-
    -C-              | | |            |              | |
    |                -C-              -C-            -C—C-
                     |                |              |
                                    -C-
                                     |
    (a)              (b)              (c)              (d)
```

For simplicity, the formula possessing the *longest horizontal*

carbon chain is preferred in such cases.

Which formula would that be in the case of methylbutane?

● (a)  →  52

● (b)  →  66

● (c)  →  64

● (d)  →  62

**46**      The *methyl group* (or methyl *residue* or methyl *radical*) is derived from methane by removal of one hydrogen atom:

```
      H                      H
      |                      |
   H-C-H                  H-C-
      |                      |
      H                      H
```

     methane                methyl

The methyl radical is not a stable compound, but the methyl group is present in numerous compounds where it is bonded to other elements or groups (e.g. $-Cl$, $-CH_3$, $-C_2H_5$, or $-OH$). These compounds will be described later.

In a similar manner, the ethyl group is obtained by removal of a hydrogen atom from ethane, propyl from propane, butyl from butane, etc.

**7**

Write down the missing structures:

methyl       $\begin{array}{c} H \\ | \\ H-C- \\ | \\ H \end{array}$

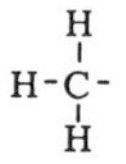

ethyl

propyl

butyl             →   53

---

**47**     Yes, propene is also called propylene.

The *alkenes* contain at least one *double bond*. The carbon-carbon double bond is particularly reactive: reaction with other atoms or molecules takes place by addition to this bond.

*Hydrogenation* is an example of addition to the double bond. In this reaction two hydrogen atoms are added.

Hydrogenation readily takes place in the presence of catalysts like platinum or nickel; hence the term *catalytic* hydrogenation.

*Problem:*

Which compounds are formed by hydrogenation of a) ethene and b) propene? Give the empirical formulas and molecular masses of the products. (Use as atomic masses: carbon = 12; H = 1.)

- a) 28    b) 42           →   25
- a) 29    b) 43           →   9
- a) 30    b) 44           →   40
- I need further details        →   4

**48**   True, branching of a three-carbon chain as in propane is not possible. (For more details, see  49  .)

Branching becomes possible in chains of four or more carbon atoms.

Give the structural formulas for n-butane and isobutane.   →   41

7

---

**49**   There are *no* isomers of propane.

Probably, you wrote the formulas

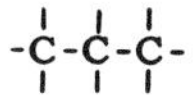

$$-\overset{|}{\underset{|}{C}}-\overset{|}{\underset{|}{C}}-\overset{|}{\underset{|}{C}}- \qquad -\overset{|}{\underset{|}{C}}-\overset{|}{\underset{|}{C}}- \qquad -\overset{|}{\underset{|}{C}}-\overset{|}{\underset{|}{C}}-$$

$$\qquad\qquad\qquad\quad -\overset{|}{\underset{|}{C}}- \qquad\qquad -\overset{|}{\underset{|}{C}}-$$

     **(a)** 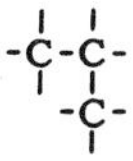        **(b)** 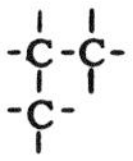     **(c)**

which are, however, identical; (b) and (c) are *not* branched molecules.

Consider that the real molecules have three-dimensional spatial structures. Our drawings are only two-dimensional planar representations of the real molecules. We draw these formulas linearly, with right angles, and symmetrically for the sake of clarity. Only the *sequence* of the atoms is justly given in these formulas; not the structure in space.

Thus, open-chain hydrocarbons do not exist in the form of rigid, stretched chains, but rather as zig-zag, wavy strings, or tangled forms, especially for the longer chains. In other words, the atoms are not bonded to each other in a rigid manner, but can be bent or rotated about the bonds.

Therefore, the angular formulas (b) and (c) above have nothing to do with *branching*. Stretching of these formulas will reproduce (a):

    **(b)**        **(a)**

    **(c)**        **(a)**

The three formulas (a), (b), and (c) do not represent *isomers,* but three different ways of drawing the same *molecule.*

Branching becomes possible only with *four* or more carbon atoms in the chain; at the same time, *isomers* can be formed.

Write down the structural formulas for n-butane and isobutane.   →   $\boxed{41}$

---

$\boxed{50}$    $C_{15}H_{32}$

**7**

(If your answer was incorrect, repeat from   $\boxed{29}$  .)

Carbon atoms can combine with each other not only to form linear chains as presented up until now, but also to form *branched* chains.

Example:

The compound to the left is the *linear* butane, which you already know. (The hydrogen atoms are omitted for clarity.)

The compound to the right also contains four carbon atoms, but in a *branched* chain (again omitting the hydrogen atoms).

Write down these two structural formulas complete with hydrogen atoms, and give the empirical formulas for the two compounds.   →   $\boxed{57}$

---

$\boxed{51}$    Acetylene: $C_2H_2$, H–C≡C–H

Acetylene is also called ethyne. In general, the hydrocarbons with at least one triple bond are known as alkynes.

Acetylene can be hydrogenated – by taking up how many hydrogen atoms?

- 2 → $\boxed{58}$

- 4 → $\boxed{80}$

- 6 → $\boxed{76}$

---

$\boxed{52}$    Correct. The pentane isomer

is a methylbutane.

How is the following pentane isomer called?

```
       -C-
        |
   -C-C-C-
        |
       -C-
```

- Tetramethylmethane → 70

- Trimethylethane → 68

- Dimethylpropane → 60

**7**

---

**53**

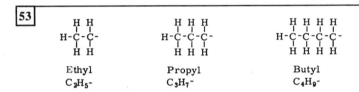

Ethyl      Propyl      Butyl
$C_2H_5-$      $C_3H_7-$      $C_4H_9-$

The names of the radicals (or residues or groups) are derived from those of the alkanes by changing the ending "ane" to "yl". (In the examples above the parent alkanes are eth*ane*, prop*ane*, and but*ane*.)

Copy and complete the following table as far as hexane:

| Methane | $CH_4$ | Methyl | $CH_3-$ |
|---|---|---|---|
| ethane | $C_2H_6$ | ethyl | $C_2H_5-$ |
| propane | $C_3H_8$ | propyl | $C_3H_7-$ |
| butane | $C_4H_{12}$ | butyl | $C_4H_9-$ |
| pentano | $C_5H_{12}$ | pentyl | $C_5H_{11}-$ |
| hexane | $C_6H_{14}$ | hexyl | $C_6H_{13}-$ |

→ 59

---

**54**     Yes, there are *four* hydrogen atoms less.

Now return to 10 .

---

**55**    $C_2H_4$           $C_3H_6$
      Ethene or ethylene     Propene or propylene

What is the structure of ethene?

We know that ethene ($C_2H_4$) has two hydrogen atoms less than ethane ($C_2H_6$).

Ethane is

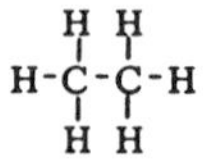

$$H-\underset{\underset{H}{|}}{\overset{\overset{H}{|}}{C}}-\underset{\underset{H}{|}}{\overset{\overset{H}{|}}{C}}-H$$

Removing two hydrogen atoms we obtain a free radical with two unpaired electrons:

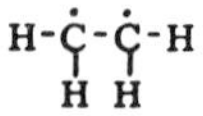

$$H-\underset{\underset{H}{|}}{\overset{\cdot}{C}}-\underset{\underset{H}{|}}{\overset{\cdot}{C}}-H$$

which we can, however, combine to form a new electron pair,

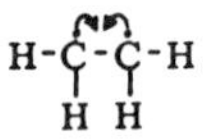

$$H-\underset{\underset{H}{|}}{\overset{\frown}{C}}-\underset{\underset{H}{|}}{\overset{}{C}}-H$$

which is the same as a new bond, symbolized by a line as usual:

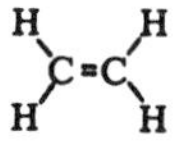

$$\underset{H}{\overset{H}{\diagdown}}C=C\underset{H}{\overset{H}{\diagup}}$$

Thus, the two carbon atoms are now joined by two lines, that is *two* bonds. This we call a *double bond*.

What is the name of the following compound?

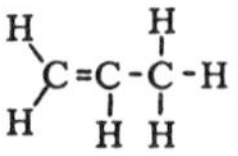

$$\underset{H}{\overset{H}{\diagdown}}C=\underset{\underset{H}{|}}{\overset{\overset{H}{|}}{C}}-\underset{\underset{H}{|}}{\overset{}{C}}-H$$

- Propane  →  [26]

- Butene  →  [74]

- Propene  →  [47]

---

[56]   You computed the molecular mass of butene, but butene has only *one* double bond; not *two*.   →  [10]

---

[57]

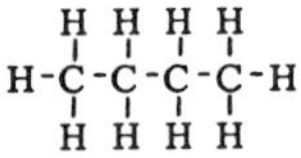

$$H-\underset{\underset{H}{|}}{\overset{\overset{H}{|}}{C}}-\underset{\underset{H}{|}}{\overset{\overset{H}{|}}{C}}-\underset{\underset{H}{|}}{\overset{\overset{H}{|}}{C}}-\underset{\underset{H}{|}}{\overset{\overset{H}{|}}{C}}-H$$

$C_4H_{10}$

$$H-\underset{\underset{H}{|}}{\overset{\overset{H}{|}}{C}}-\underset{\underset{\underset{H-C-H}{|}}{|}}{\overset{\overset{H}{|}}{C}}-\underset{\underset{H}{|}}{\overset{\overset{H}{|}}{C}}-H$$

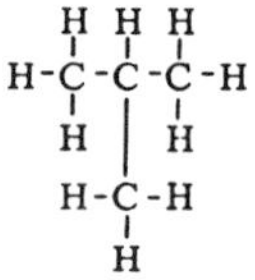

$C_4H_{10}$

As you see, there are *two* compounds with the formula $C_4H_{10}$! Compounds with identical empirical formulas but different structures are called *isomers* or *isomeric* compounds. (Pronounciation: 'isomer and iso'meric.) Thus, the two butanes shown above are *isomeric* compounds. They are different molecular entities having different melting and boiling points, for example.

Different compounds must also have different names. The linear butane (to the left) is called n-butane (n for normal). The branched butane (to the right) is called isobutane (iso for isomer).

Let us see which of the compounds presented so far have isomers.

Methane ($CH_4$), and ethane ($C_2H_6$) have *no* isomers.

Is branching possible in the case of propane, i.e. is there a propane isomer?

Write down the structure of propane before answering!

● Yes        →    49

● No         →    48

---

**58**    We can start by adding two hydrogen atoms to acetylene, $C_2H_2$, thus obtaining ethene, $C_2H_4$. However, ethene can take up a further two hydrogens, giving ethane, $C_2H_6$. Thus, a total of four hydrogen atoms can be added to acetylene.

All hydrocarbons mentioned so far are present in crude oil or can be manufactured by refining crude oil. This will be described in more detail in the next program.

However, acetylene can be obtained not only by cracking oil, but also in other ways which will now be described.

Preparation of acetylene (1).
Acetylene is formed by the addition of water to calcium carbide, $CaC_2$. Calcium hydroxide is also formed.

Formulate the reaction equation.    →    65

---

**59**

| Methane | $CH_4$ | Methyl | $CH_3-$ |
|---|---|---|---|
| ethane | $C_2H_6$ | ethyl | $C_2H_5-$ |
| propane | $C_3H_8$ | propyl | $C_3H_7-$ |
| butane | $C_4H_{10}$ | butyl | $C_4H_9-$ |
| pentane | $C_5H_{12}$ | pentyl | $C_5H_{11}-$ |
| hexane | $C_6H_{14}$ | hexyl | $C_6H_{13}-$ |

You now know the alkyl groups, e.g. the methyl and ethyl groups, etc. Write down the structural formula for the methyl group, including the hydrogen atoms.    →    45

|60|

Yes, the correct name is dimethylpropane.

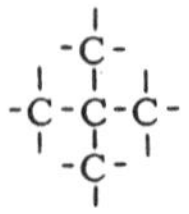

$$
\begin{array}{c}
-\overset{|}{\underset{|}{C}}- \\
-\overset{|}{\underset{|}{C}}-\overset{|}{\underset{|}{C}}-\overset{|}{\underset{|}{C}}- \\
-\overset{|}{\underset{|}{C}}-
\end{array}
$$

You now know that the existence of *isomers* depends on the possibility of having branched and unbranched carbon chains.

Try and define an isomer:

Isomers are compounds with the same .................... formula, but different .................... .

Check your answer at |67| .

---

|61| Butene, $C_4H_8$.

Ethene is also known under the name

*ethylene,*

and propene may be called

*propylene.*

Write down the empirical formulas and the *two* names for each of these compounds. → |55|

---

|62| The structure selected by you has only two carbon atoms in a horizontal row (printed in boldface below:)

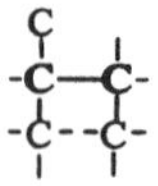

$$
\begin{array}{c}
\overset{|}{\underset{|}{C}} \\
-\mathbf{C}-\mathbf{C}- \\
-\underset{|}{C}--\underset{|}{C}-
\end{array}
$$

At |45| you will find a structure showing a longer horizontal chain. → |45|

**63**    Your number is too large!

Inspect the following:

ethane, $C_2H_6$:

$$H-\overset{\overset{\displaystyle H}{|}}{\underset{\underset{\displaystyle H}{|}}{C}}-\overset{\overset{\displaystyle H}{|}}{\underset{\underset{\displaystyle H}{|}}{C}}-H$$

ethene, $C_2H_4$:

$$\overset{H}{\underset{H}{\Large{>}}}C=C\overset{H}{\underset{H}{\Large{<}}}$$

acetylene, $C_2$... :    $-C\equiv C-$

How many hydrogen atoms are there in acetylene?    →    |51|

---

**64**    The structure selected by you has only three carbon atoms in a horizontal row:

$$-\overset{|}{C}-\overset{|}{C}-\overset{|}{C}-$$
$$-\overset{|}{C}-$$
$$-\overset{|}{C}-$$

At  |45|  you will find a structure possessing a longer horizontal chain.    →    |45|

---

**65**    $CaC_2 + 2\ H_2O \rightarrow C_2H_2 + Ca(OH)_2$

Calcium carbide is prepared technically from carbon and lime under the influence of a 2200 °C electric arc.

Preparation of acetylene (2).
Methane is passed rapidly through an electric arc at a temperature as high as 20 000 °C (in the center of the arc). Acetylene and hydrogen are formed among other products.

$$......CH_4 \rightarrow C_2H_2 + ......H_2$$

Complete this equation giving the correct numbers of molecules.    →    |73|

---

**66**    The structure selected by you has only three carbon atoms in a horizontal row:

$$-\overset{|}{C}-$$
$$-\overset{|}{C}-\overset{|}{C}-\overset{|}{C}-$$
$$-\overset{|}{C}-$$

At 45 you will find a structure possessing a longer horizontal chain. → 45

---

**67**

> Isomers are compounds with identical *empirical* formula, but different *structures*.

**7**

The number of hydrogen atoms in a branched or unbranched hydrocarbon molecule is easily calculated: it is twice the number of carbons plus two. What is the general formula expressing this?

Compare your answer at 72 .

---

**68**  The name trimethylethane for

$$
\begin{array}{c}
-\overset{|}{\underset{|}{C}}- \\
| \\
-\overset{|}{C}-\overset{|}{C}-\overset{|}{C}- \\
| \\
-\overset{|}{\underset{|}{C}}-
\end{array}
$$

is not in principle wrong, but it is not systematic and therefore not used. Compounds are named taking the longest possible horizontal carbon chain as the base. This is a three-carbon chain in the present case, i.e. it is not ethane.

Repeat from 45 .

---

**69**  Ethene, propene, and butene are *alkenes*. Alkenes are also called "olefins".

The alkenes have two hydrogen atoms less than the corresponding alkanes.

Inspect this table:

| | | | | |
|---|---|---|---|---|
| ethane | $C_2H_6$ | | ethene | $C_2H_4$ |
| propane | $C_3H_8$ | | propene | $C_3H_6$ |
| butane | $C_4H_{10}$ | | butene | .............. |

and give the formula for butene.  → 61

---

**70**  The name tetramethylmethane for

$$
\begin{array}{c}
-\overset{|}{\underset{|}{C}}- \\
| \\
-\overset{|}{C}-\overset{|}{C}-\overset{|}{C}- \\
| \\
-\overset{|}{\underset{|}{C}}-
\end{array}
$$

is not in principle wrong, but it is not systematic and therefore not used. Compounds are named taking the longest possible horizontal carbon chain as the base. This is a three-carbon chain in the present case, i.e. it is not methane.

Repeat from   45   .

---

**71**

Hexane                    Cyclohexane

The two terminal hydrogen atoms (boldface print) in hexane must be removed in order to close the chain to a ring.

The hydrocarbons discussed thus far have names ending in -ane (e.g. eth**ane,** dimethylprop**ane,** cyclohex**ane**).

The general name of these compounds is

   alkanes (or paraffins).

Note that the name "alk**ane**" also ends in *ane.*

There is another group of hydrocarbons, namely the

   alkenes (or olefins).

The names of the alkenes all end in **-ene.**

To which group do ethane, propane and butane belong? They belong to the group of ...................   →   78

---

**72**   $C_nH_{2n+2}$

Hydrocarbons are also known which do *not* have this general formula.

For example, there are hydrocarbons which have only twice as many hydrogen as carbon atoms. Such compounds are formed when an open hydrocarbon chain is closed to a ring. Therefore, they are called

   cyclic hydrocarbons

(cycle = ring).

Cyclohexane is an important example. The structure of cyclohexane is

**7**

Write down the empirical formulas for hexane and cyclohexane and compare the two. What is the difference?

Check your answer at   $\boxed{79}$ .

---

$\boxed{73}$
     $2\ CH_4 \rightarrow C_2H_2 + 3\ H_2$. A discussion of the structure of the double- and triple carbon bond, based on the orbital atomic model, is given in Program 5.

     This is the end of Program 7.

---

$\boxed{74}$
     There are *four* carbon atoms in butane.

The compound     is $C_3H_6$ = propene.

What is the alternative name for this compound?     $\rightarrow$   $\boxed{47}$

---

$\boxed{75}$
     In butane there are 4 carbons and 10 hydrogen atoms:

We have learned that *one* double bond is formed by elimination of *two* hydrogen atoms.

Butadiene has *two* double bonds and hence .................... hydrogen atoms less than butane.    $\rightarrow$   $\boxed{54}$

---

$\boxed{76}$
     If six hydrogen atoms were taken up, all three bonds between the two carbon atoms would have to be broken.

However, a single bond always remains between the two carbon atoms following hydrogenation of a double or triple bond.

Return to   $\boxed{51}$ .

---

**77**    You computed the molecular mass of butane. Butadiene has four carbon atoms but fewer hydrogen atoms than in butane since it has two double bonds.   →   10 .

**7**

**78**    Ethane, propane, and butane are *alkanes* (also known as paraffins). The names of the alkenes are quite similar, so read them carefully!

The corresponding alkenes are

      ethene, propene, and butene.

Complete the sentence:

      ethene, propene, and butene are ...................   →   69

**79**    Cyclohexane: $C_6H_{12}$;   hexane: $C_6H_{14}$

Cyclohexane has two hydrogen atoms *less* than hexane.

In order to close the hexane chain to a ring, two terminal hydrogen atoms must be removed.

Write down the structural formulas for hexane and cyclohexane, including the hydrogen atoms.   →   71

**80**    Correct.

The hydrogenation can, however, be stopped after the addition of only two hydrogen atoms.

In this case, the triple bond is converted to a double bond.

Which compound is formed by the addition of two hydrogen atoms to acetylene?

Give the name and empirical formula.   →   58

**7**

**81**     Your number is too large.

Inspect the following:

ethane $C_2H_6$:

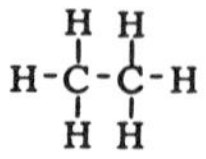

$$H-\underset{\underset{H}{|}}{\overset{\overset{H}{|}}{C}}-\underset{\underset{H}{|}}{\overset{\overset{H}{|}}{C}}-H$$

ethene $C_2H_4$:

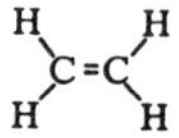

$$\underset{H}{\overset{H}{\diagdown}}C=C\underset{H}{\overset{H}{\diagup}}$$

acetylene $C_2H$......    $-C{\equiv}C-$

How many hydrogen atoms are there in acetylene?     →    51

---

**82**     Your answer is wrong.

Attachment of eight hydrogen atoms to two carbon atoms is possible only in the case of two molecules of methane: $2 \ CH_4$.

Inspect the following:

ethane $C_2H_6$:

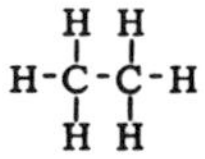

$$H-\underset{\underset{H}{|}}{\overset{\overset{H}{|}}{C}}-\underset{\underset{H}{|}}{\overset{\overset{H}{|}}{C}}-H$$

ethene $C_2H_4$:

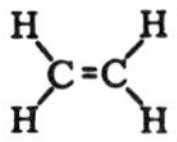

$$\underset{H}{\overset{H}{\diagdown}}C=C\underset{H}{\overset{H}{\diagup}}$$

acetylene $C_2H$ ......    $-C{\equiv}C-$

How many hydrogen atoms are there in acetylene?     →    51

# Program 8

## Hydrocarbons (II)

---

**1** The common name for open-chain and cyclic hydrocarbons is

aliphatic hydrocarbons.

These compounds were described in the preceding program. In the present
program, we shall be concerned with a particular class of cyclic hydrocarbons:
the aromatic hydrocarbons.

You already know a cyclic hydrocarbon, namely cyclohexane. Write down the
structure for $C_6H_{12}$. $\rightarrow$ $\boxed{7}$

**8**

---

**2** You have confused important phenomena.

"Polymerization" is the formation of very large molecules from many small
ones.

"Hydrogenation" is the *addition* of hydrogen.

"Dehydrogenation" is the *elimination* of hydrogen.

Return to $\boxed{14}$ .

---

**3** You have confused "hydrogenation" and "dehydrogenation".

*Hydrogenation* is the addition of hydrogen to a double bond, e.g.

$$\underset{H}{\overset{H}{\diagup}}C=C\underset{H}{\overset{H}{\diagdown}} + H_2 \longrightarrow H-\overset{\overset{H}{|}}{\underset{\underset{H}{|}}{C}}-\overset{\overset{H}{|}}{\underset{\underset{H}{|}}{C}}-H$$

*Dehydrogenation* is the opposite reaction, the elimination of hydrogen, e.g.

$$C_6H_{12} \longrightarrow C_6H_6 + 3\ H_2$$
$$\text{cyclohexane} \qquad \text{benzene}$$

The preparation of benzene from cyclohexane is a ..................... .
Complete and check at $\boxed{21}$ .

---

| 4 |

You counted *all* the bonds in the benzene molecule, including the carbon-hydrogen bonds. However, only the number of bonding electrons between the six carbon atoms was required.

Go back to    | 19 |   .

| 5 |

    1. *Aliphatic* compounds.
    2. *Aromatic* compounds.

**8**

We divide the aliphatic hydrocarbons into

a) *saturated* hydrocarbons:
   alkanes or paraffins (e.g. methane, isobutane, cyclohexane);

b) *unsaturated* hydrocarbons:
   A) with double bonds: alkenes or olefins (e.g. ethene, propene)
   B) with triple bonds: alkynes (e.g. ethyne = acetylene).

Write down the complete structures of the molecules mentioned above.

Remember that carbon is always tetravalent. Make sure that each carbon atom in your formulas has four and only four valence bonds.

Compare at    | 12 |   .

| 6 |

    Reconsider your answer.

Is *distillation* really a *chemical* process? You will probably recognize your error.

Go back to    | 15 |   .

| 7 |

Cyclohexane

Another hydrocarbon composed of six carbon atoms in a ring is

    benzene, $C_6H_6$.

Comparing the formulas for benzene and cyclohexane, you will see that benzene has *fewer* hydrogen atoms.

How many fewer?

Write down the number and check at   14   .

---

**8**     Each bond (i.e. each line) in a chemical structure represents an electron *pair,* that is: *two* bonding electrons.

Count the number of bonds (lines) between the six carbon atoms in benzene and multiply by 2 in order to obtain the number of electrons. You will find the structural formula for benzene at   19   .

**8**

---

**9**     Reconsider your answer.

Is *distillation* really a *chemical* process? You will probably recognize your error.

Go back to   15   .

---

**10**     No, they are certainly not polymers!

Polymers are very large molecules formed from many small ones.

This is evidently not the case for the xylenes.

Return to   29   .

---

**11**     Yes, the 9 bonds between the carbon atoms in benzene correspond to 18 bonding electrons.

The structural formula

gives the correct *number,* but not the precise *arrangement* of the bonding electrons in benzene. In reality, the 18 electrons are not distributed between three double and three single bonds. Rather, all the bonds between the six carbon atoms are equivalent, all having a bond order *between* that of single and double bond. The bonds being equivalent, the electrons will distribute themselves *evenly* over these bonds.

Thus, on average, there will be ............... bonding electrons between any two carbon atoms in benzene.   →   20

---

**12**    Compare your structures, and correct your errors, if any:

Aliphatic hydrocarbons

a) saturated hydrocarbons: *alkanes* or paraffins

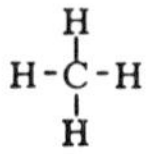

     methane 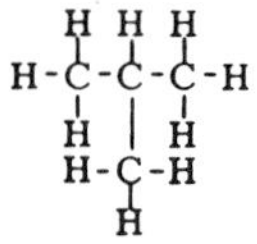           isobutane 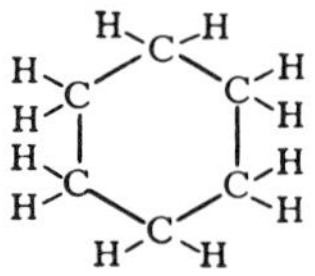           cyclohexane

b) unsaturated hydrocarbons

   A) with double bond: *alkenes* or olefins

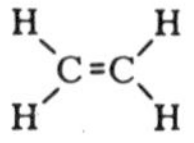

     ethene 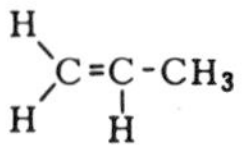           propene

   B) with triple bond: *alkynes*

$$H-C\equiv C-H$$
     ethyne or acetylene

Now define in writing alkanes, alkenes, and alkynes:

Alkanes or paraffins are .................... aliphatic hydrocarbons.
Alkenes or olefins are ....................hydrocarbons with ....................
Alkynes are .................... hydrocarbons with ....................   →   25

---

**13**    Gas oil *is* distillable.

Its boiling point was given at   74   . Answer the question again.

---

**14**    Benzene has *six* hydrogen atoms less than cyclohexane. These six hydrogens must be removed when making benzene from cyclohexane.

$$C_6H_{12} \longrightarrow C_6H_6 + 3\ H_2$$
cyclohexane     benzene

What kind of chemical reaction is this?

- Hydrogenation      →    $\boxed{3}$

- Polymerization      →    $\boxed{2}$

- Dehydrogenation    →    $\boxed{21}$

- I don't know        →    $\boxed{22}$

8

---

$\boxed{15}$     Correct; asphalt is not distillable. Hence, asphalt is the residue from the distillation of crude oil.

The fractions obtained by the initial distillation of crude oil have only limited *direct use,* except as fuels. Thus, the gasoline, kerosene, gas oil, and lubricating oil fractions are in part refined further in order to obtain basic chemicals for the chemical industry.

Three important refining methods are

1. Repeated *distillation* in order to separate mixtures into individual hydrocarbon fractions.

2. *Cracking* of larger hydrocarbons to smaller ones.

3. *Dehydrogenation* of cyclic aliphatic hydrocarbons transforms them into aromatic hydrocarbons.

One of the methods cited above is a purely physical operation; the other two are chemical processes. Pick out the *chemical* processes:

- Distillation and cracking      →    $\boxed{9}$

- Cracking and dehydrogenation      →    $\boxed{31}$

- Distillation and dehydrogenation      →    $\boxed{6}$

---

|16| Consider:

one double bond is formed by elimination of *two* hydrogen atoms.

Repeat |21| .

---

|17| Consider:

one double bond is formed by elimination of *two* hydrogen atoms.

Repeat |21| .

---

|18| Consider:

one double bond is formed by elimination of *two* hydrogen atoms.

Repeat |21| .

---

|19|

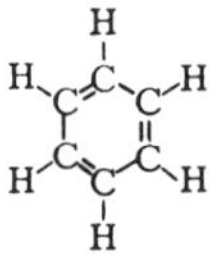

Benzene

As always, each carbon atom in benzene is tetravalent. Count the valences of each carbon atom in the formula above to verify this.

Although the above formula for benzene is generally used, only the overall arrangement of atoms is correctly depicted, not the distribution of the bonding electrons between the six carbon atoms.

Remember: each valence bond corresponds to two bonding electrons.

How many bonding electrons are there in total *between the six carbon atoms*?

● 9 　　　　　　　　　→ |26|

● 18 　　　　　　　　→ |11|

● 24 　　　　　　　　→ |4|

● I need further information → |8|

**20**    *Three.*

Thus, formally there are three bonding electrons per carbon atom. *Two* of these electrons are used for the single bonds between the carbon atom and its neighbors. The remaining electron goes to the double bond. However, these double bonds are not rigidly fixed in benzene: all six carbon-carbon bonds are identical. This can be expressed by saying that the double bonds are fluxional, or that each carbon-carbon bond has a bond order of 1.5 as shown pictorially in formula (a) below. Since the six "extra" electrons are not rigidly fixed to the individual carbon atoms, but can move around between them, we can also denote this by formula (b) where the dotted lines symbolize "half" double bonds. These two formulas are combined in formula (c), where the central circle symbolizes six electrons equally shared by the six carbon atoms:

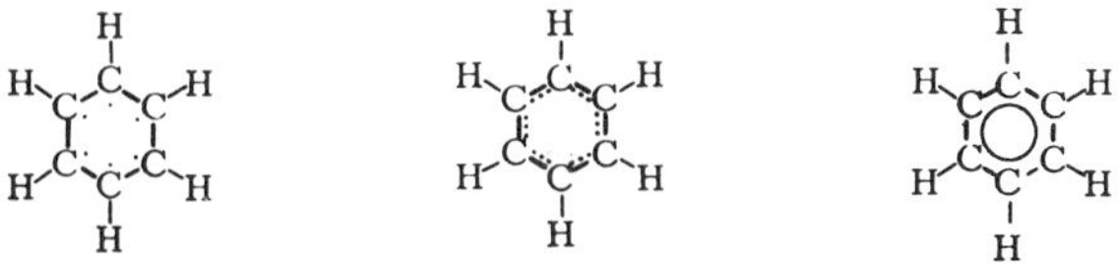

     (a)           (b)           (c)

An explanation of the structure of the benzene molecule, based on the orbital atomic model is given in Program 5. To summarize:

Of the 18 electrons, 12 are used pairwise to form six bonds between the six carbon atoms. The remaining six electrons are evenly distributed among the six carbon atoms.

This should be kept in mind when using the habitual formula for benzene with alternating single and double bonds as shown in  |19| .

Write down this formula once more. The best way to learn and memorize structural formulas is by frequent drawing!   →   |27|

---

**21**    Correct; it is a *dehydrogenation.*

Unsaturated compounds are formed by dehydrogenation of hydrocarbons. Unsaturated compounds are those containing double or triple bonds.

Consider:

Benzene is formed by dehydrogenation of cyclohexane. Six hydrogen atoms are eliminated in this process. How many double bonds are there in benzene?

● One      →    |16|

● Two      →    |17|

---

**22**

The addition of hydrogen to a double bond is a *hydrogenation,* e.g.

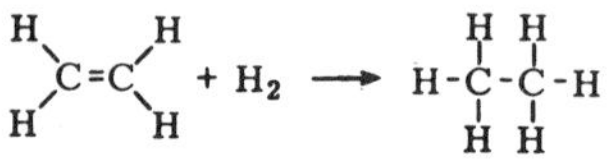

$$\underset{H}{\overset{H}{>}}C=C\underset{H}{\overset{H}{<}} \;+\; H_2 \;\longrightarrow\; H-\underset{\underset{H}{|}}{\overset{\overset{H}{|}}{C}}-\underset{\underset{H}{|}}{\overset{\overset{H}{|}}{C}}-H$$

**8**

A *dehydrogenation* is the opposite reaction, viz. elimination of hydrogen, e.g.

$$C_6H_{12} \;\longrightarrow\; C_6H_6 + 3\,H_2$$
cyclohexane      benzene

So what kind of reaction is the transformation of cyclohexane
into benzene?    →    21

---

**23**

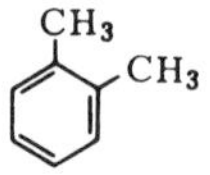

ortho−xylene 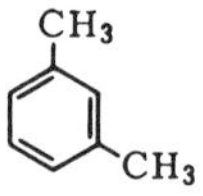         meta−xylene 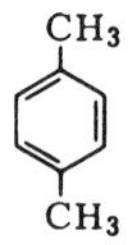         para−xylene

There are no other dimethylbenzene isomers. Any attempt to arrange the two
methyl groups in another way will only lead to formulas *equivalent* to the
above. Verify this by inspecting the following formulas (and turning them, if
necessary).

o−xylene

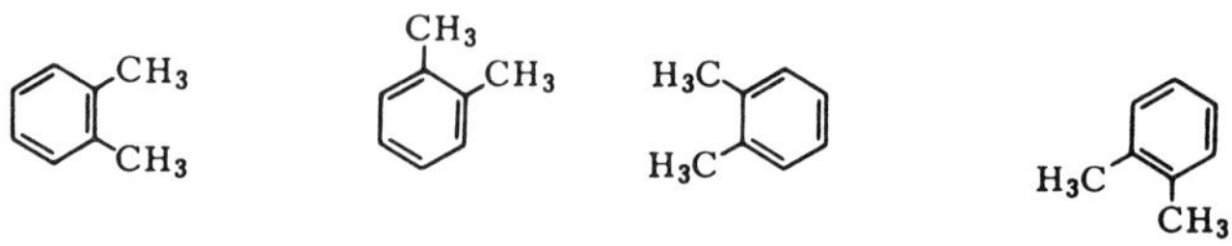

m−xylene

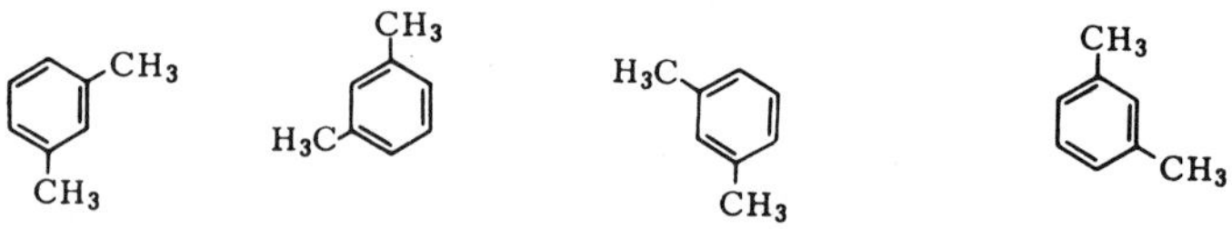

p−xylene

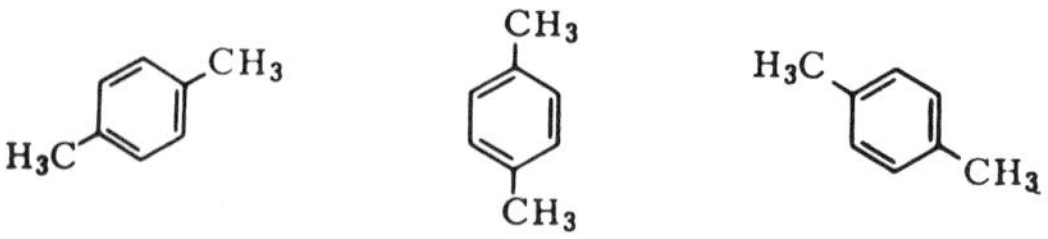

So, how many isomeric xylenes are there?    →    29

---

|24|     No, they are not isotopes.

Isotopes are not chemical *compounds,* but elements. Isotopes have the same atomic number, nuclear charge, number of electrons, and the same place in the periodic table. Only the atomic masses are different.

Repeat   |29|   .

|25|     Alkanes or paraffins are *saturated* aliphatic hydrocarbons.                     **8**

Alkenes or olefins are *unsaturated aliphatic* hydrocarbons with *double bond.*

Alkynes are *unsaturated aliphatic* hydrocarbons with *triple bond.*

Let us now look at the *properties of the hydrocarbons.*

The hydrocarbons can be gases, liquids, or solids at room temperature (20 °C) and atmospheric pressure (760 Tor; ca. 1 bar) as shown in the following table numbers in parenthesis are the molecular masses):

           gases:    e.g. methane (16), ethane (30), propane (44), ethene (28),
                     acetylene (26)

           liquids:  e.g. hexane (86), cyclohexane (84), benzene (78), toluene
                     (92)

           solids:   e.g. naphthalene (128).

Note the variation of state as a function of molecular mass!

Complete the following sentences:

a) Molecules with small mass are normally ...................

b) Molecules with medium mass are normally ...................

c) Heavier molecules are normally ...................   →   |33|

|26|     You only counted the number of bonds between the six carbon atoms in the ring. Recall that each bond corresponds to *two* electrons.

Repeat   |19|   .

27

Rather than always drawing the detailed structural formula for benzene (left), the simplified formulas (right) are often used.

**8**

A large number of derivatives of benzene is known in which one or more hydrogen atoms are replaced by $-CH_3$, $-Cl$, $-NO_2$, $-OH$, or other atoms or groups.

*Toluene* is the compound resulting from the replacement of a hydrogen atom in benzene by a methyl group. What, therefore, is the empirical formula for toluene?

- $C_7H_7$                          →      42

- $C_7H_8$                          →      34

- $C_7H_6$                          →      45

- I need further information        →      49

28        Yes, there are *three* double bonds in benzene.

We write the structure of benzene with *alternating* single and double bonds between the six carbon atoms in the ring. Thus, each carbon has one more valence which forms the bond to hydrogen.

Complete the structural formula for benzene below by inserting the bonds:

$$
\begin{array}{ccc}
 & H & \\
H & C & H \\
 C & & C \\
H & C & H \\
C & & C \\
 & H & 
\end{array}
$$

→      19

29        Correct; there are only three xylenes.

They are differentiated by the prefixes *ortho-*, *meta-*, and *para-* (or *o-*, *m-*, and *p-*).

The three xylenes have identical empirical formulas, but different structures.

Such compounds are called:

- Isotopes    → 24

- Polymers   → 10

- Isomers    → 38

- Alkanes    → 36

**8**

---

**30**    You are not yet familiar with the naming of hydrocarbons.

Repeat from   1 .

---

**31**    Correct; *cracking and dehydrogenation* are chemical reactions.

We will now consider these two methods of refining crude oil in more detail.

*Cracking* is the process of cleaving large molecules (e.g. with 10–20 carbon atoms) into smaller ones (e.g. with 2–4 carbon atoms).

The cracking of hydrocarbons is achieved by heating to high temperatures in the gaseous state for short periods of time.

To recapitulate: the transformation of larger hydrocarbon molecules into smaller ones is called ................... → 39

---

**32**    You gave the correct name of *para*-xylene only.

Repeat from   37 .

---

**33**    *Gases*
         *liquids*
         *solids*

Sharply defined borders between these states *cannot* be given (for example a particular molecular mass) since boiling and melting points are dependent *not only* on the molecular mass but on other factors as well. Nevertheless, the increasing boiling points within a homologous series are clearly seen in the following table:

|          | Molecular mass | Boiling point (°C) |
|----------|----------------|--------------------|
| Methane  | 16             | − 161              |
| ethane   | 30             | − 89               |
| propane  | 44             | − 42               |
| butane   | 58             | − 0.5              |
| pentane  | 72             | + 36               |
| hexane   | 86             | + 69               |

(You do not need to memorize these numbers!).

The molecular masses of the three compounds below are given in parenthesis:

toluene (92), xylene (106), benzene (78).

Order these compounds according to increasing *boiling point,* so that the compound with the lowest boiling point stands to the left.

................... ..................... ...................  →  $\boxed{51}$

---

$\boxed{34}$   Toluene has the empirical formula $C_7H_8$.

You will obtain the structure of toluene by replacing one hydrogen atom in benzene by a methyl group.

Write down the structure and compare at   $\boxed{43}$   .

---

$\boxed{35}$   You are not yet familiar with the naming of hydrocarbons.

Repeat from   $\boxed{43}$   .

---

$\boxed{36}$   Alkanes are *saturated* hydrocarbons. Their names end with *-ane* as in eth*ane* or hex*ane*. Xylene is *not* an alkane.

Repeat   $\boxed{29}$   .

---

$\boxed{37}$   You have probably realized that there are three ways in which the methyl groups can be attached to the benzene ring in the xylenes. Your answer is correct if you gave *one* of the following structures:

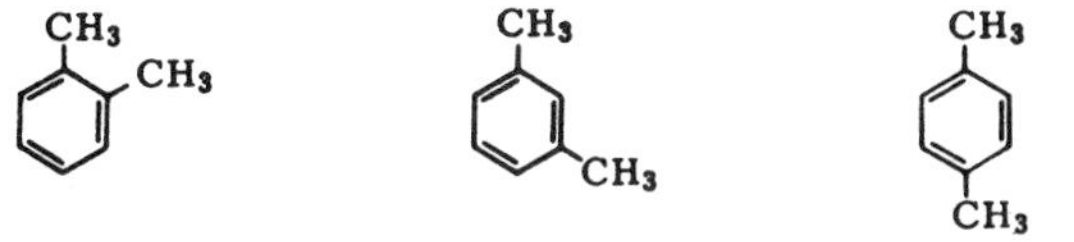

name:  *ortho*-xylene        *meta*-xylene        *para*-xylene
or     *o*-xylene           *m*-xylene           *p*-xylene

Are there other xylenes, i.e. are there other ways of attaching two methyl groups to a benzene ring leading to compounds different from the above?

● Yes        →    23

● No         →    29

---

**38**      Correct, they are isomers. The *isomeric* xylenes all have the same empirical formula, $C_8H_{10}$, but different structures.

Write down the structures of the three xylenes using the simplified benzene formula.   →    46

---

**39**    *Cracking*

Ethene, propene, and the isomeric butenes are some of the most important products formed by cracking. You will notice that *unsaturated* hydrocarbons predominate.

The possible cracking of a ten-carbon molecule is illustrated below. The molecules formed are only examples; smaller hydrocarbon molecules could be formed as well:

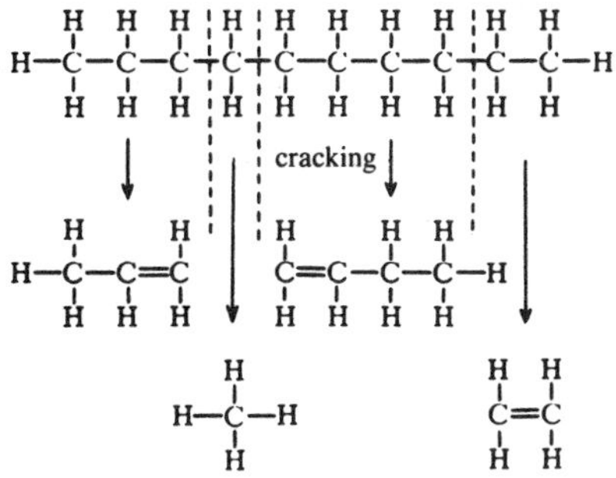

Write down the names of the four products shown.   →    56

---

| 40 |

You gave the correct name of *ortho*-xylene only.

Repeat from   | 37 |   .

---

| 41 |

You are not yet familiar with the naming of hydrocarbons.

Repeat from   | 43 |   .

**8**

---

| 42 |   $C_7H_7$ is not the formula.

Benzene is $C_6H_6$.

Replace one hydrogen atom (H) in benzene by a methyl residue ($CH_3$) in order to derive toluene.

What is the empirical formula for toluene?    →   | 34 |

---

| 43 |

Toluene: 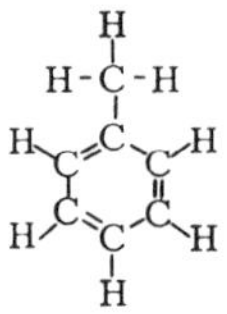 or simplified:

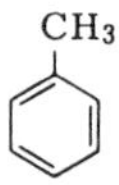

*Xylene* is similarly obtained by replacing *two* hydrogen atoms in benzene by methyl groups.

Which empirical formula does this give for xylene?

- $C_7H_8$                 →   | 54 |

- $C_8H_{10}$               →   | 50 |

- $C_8H_{12}$               →   | 57 |

---

| 44 |

You are not yet familiar with the naming of hydrocarbons.

Repeat from   | 43 |   .

---

**45**     $C_6H_7$ is not the formula.

Benzene is $C_6H_6$.

Replace one hydrogen atom (H) in benzene by a methyl residue ($CH_3$) in order to derive toluene.

What is the empirical formula for toluene?     →    34

**8**

---

**46**

          (a)                        (b)                        (c)

Give the names of these three xylenes.

● a) *ortho-*       b) *para-*        c) *meta*-xylene   →    40

● a) *meta-*       b) *ortho-*      c) *para*-xylene   →    32

● a) *para-*       b) *meta-*       c) ortho-xylene   →    65

● a) *ortho-*       b) *meta-*       c) *para*-xylene   →    55

---

**47**     No. A detached structure of styrene is given below in order to help you identify the components:

      H-C=CH₂

Repeat the question at    69   .

---

**48**     You are not yet familiar with the naming of important hydrocarbons.

Repeat from    34   .

---

---

**49**    Benzene is $C_6H_6$.

Replacing one H in benzene by $CH_3$ gives toluene.

What is the empirical formula for toluene?    →    34

---

**50**    Correct; xylene is $C_8H_{10}$.

Now construct the xylene molecule starting from the benzene structure by replacing two hydrogen atoms by methyl groups.    →    37

---

**51**    Benzene (80 °C), toluene (111 °C), xylene (mixture of isomers) (140 °C).

(The boiling points are given in parentheses).

Note that the boiling points increase with increasing molecular mass.

The chemical properties of the hydrocarbons, e.g. reaction witth chlorine or nitric acid, will be treated in subsequent programs. At this point we will describe the process of combustion only.

Hydrocarbons will burn, sometimes explosively, when ignited in the presence of sufficient oxygen. They are completely burnt to $CO_2$ and $H_2O$ in the process.

$CH_4$ + .................... → ............... + ...............

Complete the equation for the combustion of methane.    →    58

---

**52**    You erroneously gave twice the empirical formula for benzene as the composition of naphthalene. Note, however, that the two benzene rings in naphthalene have two carbon atoms in common. Count the carbon atoms!    →    62

---

**53**    You ignored the fact that the two carbon atoms through which the two benzene rings are joined have no free valences. Hence, they cannot carry any hydrogen atoms.

What is the correct empirical formula for naphthalene?    →    69

**54** No, it is not $C_7H_8$.

You removed *two* hydrogen atoms from benzene, but added only *one* methyl group. Both hydrogens should be replaced by methyl groups.

Go back to 43 and try again.

**8**

**55** Your answer is correct.

The name "xylene" without any further specification denotes a mixture of the three xylene isomers.

When naming a particular xylene, the appropriate prefix must be used.

Write down the names of the three xylenes, giving the complete as well as the abbreviated prefixes. → 62

**56** Propene (= propylene), methane, butene, ethene (= ethylene).

As already mentioned, these are not the only hydrocarbons formed in cracking reactions. For example, ethane, propane, butane, and isobutane may also be formed. Thus, branched and unbranched, saturated and unsaturated products are possible.

The *unsaturated* products are of special importance.

The *mixtures* of hydrocarbons obtained by cracking must – depending on their future use – be separated more or less completely, e.g. by distillation.

Another chemical procedure for refining crude oil is *dehydrogenation,* e.g.:

Write down the completed equation together with the names of all the compounds. → 64

**57**

$C_8H_{12}$ is not the formula.

You simply added two $CH_3$-groups to benzene without removing the two hydrogen atoms.

Repeat  43 .

**8**

**58**

$$CH_4 + 2 O_2 \rightarrow CO_2 + 2 H_2O$$

The danger of explosion is particularly pronounced for acetylene. Acetylene-air mixtures easily detonate. Therefore: **extreme caution should be taken when working with acetylene.**

Write down the equation for the complete combustion of acetylene. → 66

**59**

Acetylene has a triple bond!

$$H–C≡C–H$$

Repeat  69 .

**60**

Wrong! The boiling points of the fractions were given in the text. Study this carefully. → 74

**61**

Toluene

benzyl
(radical or residue)

Unfortunately, the benzyl and phenyl radical or residue or group are easily confused, since "benzyl" could logically have been derived from "benzene". You should carefully note the interrelationship between these names, completing the following sentences:

a) The phenyl residue is derived from ...................

b) The benzyl residue is derived from ................... → 70

**62**

| ortho-xylene | | o-xylene |
|---|---|---|
| meta-xylene | or | m-xylene |
| para-xylene | | p-xylene |

*Naphthalene* is another important compound containing benzene rings.
(Naphthalene is pronounced "naftaleene".) The following is an incomplete
formula,

showing that naphthalene consists of two "condensed" (fused together)
benzene rings. The hydrogen atoms are omitted in this formula.

You should now redraw this formula completed with hydrogen atoms.
Make sure that each carbon atom is tetravalent! The empirical formula for
naphthalene is:

- $C_{10}H_{10}$  →  53

- $C_{10}H_8$  →  69

- $C_{12}H_{12}$  →  52

---

**63**

You have confused

the ethyl and vinyl groups, and
the benzyl and phenyl groups.

Repeat from  83  .

---

**64**

cyclohexane      dehydrogenation      benzene      +      3 H₂      hydrogen

Thus, aromatic hydrocarbons can be obtained from alicyclic ones
(= cyclic aliphatic hydrocarbons) by dehydrogenation.

Describe briefly for which *purpose* the following methods are used as regards
crude oil refining:

a) Distillation

b) Cracking

c) Dehydrogenation  →  |73|

---

|65|  You only named *meta*-xylene correctly.

Repeat from  |37|  .

---

|66|  $2 C_2H_2 + 5 O_2 \rightarrow 4 CO_2 + 2 H_2O$

Care should be exercised not only with acetylene but with all hydrocarbons:

> All hydrocarbons are easily inflammable and may form explosive mixtures with air.

The numbers of carbon and hydrogen atoms in a given compound can be experimentally determined by controlled combustion, whereby the amounts of $CO_2$ and $H_2O$ formed are quantitatively measured.

Where are the hydrocarbons found, and how are they produced?

They occur in natural gas, in mineral oil, and in tar. Natural gas is mainly methane, and its predominant use is for heating purposes.

Complete the sentence:

Natural gas is largely .................... and therefore it is highly ....................  →  |72|

---

|67|  You only named styrene correctly.

Repeat from  |76|  .

---

|68|  *Toluene*

When a hydrogen atom is removed from the methyl group in toluene, the benzyl radical is obtained. Draw the structural formulas for toluene and the benzyl radical and write their names underneath.  →  |61|

**69**     Naphthalene: empirical formula $C_{10}H_8$.

Structure:     or simpler

There are higher aromatics known in which three, four, five or more benzene rings are fused together. However, naphthalene will suffice as an example.

Let us now consider another derivative of benzene, namely

styrene

which is an important starting material for obtaining synthetic polymers (see Program 23).

The structure of styrene is:

$HC=CH_2$

Here you should recognize parts from two familiar compounds. Which ones?

● Benzene and acetylene     →     59

● Toluene and methane     →     47

● Benzene and ethene     →     76

● I don't understand the question     →     77

**70**     a) The phenyl residue is derived from *benzene*.

b) The benzyl residue is derived from *toluene*.

Let us briefly return to styrene. Remember the polymerization of ethene (ethylene). Styrene can polymerize in a similar manner:

Individual styrene molecules

polymerize to a
polymer,
polystyrene:

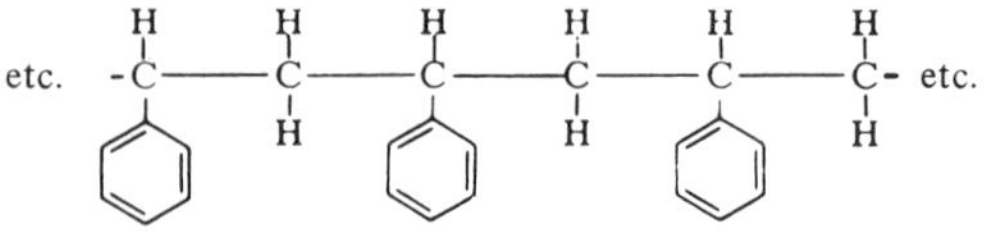

etc. $-\overset{\overset{\displaystyle H}{|}}{\underset{\underset{\displaystyle \bigcirc}{|}}{C}}\!-\!\overset{\overset{\displaystyle H}{|}}{\underset{\underset{\displaystyle H}{|}}{C}}\!-\!\overset{\overset{\displaystyle H}{|}}{\underset{\underset{\displaystyle \bigcirc}{|}}{C}}\!-\!\overset{\overset{\displaystyle H}{|}}{\underset{\underset{\displaystyle H}{|}}{C}}\!-\!\overset{\overset{\displaystyle H}{|}}{\underset{\underset{\displaystyle \bigcirc}{|}}{C}}\!-\!\overset{\overset{\displaystyle H}{|}}{\underset{\underset{\displaystyle H}{|}}{C}}\!-$ etc.

Thus, polystyrene consists of long carbon chains in which every second carbon atom carries a phenyl group.

The styrene molecule can be constructed by the union of two radicals. What are they called?

- Ethyl and benzyl radicals　　$\rightarrow$　　$\boxed{63}$

- Ethyl and phenyl radicals　　$\rightarrow$　　$\boxed{80}$

- Vinyl and  phenyl radicals　　$\rightarrow$　　$\boxed{82}$

- Vinyl and benzyl radicals　　$\rightarrow$　　$\boxed{86}$

---

$\boxed{71}$　　You gave the correct name of styrene only.

Repeat from　$\boxed{76}$　.

---

$\boxed{72}$　　The sentence should read:

"Natural gas is largely methane, and therefore it is highly inflammable".

Mineral oil (also called petroleum, crude oil, or simply "crude") is a mixture of many different hydrocarbons. All the hydrocarbons mentioned so far, and many more, are found in crude oil.

Today, crude oil is the most important source of hydrocarbons. Therefore, it is an essential raw material for the production of organic chemicals.

As we shall see later, most organic compounds are prepared directly or indirectly from chemicals produced from mineral oil.

What is crude oil composed of? Write down your answer and compare at　$\boxed{79}$　.

---

$\boxed{73}$　　The answers should read:

a) Distillation:　　　　separation of hydrocarbons.

b) Cracking:         production of small (especially unsaturated) hydrocarbon molecules from large ones.

c) Dehydrogenation: production of aromatic hydrocarbons.

These and other methods are widely used today in the production of most of the raw materials for the organic chemical industry.

Tar, obtained by the dry distillation of coal (see "Chemistry made easy – a Programmed Course for Self-Instruction I") was previously the major source of organic starting materials.

**8**

The distillation of tar yields mainly aromatic hydrocarbons, such as benzene, toluene, xylene, and naphthalene.

Coal and coal tar are still used today for this purpose, and their importance is rapidly increasing.
Still, the most important raw material for organic chemistry at present is
...............

→    84

---

**74**

Gasoline, kerosene, gas oil, lubricating oils, asphalt.

These mineral oil fractions are also mixtures of various hydrocarbons.

*Gasoline* is the hydrocarbon mixture with boiling point not higher than 150 °C.

*Kerosene* is the hydrocarbon mixture boiling between ca 150 °C and 250 °C.

*Gas oil* is the fraction boiling between ca. 250 °C and 350 °C.

Which of the five major constituents of crude oil is *not* distillable?

● Lubricating oils     →    91

● Gas oil            →    13

● Kerosene         →    97

● Gasoline          →    60

● Asphalt           →    15

|75|    Compare:

| Methane | $CH_4$ | Methyl radical | $-CH_3$ |
| ethane | $C_2H_6$ | ethyl radical | $-C_2H_5$ |
| propane | $C_3H_8$ | propyl radical | $-C_3H_7$ |
| benzene | $C_6H_6$ | phenyl radical | $-C_6H_5$ |
| ethene | $C_2H_4$ | vinyl radical | $-C_2H_3$ |

**8**

As you will have noticed, particular care is necessary with "vinyl" and "phenyl" since these radical names bear no resemblance to the hydrocarbons from which they are derived.

Another complication is the *benzyl radical:*

which is *not* derived from benzene, but from ..................... . Write down the missing word.  →  |68|

---

|76|    Yes, the styrene molecule can be thought of as arising from the union of benzene with ethene.

Removing one hydrogen atom from ethene we obtain the radical

$$H-\overset{|}{C}=CH_2$$

Similarly, removing one H from benzene we get

   or simpler   

By combining the two radicals derived from ethene and benzene, one obtains the styrene molecule.

Write down both the structure and the empirical formula for styrene.  →  |83|

---

|77|    The following example should help you understand and answer the question posed at |69| .

Naphthalene is   

You will see that this molecule is composed of two familiar rings. What is the compound with *one* such ring?

Answer:                benzene

Now try and answer the question at    69  .

---

**78**    Your answer is correct.

**8**

What are the names of the radicals and molecules given below?

H-C=CH₂

                                         H-C=CH₂

  (a)              (b)              (c)

● Benzene        styrene          ethene        →    71

● Phenyl         styrene          vinyl         →    85

● Vinyl          styrene          phenyl        →    67

● Phenyl         toluene          vinyl         →    87

---

**79**    A correct answer is "crude oil is a mixture of several different hydrocarbons".

Crude oil cannot be used directly for chemical reactions. The hydrocarbon mixture must first be separated by distillation or other means.

The following fractions are isolated sequentially by distillation at increasing temperatures:

        gasoline
        kerosene
        gas oil
        lubricating oils.

*Gasoline* distills off first; then comes *kerosone* followed by *gas oil* at higher temperatures. The last fraction composed of *lubricating oils* is isolated by distillation in vacuum. The distillation residue is *asphalt.*

You surely know some of the uses of these products.

Complete the sentences:

.................... is used as fuel in cars.

.................... is used for making roads.

.................... is used as a jet fuel and was previously used for illumination.

.................... are used as lubricants.

.................... is used in diesel engines and furnaces.    →    88

---

**80**    You are still confusing the ethyl and vinyl radicals.

Repeat from    83    .

---

**81**    The name "xylene" is not complete. You must use a prefix to specify the positions of the methyl groups.    →    89

---

**82**    Your answer is correct.

You now know the names, structures, and formulas of the most important hydrocarbons. Before continuing with their properties, occurrence, and production, let us quickly recapitulate.

It is important that you are familiar with the names and structures. (Knowing the structures, you can easily derive the empirical formulas.) Due to the large number of new structures presented in this program, it is necessary to exercise them thoroughly.

What are the names of the following compounds?

|  | (a) | (b) | (c) |  |
|---|---|---|---|---|
| ● | Toluene | benzene | styrene | → 30 |
| ● | Benzene | xylene | toluene | → 48 |
| ● | Benzene | phenyl | o-xylene | → 90 |
| ● | Benzene | toluene | o-xylene | → 89 |
| ● | Benzene | toluene | xylene | → 95 |

---

**83**

Styrene: $C_8H_8$,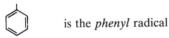

The radicals derived from ethene and benzene by removal of one hydrogen atom have special names:

$$-\overset{\text{H}}{\underset{}{\text{C}}}=\text{CH}_2 \quad \text{is the } \textit{vinyl} \text{ radical}$$

is the *phenyl* radical

Since these names will be used frequently, you should memorize them!

The ending (suffix) -yl always denotes a *radical* (= residue or group).

Write down and complete the following table. Take care to give the correct names of the radicals derived from benzene and ethene.

| | | | |
|---|---|---|---|
| Methane | $CH_4$, | methyl radical | $-CH_3$ |
| ethane | .............., | ethyl radical | $-C_2H_5$ |
| propane | .............., | ........... radical | ............ |
| benzene | .............., | ........... radical | ............ |
| ethene | .............., | ........... radical | ............  → 75 |

---

**84**     Crude oil.

Give the names of at least two aromatics which can be isolated from
*tar.*   →   98

---

**85**      Your answer is correct.

We will now review the classification of hydrocarbons. There are two major groups:

1. Aliphatic hydrocarbons (or aliphatics).
2. Aromatic hydrocarbons (or aromatics).

All compounds mentioned in Program 8 were aliphatic hydrocarbons.

8

The compounds described so far in this program are aromatics.

Thus, the *aliphatic hydrocarbons* are all the straight-chain, branched, or cyclic hydrocarbons *except* benzene and its derivatives.

Benzene and its derivatives are the *aromatic hydrocarbons* (that is, all compounds containing one or more benzene rings.).

Pick out the line below which contains *aliphatic* hydrocarbons *only:*

**8**

● Methane, toluene, styrene, acetylene, cyclohexane   →   93

● Cyclohexane, pentane, ethene, *o*-xylene, ethane   →   94

● Isobutane, cyclohexane, ethene, hexane, acetylene   →   96

● Styrene, *p*-xylene, ethane, propane, butane   →   92

---

**86**   You are still confusing the phenyl and benzyl radicals!

Repeat from   83   .

---

**87**   You gave the correct names of the phenyl and vinyl radicals.

However, the compound

H–C=CH₂

is styrene.

Toluene has the structure   CH₃

Be sure to note the difference between these two compounds in order not to repeat your mistake.

Go now to   85   .

---

**88**   Gasoline, asphalt, kerosene, lubricating oils, gas oil.

These are the fractions obtained by distillation of crude oil.

In which *order* are the fractions abtained upon the distillation of crude oil?

Write down the names in the right order:

     1. Fraction:     ....................

     2. Fraction:     ....................

     3. Fraction:     ....................

     4. Fraction:     ....................

     Residue:      ....................  →  |74|

8

---

**89**   Correct; the compounds are benzene, toluene, and *ortho*-xylene.

What are the following compounds called?

     **(a)**          **(b)**          **(c)**

| | | | | |
|---|---|---|---|---|
| ● | Xylene | naphthalene | xylene | → |81| |
| ● | Styrene | naphthalene | phenyl | → |41| |
| ● | p-Xylene | styrene | o-xylene | → |44| |
| ● | p-Xylene | naphthalene | o-xylene | → |78| |
| ● | m-Xylene | naphthalene | p-xylene | → |35| |

---

**90**   You are still not familiar with the names of the important hydrocarbons.

Repeat from   |83|   .

---

**91**   Lubricating oil is not distillable at ordinary pressure, having a boiling point above 350 °C. It would decompose at this high temperature.

Therefore, lubricating oil is distilled in a vacuum, the advantage being a lower boiling temperature.

Go back to   |74|   .

---

---

**92**      Styrene and *p*-xylene are *not* aliphatic compounds. Write down the structural formulas for these two molecules.    →    99

---

**93**      Toluene and styrene are *not* aliphatic compounds. Write down the structural formulas for these two molecules.    →    99

---

**8**    **94**      *o*-Xylene is *not* an aliphatic compound.

Write down the structural formula for *o*-xylene.    →    99

---

**95**      You gave the proper names of benzene and toluene. However, the word "xylene" denotes a *mixture* of the three xylene isomers.

A particular xylene isomer was asked for, namely

What is the *complete* name of this material?    →    89

---

**96**      Your answer is correct.

Now complete the following sentences:

1. Isobutane, cyclohexane, ethene, hexane, and acetylene are ................... compounds.

2. Benzene, toluene, the xylenes, and naphthalene are ..................... compounds.    →    5

---

**97**      Kerosene is in fact distillable. Its boiling range was given in the last section. Read   74   again.

---

**98**      Benzene, toluene, xylene, naphthalene.

This is the end of Program 8.

---

| toluene | styrene | $o$-xylene | $p$-xylene |

Each of these compounds possesses a benzene ring. Hence, they are *aromatic* compounds.

Go back to　85 .

8

# Program 9

## Halogen Compounds (I)

---

**1**    The *hydrocarbons* were presented in the two preceding programs. These organic compounds are composed of two elements only: carbon and hydrogen.

We shall now be concerned with organic compounds containing *other elements* in addition to carbon and hydrogen. First the halogens, *fluorine, chlorine, bromine,* and *iodine.*

The *chlorine* compounds are the most important organic halogen compounds.

A chlorine compound is obtained by the replacement of one or more hydrogen atoms in a hydrocarbon by chlorine atoms. (The practical execution of this reaction will be discussed below.)

What is the empirical formula for the compound obtained by the replacement of *one hydrogen atom* in the simplest of hydrocarbons (methane) by *one chlorine atom?*

- $CH_4Cl$    $\rightarrow$    8

- $C_2H_5Cl$    $\rightarrow$    13

- $CH_3Cl$    $\rightarrow$    11

---

**2**

$$H-\overset{\overset{\displaystyle H}{|}}{\underset{\underset{\displaystyle H}{|}}{C}}-Cl \qquad \text{methyl chloride}$$

Not only one, but also two, three, or all four hydrogens in methane can be replaced by chlorine atoms.

Write down the empirical formulas for the compounds obtained by replacement of two, three, or four hydrogen atoms in methane by chlorine.    $\rightarrow$    12

---

**3**    Your names for the four compounds are all wrong.

Repeat from 1 .

---

---

**4**     Your chosen reaction conditions would *not* lead to the formation of *o*-chlorotoluene.

The preparation of *o*-chlorotoluene by chlorination of toluene is a

- ring chlorination      →    $\boxed{61}$

- side-chain chlorination      →    $\boxed{54}$

---

**5**

$$CH_3\text{-benzene} + Cl_2 \xrightarrow[\text{catalyst}]{20°C} CH_3\text{-benzene-Cl} + HCl$$

This is a *ring* chlorination. Characteristically, such chlorinations are carried out at relatively low temperatures and in the presence of a catalyst.

Write down the characteristic features of this type of chlorination.    →    $\boxed{14}$

---

**6**     No, 92 g of benzyl chloride are *not* formed.

The chlorination of 92 g (1 mol) of toluene gives *1 mol* of benzyl chloride.

Go back to    $\boxed{43}$   .

---

**7**     The *ring* chlorination is catalyzed by *iron* or *iodine*.

Which temperature is used for this reaction?

- Boiling point temperature      →    $\boxed{38}$

- Room temperature      →    $\boxed{45}$

---

**8**     You just *added* a chlorine atom to the formula for methane, $CH_4$.

However, a hydrogen atom should be *replaced* by a chlorine atom. Remove a hydrogen atom from $CH_4$ and replace it by Cl.    →    $\boxed{1}$

---

**9**     The *ring* chlorination is catalyzed by *iron* or *iodine*.

Which temperature is used for this reaction?

● Boiling point temperature      →      38

● Room temperature               →      45

---

**10**      Your chosen reaction conditions would *not* lead to the formation of *o*-chlorotoluene.

The preparation of *o*-chlorotoluene by chlorination of toluene is a

● ring chlorination          →      61

● side-chain chlorination     →      54

**9**

---

**11**      Yes, $CH_3Cl$.

This compound is called methyl chloride. Note the presence of a *methyl group*.

Write down the structural formula for this compound.    →    2

---

**12**      $CH_2Cl_2$                    $CHCl_3$                         $CCl_4$
      methylene chloride        chloroform          carbon tetrachlorine

Memorize these names. You probably know chloroform already, also perhaps carbon tetrachloride.

Write down the structural formulas for the three compounds together with their names.    →    19

---

**13**      You started from ethane, $C_2H_6$. *Methane* is required in order to obtain the product of replacement of a hydrogen atom by chlorine.

Go back to   1  .

---

**14**      Low temperature and catalysts favor ring chlorination.

Iron (in the form of iron filings or an iron salt) or iodine are used as catalysts.

What are the characteristic reaction conditions for side-chain chlorination?    →    34

**15**

No, [structure: benzene ring with Cl at top and Cl at lower right] is *m*-dichlorobenzene

Remember the formula for *o*-xylene: the two methyl groups are on neighboring carbon atoms:

[structure: benzene ring with two adjacent CH₃ groups]

The same is true for *o*-dichlorobenzene: [structure: benzene ring with two adjacent Cl groups]

In *m*-xylene the two substituted carbon atoms are *not* neighbors, but separated by one carbon:

[structure: benzene ring with two CH₃ groups in meta position]

Similarly in *m*-dichlorobenzene: [structure: benzene ring with two Cl groups in meta position]

In *p*-xylene the two methyl groups are opposite: [structure: benzene ring with two CH₃ groups in para position]

and likewise in *p*-dichlorobenzene: [structure: benzene ring with two Cl groups in para position]

Now write down the structural formulas for all three dichlorobenzenes and compare at  **53**

**16**  The *ring* chlorination is catalyzed by *iron* or *iodine*.

Which temperature is used for this reaction?

● Boiling point temperature  →  **38**

● Room temperature  →  **45**

**17**     "Carbon tetrachloride" is a correct and common name for $CCl_4$.
However, the strictly systematic name is a different one.    →   19

**18**     Your answer is correct!

We will now introduce an important term before continuing with chlorine compounds.

We have seen that chlorine compounds are derived from hydrocarbons by the replacement of hydrogen atoms by chlorine. The chemical procedure of replacing hydrogen atoms by other atoms or groups is called *substitution*.

**9**

Thus, methyl chloride is the product of the *substitution* of chlorine for a hydrogen atom in methane. We can also say that chlorine has .................... a hydrogen atom.

Complete this sentence.    →   25

**19**

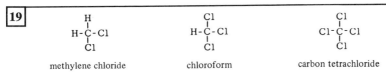

| methylene chloride | chloroform | carbon tetrachloride |

Unfortunately, many organic compounds have more than one common name which it may be necessary to learn.

By international convention each molecule can be named *systematically* in a unique fashion. Nevertheless, other, non-systematic, names are in current use, largely for historical reasons.

This is the case, for example, for the four chlorine compounds which we have just seen. The names "methyl chloride", "methylene chloride", "chloroform", and "carbon tetrachloride" are *not* systematic ones. Yet, they are commonly used in the laboratory and in the chemical industry.

The correct systematic names are:

       chloromethane,
       dichloromethane,
       trichloromethane, and
       tetrachloromethane.

Write down these four systematic names together with the appropriate empirical formulas for the compounds.    →   26

**20**

No, [structure] is *m*-dichlorobenzene.

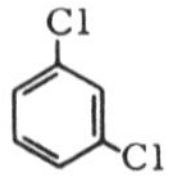

Remember the formula for *o*-xylene: the two methyl groups are on neighboring carbon atoms:

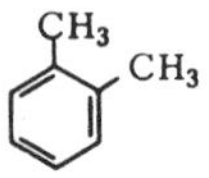

[structure: CH₃ CH₃]

**9**

The same is true for *o*-dichlorobenzene: [structure: Cl Cl]

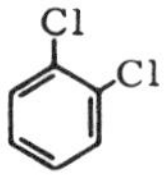

In *m*-xylene the two substituted carbon atoms are *not* neighbors, but separated by one carbon:

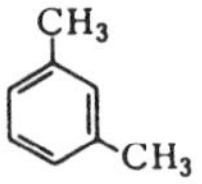

[structure: CH₃ CH₃]

Similarly in *m*-dichlorobenzene: [structure: Cl Cl]

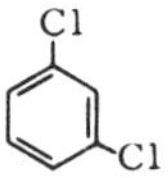

In *p*-xylene the two methyl groups are opposite: [structure: CH₃ CH₃]

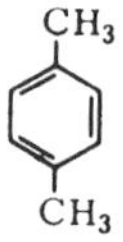

and likewise in *p*-dichlorobenzene: [structure: Cl Cl]

Now write down the structural formulas for all three dichlorobenzenes and compare at   53

---

**21**      Correct; toluene reacts with chlorine gas at *room temperature* and in the presence of a *catalyst,* forming the product of *ring chlorination,* i.e. *o*-chlorotoluene. (Some *p*-chlorotoluene is formed as well; this will be explained later.)

Complete the equation for this reaction:

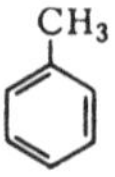

[structure: CH₃] +     $\xrightarrow[\text{catalyst}]{20\,°C}$     +      →   5

**22**     Your reaction conditions do *not* favor the formation of *o*-chlorotoluene.

The production of *o*-chlorotoluene by chlorination of toluene is a

● chlorination of the ring     →    61

● side-chain chlorination     →    54

**23**

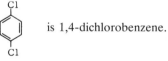

toluene          benzyl chloride

The formation of benzyl chloride from toluene is a side-chain chlorination.

Remember the name "benzyl" for the radical

The side-chain chlorination is carried out in practice by addition of chlorine gas to *boiling* toluene under UV-irradiation. The chlorination takes place exclusively in the *side-chain* and not in the ring under these conditions.

The two requirements for a successful *side-chain* chlorination of toluene are

       1. The toluene should be ....................
       2. It should be irradiated with ....................

Complete these sentences.    →    30

**24**      is 1,4-dichlorobenzene.

As you see, *p*-dichlorobenzene can also be called 1,4-dichlorobenzene.

Copy and complete the following:

*o*-dichlorobenzene or      ....,.... dichlorobenzene
*m*-dichlorobenzene or      ....,.... dichlorobenzene
....................     or     1,4-dichlorobenzene     →    36

25

*Substituted.*

*Thus, we may say either*

"*a hydrogen atom is replaced* by a chlorine atom",

or

"a chlorine atom is *substituted* for a hydrogen atom".

In like manner, chlorine can be substituted for hydrogen atoms in benzene.

**9**

Substitution of *one* hydrogen atom gives

chlorobenzene.

Substitution of *two* hydrogen atoms gives

...................

Give the name of the product.    →    32

---

26

Chloromethane,        $CH_3Cl$
dichloromethane,      $CH_2Cl_2$
trichloromethane,     $CHCl_3$
tetrachloromethane,   $CCl_4$.

There are other common names for these compounds. Which ones?

| $CH_3Cl$ | $CH_2Cl_2$ | $CHCl_3$ | $CCl_4$ | | |
|---|---|---|---|---|---|
| ● Methylene chloride | chloroform | carbon tetra-chloride | methyl chloride | → | 3 |
| ● Methylene chloride | methyl chloride | chloroform | carbon tetrachloride | → | 33 |
| ● Methyl chloride | methylene chloride | chloroform | carbon tetrachloride | → | 18 |
| ● Methyl chloride | chloroform | methylene chloride | carbon tetrachloride | → | 35 |

---

27

Methyl chloride, $CH_3Cl$
methylene chloride, $CH_2Cl_2$
chloroform, $CHCl_3$

What is the correct, systematic name of CCl$_4$?

● Carbon tetrachloride       →      17

● Tetrachloromethane        →      18

---

**28**      You have forgotten the difference between the notations *ortho-*, *meta-*, and *para-*.

This was thoroughly exercised in the previous program, using *o-*, *m-*, and *p*-xylene as examples.

These compounds are repeated below:

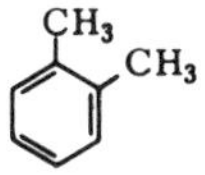

| | | |
|---|---|---|
| *o*−xylene | *m*−xylene | *p*−xylene |

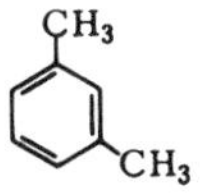

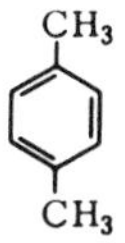

Give the name of the compound:

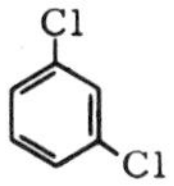

● *o*-Dichlorobenzene       →      15

● *m*-Dichlorobenzene       →      44

● *p*-Dichlorobenzene       →      20

---

**29**      You have forgotten the difference between the notations *ortho-*, *meta-*, and *para-*.

This was thoroughly exercised in the previous program, using *o-*, *m-*, and *p*-xylene as examples.

These compounds are repeated below:

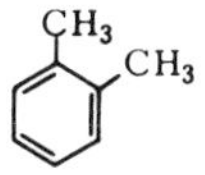

| | | |
|---|---|---|
| *o*−xylene | *m*−xylene | *p*−xylene |

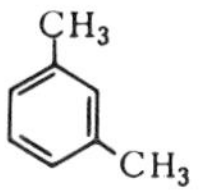

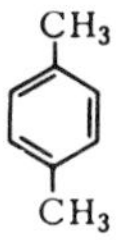

Give the name of the compound:

- *o*-Dichlorobenzene       →     15

- *m*-Dichlorobenzene       →     44

- *p*-Dichlorobenzene       →     20

**9**

---

**30**     The conditions for side-chain chlorination are

a) The toluene should be boiling.
b) It should be irradiated with UV-light.

UV-light (*u*ltraviolet light) is present in sunlight, especially at high altitudes.

In contrast to this, the *ring chlorination* is carried out at *room temperature* in the presence of a *catalyst*.

In summary,

boiling point temperature and UV-irradiation give side-chain reaction;

low temperature and a catalyst give reaction in the ring.

Thus, by proper choice of reaction conditions, the desired product can be obtained.

Specify the conditions necessary for the preparation of *o*-chlorotoluene:

- UV-irradiation, room temperature     →     4

- UV-irradiation, boiling     →     10

- Catalyst, room temperature     →     21

- Catalyst, boiling     →     22

---

**31**    Correct.

Now give the equation for *side-chain chlorination* of toluene:

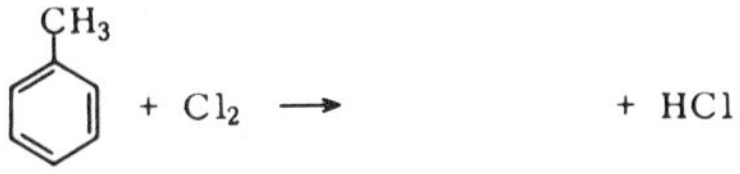

$$\text{CH}_3\text{-}C_6H_5 + Cl_2 \longrightarrow \qquad + HCl$$

toluene                          $\rightarrow$   **23**

**9**

---

**32**    *Dichlorobenzene*

Write the empirical formulas for chlorobenzene and dichlorobenzene.    $\rightarrow$   **39**

---

**33**    You have confused *methyl chloride* and *methylene chloride*.

The systematic name for methyl chloride is chloromethane.

The systematic name for methylene chloride is dichloromethane.

Copy the following names and give the empirical formulas for the compounds:

Methyl chloride has the formula ...................
Methylene chloride has the formula ...................
Chloroform has the formula ......................    $\rightarrow$   **27**

---

**34**    UV- or sunlight and boiling point temperature.

The preparation of benzyl chloride from toluene is a side-chain chlorination:

$$\text{CH}_3\text{-}C_6H_5 + Cl_2 \xrightarrow[\text{boiling}]{\text{UV-light}} \text{CH}_2Cl\text{-}C_6H_5 + HCl$$

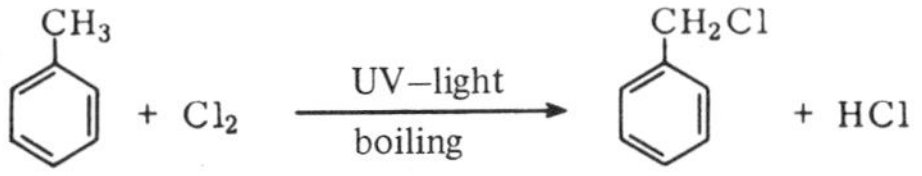

Other products may also be formed in this reaction. Toward the end of the reaction, we will have a large proportion of benzyl chloride and only little toluene. At this point, the benzyl chloride formed may react further with chlorine, giving the product of *double* side-chain chlorination:

$$\text{CH}_2Cl\text{-}C_6H_5 + Cl_2 \xrightarrow[\text{boiling}]{\text{UV-light}} \text{CHCl}_2\text{-}C_6H_5 + HCl$$

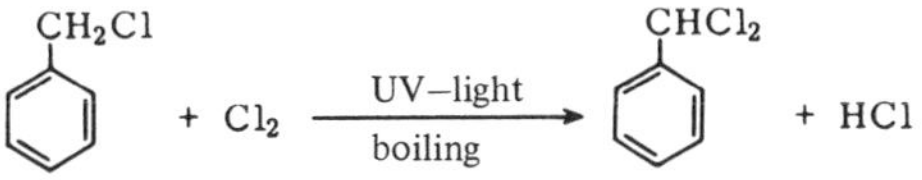

benzyl chloride

and finally, a *third* chlorine atom may be introduced:

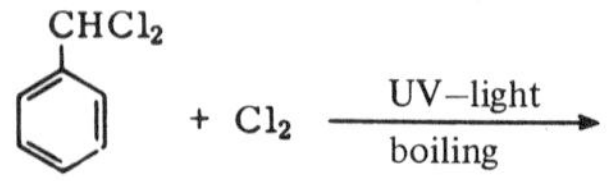

$$\underset{\text{CHCl}_2}{\bigcirc} + Cl_2 \xrightarrow[\text{boiling}]{\text{UV--light}}$$

Complete the latter equation.    →   $\boxed{43}$

---

$\boxed{35}$    You have confused *chloroform* with *methylene chloride*.

The systematic name for methylene chloride is dichloromethane.

The systematic name for chloroform is trichloromethane.

Copy the following names and give the empirical formulas for the compounds:

Methyl chloride has the formula ....................
Methylene chloride has the formula ....................
Chloroform has the formula ..................    →   $\boxed{27}$

---

$\boxed{36}$    *o*-Dichlorobenzene or 1,2-dichlorobenzene.
*m*-Dichlorobenzene or 1,3-dichlorobenzene.
*p*-Dichlorobenzene or 1,4-dichlorobenzene.

Xylene may also be called dimethylbenzene. Thus,

       *o*-xylene = 1,2 dimethylbenzene
       *m*-xylene = ..,..-dimethylbenzene
       *p*-xylene = ..,..-dimethylbenzene

Complete the names.    →   $\boxed{42}$

---

$\boxed{37}$

| | Reaction time | Completeness |
|---|---|---|
| Inorganic reactions | short | complete |
| Organic reactions | long | incomplete |

The above table is only intended as a rough rule of thumb. Naturally, exceptions are possible.

It is typical for organic reactions that they are rather slow. Slow reactions can be accelerated by

       a) a temperature increase
       b) use of a catalyst.

What are the catalysts used to accelerate the ring chlorination of aromatics?

- Iodine and iron         →   45

- Manganese dioxide       →   7

- Platinum                →   9

- I have forgotten        →   16

---

**38**     Have you forgotten the criteria for *ring* substitution?

They are: low temperature and the use of catalysts. What are the criteria for side-chain substitution?   →   62

---

**39**     Chlorobenzene,  $C_6H_5Cl$.
       Dichlorobenzene,  $C_6H_4Cl_2$.

Give the structural formula for chlorobenzene.   →   46

---

**40**     *Benzene ring.*
       *Side-chains.*

*The side-chain on an aromatic compound can be methyl, ethyl, propyl, or another radical.*

*Ring chlorination*
        means replacement of a *ring*-hydrogen atom by a chlorine atom.

*Side-chain chlorination*
        means replacement of a hydrogen atom in the *side-chain* by a chlorine atom.

How can the following compound be prepared?

- Ring chlorination of toluene          →   55

- Side-chain chlorination of xylene     →   50

- Side-chain chlorination of toluene    →   31

41    No, the use of UV-light and boiling would lead to *side-chain chlorination*.

Repeat from   30  .

42    1,3-Dimethylbenzene (*m*-xylene).
      1,4-Dimethylbenzene (*p*-xylene).

Let us now continue with the chlorine compounds. The introduction of one or more chlorine atoms into a molecule is called *"chlorination"*.

We have already described the chlorination of methane and benzene. In each case, one or more hydrogen atoms can be replaced by chlorine. Other hydrocarbons can be chlorinated in the same manner.

The chlorination of *aromatic* compounds follows. The simplest example is the chlorination of benzene to chlorobenzene:

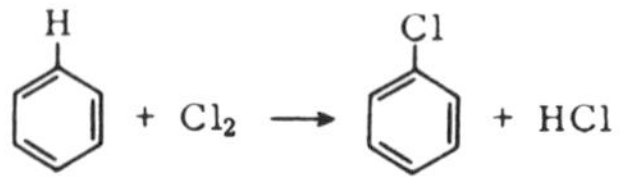

The ring hydrogen atom which is replaced by chlorine is drawn in for clarity. This is not necessary, however, and normally such substitution reactions are given as:

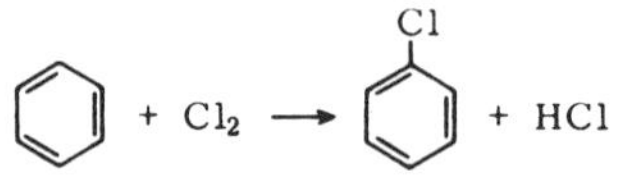

As you will notice, *one* of the atoms in the chlorine molecule is engaged in *substitution* on the benzene ring; *the other* recombines with the hydrogen atom which was replaced, thus forming a molecule of HCl.

Other *halogens*, e.g. *bromine* or *iodine* can substitute on the aromatic ring in the same way as *chlorine*.

Write the equations for the reactions of

a) benzene with bromine
b) benzene with iodine     →     49

**43**

$CHCl_2$ + $Cl_2$ $\xrightarrow[\text{boiling}]{\text{UV-light}}$ $CCl_3$ + HCl

This example illustrates the difficulty of achieving specific reactions in organic chemistry. Unwanted *side-reactions* often lower the yields.

The yield of the desired compound is always expressed in % of the theoretically possible yield.

*Example:*

**9**

The chlorination of 92 g of toluene (1 mol) according to the equation

$CH_3$ + $Cl_2$ $\xrightarrow[\text{boiling}]{\text{UV-light}}$ $CH_2Cl$ + HCl

should theoretically give .................... g of benzyl chloride.

How many grams should be formed?

- 253 g      →   57

- 126.5 g      →   51

- 92 g      →   6

- I don't know      →   63

(Atomic masses: C = 12, H = 1, Cl = 35.5).

---

**44**     Your answer is correct.

The systematic nomenclature (= naming) of the dichlorobenzenes is a different one.

The six carbon atoms in benzene are numbered 1 to 6:

is therefore 1,2-dichlorobenzene
(pronounced: "one-two-dichlorobenzene").

is 1,3-dichlorobenzene
(pronounced: "one-three-dichlorobenzene").

The substituents (here chlorine atoms) are located by giving the numbers of the carbon atoms to which they are attached.

Give the corresponding name of the compound     →  24

---

**45**     Yes, ring chlorination is carried out at *room temperature* and catalyzed by *iron* or *iodine*.

Organic reactions can, in general, be accelerated by

1. ....................
2. ....................   →   60

---

**46**

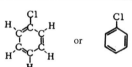    or

chlorobenzene

The simplified formula for the benzene ring shown on the right will be used from here on. A different simiplified formula for the benzene ring is

Draw the structural formula for dichlorobenzene. You will realize that there are *three* ways of arranging the two chlorine atoms on the ring. Remember *o*-xylene, *m*-xylene, and *p*-xylene!

Give the structures of the three dichlorobenzenes.   →   53

---

**47**     a) Side-reactions.
b) Incomplete conversion.
c) Loss during purification.

These points differentiate many *organic* reactions from *inorganic* ones.

Inorganic reactions are often complete and free from side-reactions.

An example is the quantitative precipitation of silver chloride by treatment of an aqueous solution of silver nitrate with hydrochloric acid.

Organic and inorganic reactions also differ in the *speed* of reaction. *Inorganic* reactions are often very fast (example: the precipitation of AgCl is spontaneous).

*Organic* reactions, in contrast, may require long reaction times.

**9**

Copy the following table and complete column A with "long/short", column B with "complete/incomplete" as appropriate.

|                     | a) Reaction time | b) Completeness |       |
|---------------------|:----------------:|:---------------:|-------|
| Inorganic reactions | .................. | .................. |       |
| Organic reactions   | .................. | .................. | →  37 |

---

**48**

*Fluorine F*
*Chlorine Cl*
*Bromine Br*
*Iodine I*

The halogenation of an aromatic compound, i.e. the chlorination of toluene, can take two different courses:

1. *ring* chlorination
2. *side-chain* chlorination

The expressions "ring" and "side-chain" denote the benzene ring and an aliphatic substituent, respectively.

Thus, toluene,  is composed of a *benzene ring* and a *methyl side-chain*.

Similarly, *o*-xylene,  is composed of a .....................

and two methyl ....................

Complete this sentence.    →    40

**49**

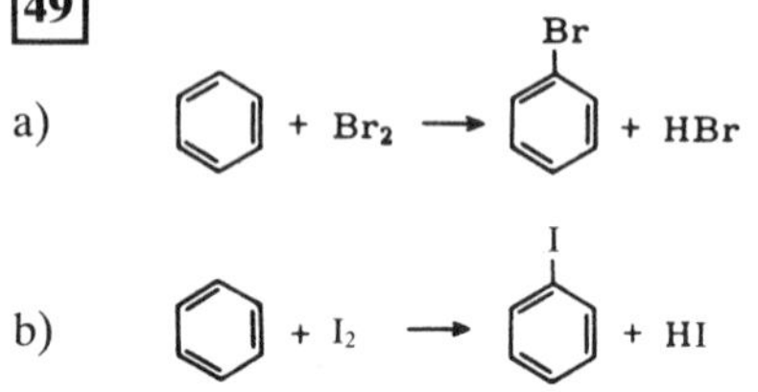

a)

+ Br$_2$ → + HBr

b)

+ I$_2$ → + HI

a) is a *bromination* of benzene;
b) is an *iodination* of benzene.

The introduction of *chlorine* into an organic compound by treatment with Cl$_2$ is a ....................

The introduction of *bromine* into an organic compound by treatment with Br$_2$ is a ....................

The introduction of *iodine* into an organic compound by treatment with I$_2$ is an ....................

Write the missing words.    →    56

---

**50**    Xylene has *two* side-chains and cannot, therefore, be the starting material.

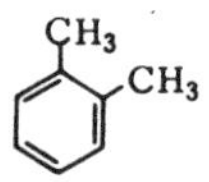

o−xylene

Return to    40  .

---

**51**    Correct, 92 g of toluene (1 mol) must theoretically give 126.5 g of benzyl chloride (1 mol).

These 126.5 g correspond to a 100% yield.

In practice, only ca. 90 g of benzyl chloride can be obtained from the 92 g of toluene.

These 90 g of benzyl chloride correspond to a yield of 71%.

(If you wish to try another example of the calculation of yields, go to    52   .)

The theoretical yield of 100% is *not* usually obtained in organic reactions.

You already know one reason for this: the occurrence of side-reactions.

Another reason could be that the starting materials are not completely used up in the reaction; in other words, the transformation is incomplete.

Benzyl chloride is formed by side-chain chlorination of toluene, but the crude product will be *impure* because of the side-reactions and the incomplete transformation of the starting compound.

What are the possible impurities?

**9**

a) impurities due to side-reactions: ....................
b) impurities due to incomplete transformation: ....................

Give the structural formulas for the four compounds. → |58|

---

|52|    Bromobenzene is prepared by the bromination of benzene:

$$\text{+ Br}_2 \xrightarrow[20°C]{\text{catalyst}} \text{+ HBr}$$

151 g of bromobenzene are obtained from 100 g of benzene. What is the yield in %? (theoretical yield = 100%) (atomic mass of bromine = 80). → |59|

---

|53|

Give the name of the compound on the left only!

● *o*-Dichlorobenzene    → |44|

● *m*-Dichlorobenzene    → |29|

● *p*-Dichlorobenzene    → |28|

---

|54|    The distinction between the *benzene ring* and the *side-chain* should be restudied from |48| .

---

**55**    No; study the following scheme closely:

side-chain:    $CH_3$ $\xleftarrow{+\,Cl_2}$    *side chain* chlorination

ring:     $\xleftarrow{+\,Cl_2}$    *ring* chlorination

Return to   40   .

**9**

**56**    Chlorination
bromination
iodination
In general: halogenation.

The term *halogenation* denotes introduction of any given halogen atom into an organic molecule.

Thus, a *halogenation* could be a *fluorination,* a *chlorination,* a *bromination,* or an *iodination.* The abbreviation *Hal* for halogen atom is often used.

   is a general formula which can represent fluorobenzene, chlorobenzene, bromobenzene, or iodobenzene.

Give the names and chemical symbols of the four elements denoted by
*Hal.*   →   48

**57**    Your calculation refers to the formation of two moles of benzyl chloride. However, only *one* mole of benzyl chloride can be formed from *one* mole of toluene.

Return to   43   .

**58**

a) Impurities due to side reactions:
    These compounds arise due to further chlorination
    of the side-chain.

b) Impurities due to incomplete transformation:
    These are the unchanged starting materials.

In order to isolate the desired product, the impurities must be removed. Benzyl chloride is best purified by distillation. (Other procedures may be advantageous for other compounds, e.g. recrystallization.)

Loss of the desired material can also occur during the purification procedure (= work-up).

Thus, there are three reasons why a 100% yield is seldom obtained. Give these reasons:

a) ..................

b) ..................

c) ..................  →  47

9

---

59    Solution:

1 mol benzene (78 g) theoretically gives 1 mol (157 g) bromobenzene.

100 g benzene theoretically gives $\dfrac{157 \cdot 100}{78} = 201$ g bromobenzene

Thus, 201 g corresponds to a 100% yield.
However, only 151 g bromobenzene was obtained. This corresponds to a yield of 75%.

Return to    51  .

---

60    1. Temperature increase.
      2. Catalysts.

Up until now we have only dealt with the halogenation of *aromatic* hydrocarbons.

*Aliphatic* hydrocarbons can also be halogenated. Thus, the direct reaction with chlorine causes chlorination. These reactions are complicated by the fact that mixtures of differently substituted and often inseparable reaction products are formed. For this reason, indirect methods are often used to introduce a halogen atom into a specific position in the molecule. These methods will be the subject of the following program.

This is the end of Program 9.

---

**61**     Correct; *o*-chlorotoluene is the product of ring chlorination of toluene.

Which reaction conditions are required for *ring chlorination?*

- UV-light, boiling            →    41

- Catalyst, room temperature       →    21

**62**     The necessary reaction conditions for side-chain substitution are UV-irradiation and high temperature (boiling).

Answer the following questions in key-word fashion:

Which catalysts accelerate ring chlorination?

Which temperature is used for ring chlorination?    →    45

**63**     One mole of toluene (92 g) gives, on chlorination, *one* mole of benzyl chloride.

You can easily calculate the molecular mass of benzyl chloride, adding up:

| | |
|---|---|
| molecular mass of toluene: | 92 |
| minus a hydrogen atom: | −1 |
| plus a chlorine atom: | +35,5 |
| | |
| molecular mass of benzyl chloride: | .......... |

Check your answer at    51    .

# Program 10

## Halogen Compounds (II)

**1**    At the end of the last program we mentioned that the direct chlorination of aliphatic hydrocarbons (aliphatics) often leads to a mixture of substitution products which are hard to separate.

This program will start with a method allowing the introduction of halogen atoms into aliphatic compounds in such a way that unique products – not mixtures – are formed.

Bromine is a red-brown liquid which dissolves in chloroform to give a brown solution. The chloroform only serves as solvent; no reaction takes place. When ethene is passed into the solution, it will decolorize due to the following reaction taking place:

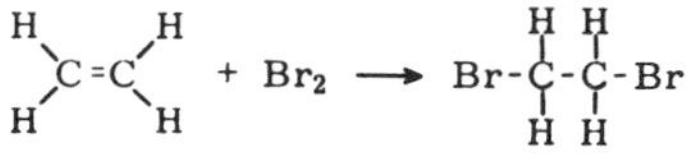

$$
\begin{array}{c}
\text{H} \\ \diagdown \\ \end{array} \text{C} = \text{C} \begin{array}{c} \diagup \text{H} \\ \diagdown \text{H} \end{array} + \text{Br}_2 \longrightarrow \text{Br} - \overset{\overset{\text{H}}{|}}{\underset{\underset{\text{H}}{|}}{\text{C}}} - \overset{\overset{\text{H}}{|}}{\underset{\underset{\text{H}}{|}}{\text{C}}} - \text{Br}
$$

What kind of reaction has taken place?

- A substitution reaction of ethene, bromine replacing hydrogen.    →    21

- A substitution reaction of ethane, bromine replacing hydrogen    →    14

- An addition of bromine to the double bond in ethene.    →    8

**2**    The name 1,1-dichloroethane signifies that two chlorine atoms are bonded to the same carbon atom (the one carrying number 1).

This was not the case in structure (b), however:

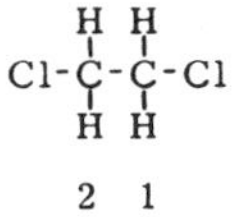

$$
\text{Cl} - \overset{\overset{\text{H}}{|}}{\underset{\underset{\text{H}}{|}}{\underset{2}{\text{C}}}} - \overset{\overset{\text{H}}{|}}{\underset{\underset{\text{H}}{|}}{\underset{1}{\text{C}}}} - \text{Cl} \quad \text{(b)}
$$

The two carbon atoms are numbered 1 and 2. The name of the compound is obtained by specifying the *numbers* of the carbon atoms to which the chlorines are attached.

Give the name of compound (b) once more.    →    12

3 │    Your answer is wrong.

The addition of Br$_2$ to acetylene first gives 1,2-dibromo-*ethene*; further addition of Br$_2$ gives 1,1,2,2-tetrabromoethane as the final product.

However, the problem was how to prepare 1,2-dibromo*ethane*.

This is described at  8 . Please repeat from there.

---

4 │    The compound you chose is in fact flammable. It contains too much hydrogen.

**10**

Return to  24  and write down the structures of the three compounds mentioned before answering again.

---

5 │    Compare:

a) 1,2-diiodoethane

$$
\begin{array}{cc}
\text{H} & \text{H} \\
| & | \\
\text{I}-\text{C}-\text{C}-\text{I} \\
| & | \\
\text{H} & \text{H}
\end{array}
$$

b) 1,1-dibromoethane

$$
\begin{array}{cc}
\text{H} & \text{Br} \\
| & | \\
\text{H}-\text{C}-\text{C}-\text{Br} \\
| & | \\
\text{H} & \text{H}
\end{array}
$$

c) 2-chloropropane

$$
\begin{array}{ccc}
\text{H} & \text{H} & \text{H} \\
| & | & | \\
\text{H}-\text{C}-\text{C}-\text{C}-\text{H} \\
| & | & | \\
\text{H} & \text{Cl} & \text{H}
\end{array}
$$

The *lowest* possible numbers should be used when naming a compound.

Example:

$$
\begin{array}{cc}
\text{H} & \text{Br} \\
| & | \\
\text{H}-\text{C}-\text{C}-\text{Br} \\
| & | \\
\text{H} & \text{H}
\end{array}
$$

is called 1,1-dibromoethane. The first carbon atom (the one to the right) is given number 1. The one to the left is number 2.

We could also have used the opposite numbering: 1 for the carbon on the left, 2 for the one on the right. The molecule would then be called 2,2-dibromoethane.

We cannot have two names for the same compound, hence the rule that the *lowest* possible numbers should be used. Thus, only *1,1-dibromoethane is correct*.

However, there is no rule which stipulates at which end of the molecule the numbering should start. The formula

$$
\begin{array}{c}
\text{Br} \quad \text{H} \\
| \quad\ | \\
\text{Br}-\text{C}-\text{C}-\text{H} \\
| \quad\ | \\
\text{H} \quad \text{H}
\end{array}
$$

also denotes 1,1-dibromoethane.

Give the name of the following compound:

$$
\begin{array}{c}
\text{H} \ \ \text{H} \ \ \text{Cl} \\
| \ \ | \ \ | \\
\text{H}-\text{C}-\text{C}-\text{C}-\text{Cl} \\
| \ \ | \ \ | \\
\text{H} \ \ \text{H} \ \ \text{H}
\end{array}
$$

● 3,3-Dichloropropane          →          |27|

● 1,2-Dichloropropane          →          |20|

● 1,1-Dichloropropane          →          |13|

**10**

---

**6**          The reaction of halogens with double or triple bonds leads to products in which the two halogen atoms are *not* attached to the same carbon atom.

Give the equation for the addition of chlorine to propene.          →          |19|

---

**7**          Wrong.

Any compound with a density higher than that of water will sink to the bottom.

Since halogenated hydrocarbons have a higher density than water, they form the .............. (upper/lower) layer.

Complete this sentence.          →          |24|

---

**8**          Yes, it was an *addition* of bromine to the double bond in ethene.

The term *addition* has much the same meaning in chemistry as in mathematics.

Bromine *adds* to ethene. The empirical formula of the reaction product is obtained by *addition* of those for the reactants:

$$C_2H_4 + Br_2 \rightarrow C_2H_4Br_2$$

Write down this equation using structural formulas, and also give the name of each compound.          →          |15|

---

**9**    You added *2* mol of chlorine to *1* mol of acetylene. However, the addition of only *1* mol of chlorine was asked for.

Give the name of the compound formed by the addition of *1* mol of chlorine to *1* mol of acetylene.    →    17

**10**    Wrong.

The rule is:

> in the addition of hydrogen halides to olefins, the halogen atom adds to the more highly substituted carbon atom (i.e. the one carrying the fewest hydrogen atoms).

Specify this carbon atom in the compound

$$H-\overset{H}{\underset{H}{C}}-\overset{H}{\underset{H}{C}}-\overset{H}{\underset{}{C}}=\overset{H}{\underset{H}{C}}$$

4   3   2   1    →    18

**11**    The compound you chose is in fact flammable. It contains too much hydrogen.

Return to   24   and write down the structures of the three compounds mentioned before answering again.

**12**    1,2-Dichloroethane is correct.

The name 1,2-dichloroethane signifies that one chlorine atom is bonded to carbon atom number 1, the other to number 2.

$$Cl-\overset{H}{\underset{H}{C}}-\overset{H}{\underset{H}{C}}-Cl$$

2   1

Now write structural formulas for:

a) 1,2-diiodoethane

b) 1,1-dibromoethane

c) 2-chloropropane    →    5

| **13** | Correct; the compound |
|---|---|

$$H-\underset{\underset{H}{|}}{\overset{\overset{H}{|}}{C}}-\underset{\underset{H}{|}}{\overset{\overset{H}{|}}{C}}-\underset{\underset{H}{|}}{\overset{\overset{Cl}{|}}{C}}-Cl$$

is called 1,1-dichloropropane.

(3,3-Dichloropropane would be wrong since the *lowest* possible numbers should be used.)

Having exercised the nomenclature, let us continue with addition reactions.

*Hydrogen halides* add to olefins (alkenes) much like the free halogens. Write down the equation for the addition of hydrogen bromide (HBr) to ethene using structural formulas.   →   23

**10**

| **14** | The reaction product |
|---|---|

$$Br-\underset{\underset{H}{|}}{\overset{\overset{H}{|}}{C}}-\underset{\underset{H}{|}}{\overset{\overset{H}{|}}{C}}-Br$$

could be thought of as being derived from ethane by substituting two bromine atoms for hydrogen. However, this is *not* the reaction described by the equation given at   1  . Have a closer look at this equation!   →   1

| **15** |
|---|

$$\underset{H}{\overset{H}{>}}C=C\underset{H}{\overset{H}{<}} \;+\; Br-Br \;\longrightarrow\; Br-\underset{\underset{H}{|}}{\overset{\overset{H}{|}}{C}}-\underset{\underset{H}{|}}{\overset{\overset{H}{|}}{C}}-Br$$

   ethene      bromine      dibromoethane

Note that there is only *one product* of this reaction. In contrast, when a halogenation is carried out by *substitution,* one of the reaction products will be a *hydrogen halide,* e.g.

$CH_3$ benzene ring $\;+\; Cl_2 \;\longrightarrow\;$ $CH_2Cl$ benzene ring $\;+\; HCl$

| toluene | chlorine | benzyl chloride | hydrogen chloride |

Let us return to the addition reactions:

chlorine or iodine can be added to an alkene just like bromine.

Give reaction equations using structural formulas for:

a) chlorine + ethene
b) iodine + ethene

and also write the names of the reaction products.    →    $\boxed{22}$

---

$\boxed{16}$    The compounds $CCl_2F_2$ and $CCl_4$ are regarded as *organic* compounds even though they have no hydrogen atoms. The chlorine atoms in $CCl_2F_2$ and $CCl_4$ are *not* ionizable.    →    $\boxed{39}$

---

**10**

$\boxed{17}$    Correct; the reaction product is called 1,2-dichloroethene:

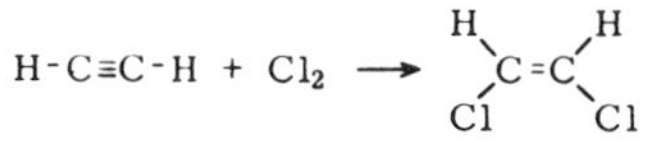

$$H-C\equiv C-H \; + \; Cl_2 \; \longrightarrow \; \begin{array}{c} H \\ \diagdown \\ C=C \\ \diagup \quad \diagdown \\ Cl \qquad Cl \end{array} \begin{array}{c} H \\ \diagup \\ \\ \end{array}$$

1,2–dichloroethene

Since 1,2-dichloroethene possesses a double bond, it can react further with halogens or hydrogen halides. Write down the equations for the addition of:

a) chlorine
b) hydrogen chloride

to 1,2-dichloroethene.    →    $\boxed{25}$

---

$\boxed{18}$    Correct; carbon atom number 2 is poorest in hydrogen.

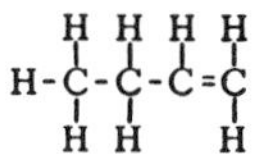

$$\begin{array}{c} H \; H \; H \; H \\ | \quad | \quad | \quad | \\ H-C-C-C=C \\ | \quad | \qquad | \\ H \; H \qquad H \\ 4 \; 3 \; 2 \; 1 \end{array}$$

Thus, when HCl is added to 1-butene (the compound shown above), the chlorine atom goes to carbon number 2; the hydrogen atom goes to carbon number 1. Give the name and structural formula of the reaction product.    →    $\boxed{28}$

---

**19**

$$H-\overset{\overset{\displaystyle H}{|}}{\underset{\underset{\displaystyle H}{|}}{C}}-\overset{\overset{\displaystyle H}{|}}{C}=\overset{\overset{\displaystyle H}{|}}{C}-H \;+\; Cl_2 \;\longrightarrow\; H-\overset{\overset{\displaystyle H}{|}}{\underset{\underset{\displaystyle H}{|}}{C}}-\overset{\overset{\displaystyle H}{|}}{\underset{\underset{\displaystyle Cl}{|}}{C}}-\overset{\overset{\displaystyle H}{|}}{\underset{\underset{\displaystyle Cl}{|}}{C}}-H$$

1,2–dichloropropane

You will notice that the two halogen atoms end up at *different* carbon atoms in an addition reaction.

Return to [28] .

---

**20**      The two chlorine atoms in 1,2-dichloropropane are at *different* carbon atoms. Repeat carefully from [22] .

---

**21**      Have a closer look at the equation given at [1] .

The starting material (ethene) has 4 hydrogen atoms. The reaction product also has 4 hydrogen atoms.

Hence, no *substitution* has taken place in this reaction   →   [1]

---

**22**

$$\overset{\displaystyle H}{\underset{\displaystyle H}{>}}C=C\overset{\displaystyle H}{\underset{\displaystyle H}{<}} \;+\; Cl-Cl \;\longrightarrow\; Cl-\overset{\overset{\displaystyle H}{|}}{\underset{\underset{\displaystyle H}{|}}{C}}-\overset{\overset{\displaystyle H}{|}}{\underset{\underset{\displaystyle H}{|}}{C}}-Cl$$

dichloroethane

$$\overset{\displaystyle H}{\underset{\displaystyle H}{>}}C=C\overset{\displaystyle H}{\underset{\displaystyle H}{<}} \;+\; I-I \;\longrightarrow\; I-\overset{\overset{\displaystyle H}{|}}{\underset{\underset{\displaystyle H}{|}}{C}}-\overset{\overset{\displaystyle H}{|}}{\underset{\underset{\displaystyle H}{|}}{C}}-I$$

diiodoethane

Names like "dichloroethane", "dibromoethane", and "diioodethane" are derived in the correct way but are not yet complete. This becomes clear from the fact that the compound

$$H-\overset{\overset{\displaystyle H}{|}}{\underset{\underset{\displaystyle H}{|}}{C}}-\overset{\overset{\displaystyle Cl}{|}}{\underset{\underset{\displaystyle H}{|}}{C}}-Cl$$

(a)

is also a dichloroethane.

As you see, the two chlorine atoms can be bonded to *the same* carbon atom – as in (a) – or to two *different* carbon atoms – as in (b).

$$\text{Cl-}\overset{\displaystyle \overset{H}{|}\;\overset{H}{|}}{\underset{\displaystyle \underset{H}{|}\;\underset{H}{|}}{C-C}}\text{-Cl}$$

(b)

In order to distinguish between these two situations, the carbon atoms are numbered, and the numbers joined to the name as prefixes.

Thus, (a) is called 1,1-dichloroethane.

(Both chlorine atoms are joined to carbon atom number 1.)

**10**

What is the name of compound (b)?

- 2,2-Dichloroethane     →    29

- 1,2-Dichloroethane     →    12

- 1,1-Dichloroethane     →    2

---

**23**

$$\overset{H}{\underset{H}{\phantom{.}}}C=C\overset{H}{\underset{H}{\phantom{.}}} + \text{H-Br} \longrightarrow \text{H-}\overset{\displaystyle \overset{H}{|}\;\overset{H}{|}}{\underset{\displaystyle \underset{H}{|}\;\underset{H}{|}}{C-C}}\text{-Br}$$

bromoethane

Thus, the addition of hydrogen halides to alkenes (olefins) leads to the formation of *mono*halogen compounds.

In contrast, the addition of halogens gives rise to ...................

Complete this sentence.     →    30

---

**24**     Yes, the halogenated hydrocarbon forms the lower layer.

Halogenated hydrocarbons are good solvents for many organic compounds, e.g. fats. Therefore, they are often used as solvents or cleansing agents – e.g. in the removal of grease stains.

A compound has less tendency to burn when it contains many halogen atoms and few or no hydrogen atoms. Some halogenated hydrocarbons are even used as fire extinguishers.

Judging from the above, which compound would you expect to be the best fire extinguisher?

- 2-Chloropropane        →   ☐ 4

- Vinyl chloride           →   ☐ 11

- Carbon tetrachloride     →   ☐ 36

**10**

---

**25**

a)

$$\underset{Cl}{\overset{H}{\diagdown}} C = C \underset{Cl}{\overset{H}{\diagup}} + Cl_2 \longrightarrow Cl-\underset{\underset{Cl}{|}}{\overset{\overset{H}{|}}{C}}-\underset{\underset{Cl}{|}}{\overset{\overset{H}{|}}{C}}-Cl$$

b)

$$\underset{Cl}{\overset{H}{\diagdown}} C = C \underset{Cl}{\overset{H}{\diagup}} + HCl \longrightarrow H-\underset{\underset{Cl}{|}}{\overset{\overset{H}{|}}{C}}-\underset{\underset{Cl}{|}}{\overset{\overset{H}{|}}{C}}-Cl$$

Write down the names of the two reaction products. Do not forget the prefixes specifying the positions of the chlorine atoms, and remember that the lowest possible numbers should be used.   →   ☐ 32

---

**26**     Your answer is wrong.

The carbon atoms numbers 3 and 4 are out of the question since they are not at the double bond.

Repeat  ☐ 37  carefully.

---

**27**     Remember:

the lowest possible numbers should be used when numbering the carbon atoms in a compound.

You violated this principle. The compound

$$H-\underset{\underset{H}{|}}{\overset{\overset{H}{|}}{C}}-\underset{\underset{H}{|}}{\overset{\overset{H}{|}}{C}}-\underset{\underset{H}{|}}{\overset{\overset{H}{|}}{C}}-Cl$$

*could* be called either 1-chloropropane or 3-chloropropane, since we could number from the right to the left, or from the left to the right:

$$
\begin{array}{ccc}
3 \ 2 \ 1 & & 1 \ 2 \ 3 \\
C\text{-}C\text{-}C & \text{or} & C\text{-}C\text{-}C
\end{array}
$$

*Only* 1-chloropropane is correct since the lowest possible number should be used.

Return to  [5] .

---

**28**    Yes, 2-chlorobutane is formed:

$$\begin{array}{cccc} H & H & H & H \\ | & | & | & | \\ H-C-C-C-C-H \\ | & | & | & | \\ H & H & Cl & H \end{array}$$

Halogens and hydrogen halides can be added not only to double bonds, but also to triple bonds. What is the product of the reaction of *1* mol of chlorine with *1* mol of acetylene? Formulate the equation!

Which product do you obtain?

- Chloroethene                    →    [42]

- 1,1-Dichloroethene              →    [6]

- 1,2-Dichloroethene              →    [17]

- 1,1,2,2-Tetrachloroethane       →    [9]

- I need further information       →    [35]

---

**29**    The name given by you, 2,2-dichloroethane, signifies that the two chlorine atoms are bonded to the same carbon atom (since the same number occurs twice). However, such was not the case in formula (b), which is repeated below:

$$\begin{array}{cc} H & H \\ | & | \\ Cl-C-C-Cl \\ | & | \\ H & H \end{array} \quad (b)$$

$$\begin{array}{cc} 2 & 1 \end{array}$$

The two carbon atoms are numbered 1 and 2. The name of the compound is obtained by prefixing these numbers (i.e. putting them in front of the name).

Hence, the name of compound (b) is .....................

Write down the name!   →    [12]

**30**          Dihalogen compounds (or dihalogenoalkanes).

The addition of HCl to propene (propylene) gives 2-chloropropane.

Write down the equation for this reaction using structural
formulas.   →   |37|

**31**          Your answer is wrong.

It was explicitly mentioned that organic halogen compounds do *not dissociate*.

Continue at   |74|   .

**10**

**32**

a)

$$\underset{Cl}{\overset{H}{\diagdown}} C = C \underset{Cl}{\overset{H}{\diagup}} + Cl_2 \longrightarrow Cl - \underset{\underset{Cl}{|}}{\overset{\overset{H}{|}}{C}} - \underset{\underset{Cl}{|}}{\overset{\overset{H}{|}}{C}} - Cl$$

1, 1, 2, 2–tetrachloroethane

b)

$$\underset{Cl}{\overset{H}{\diagdown}} C = C \underset{Cl}{\overset{H}{\diagup}} + HCl \longrightarrow H - \underset{\underset{Cl}{|}}{\overset{\overset{H}{|}}{C}} - \underset{\underset{Cl}{|}}{\overset{\overset{H}{|}}{C}} - Cl$$

1, 1, 2–trichloroethane

Since the lowest possible numbers should be used, the latter compound is
1,1,2-trichloroethane and *not* 1,2,2-trichloroethane.

As you will see, the addition of halogens or hydrogen halides to alkenes or
alkynes allows the synthesis of a variety of halogenated hydrocarbons.

The industrially important compound, *vinyl chloride,* is prepared technically in
such a reaction, namely the addition of 1 mol of hydrogen chloride to 1 mol of
acetylene.

Give the equation for the addition of HCl to acetylene.   →   |40|

**33**          Your answer is wrong.

The addition of HBr to ethene gives bromoethane.

Write down the reaction equation using structural formulas, and repeat
from   |23|   .

34

$$
\begin{array}{c} H \\ | \\ H-C-Cl \\ | \\ H \end{array}
$$

chloromethane
(methyl chloride)
gaseous

$$
\begin{array}{c} H \\ | \\ Cl-C-Cl \\ | \\ H \end{array}
$$

dichloromethane
(methylene chloride)
liquid

$$
\begin{array}{c} H \\ | \\ H-C-Br \\ | \\ H \end{array}
$$

bromomethane
(methyl bromide)
liquid

$$
\begin{array}{c} Cl \\ | \\ H-C-Cl \\ | \\ Cl \end{array}
$$

trichloromethane
(chloroform)
liquid

$$
\begin{array}{c} Cl \\ | \\ Cl-C-Cl \\ | \\ Cl \end{array}
$$

tetrachloromethane
(carbon tetrachloride)
liquid

$$
\begin{array}{c} Br \quad\quad Br \\ \diagdown C=C \diagup \\ Br \diagup \quad\quad \diagdown Br \end{array}
$$

tetrabromoethene
(tetrabromoethylene)
solid

**10**

Many halogenated hydrocarbons have a sweetish odor and narcotic action (e.g. chloroform!).

The halogenated hydrocarbons are immiscible with water, and their densities are higher than that of water.

Thus, two phases will form when a liquid halogenated hydrocarbon is added to water. Which will form the lower layer?

● The water                                → 7

● The halogenated hydrocarbon        → 24

---

35   When a halogen is added to a *double bond,* one halogen atom goes to each end of the double bond:

$$
\begin{array}{c} H \; H \\ | \; | \\ C=C \\ | \; | \\ H \; H \end{array}
+ Hal_2 \longrightarrow
\begin{array}{c} H \; H \\ | \; | \\ Hal-C-C-Hal \\ | \; | \\ H \; H \end{array}
$$

The addition of one mole of a halogen to one mole of a compound possessing a triple bond takes the following course:

$$
H-H\equiv C-H + Hal_2 \longrightarrow
\begin{array}{c} H \quad\quad H \\ \diagdown C=C \diagup \\ Hal \diagup \quad\quad \diagdown Hal \end{array}
$$

Now try and solve the problem at   28   once more.

| **36** | Correct; carbon tetrachloride is used in fire extinguishers.

The extinguishing of a fire by spraying with $CCl_4$ is due to the evaporation of the $CCl_4$. The gaseous carbon tetrachloride prevents air from reaching the fire. Hence the fire goes out.

---

Fire extinguishers containing $CCl_4$ should not be used in closed rooms because poisonous gases (e.g. *phosgene*) are formed by the pyrolysis of $CCl_4$.

---

(Phosgene will be described in Program 17)

There are numerous other uses of halogenated hydrocarbons. For example, many pesticides are halogenated compounds.

**10**

The chemical inertia and good solubilizing properties of $CH_2Cl_2$ and $CCl_4$ also lead to many applications in the chemical industry.

Before we come to the chemical properties of these compounds, we have to familiarize ourselves with a couple of commonly used terms.

The general expression "halogenated hydrocarbon" covers all compounds made up of a hydrocarbon radical and one or more halogen atoms. One also speaks of "chlorocarbon", "fluorocarbon", etc. for the fully halogenated hydrocarbons.

Give answers to the questions:

a) Which four elements are known as "halogens"?
b) What is the abbreviation for "halogen"?     →     | 44 |

---

| **37** |

$$H-\overset{\overset{H}{|}}{\underset{\underset{H}{|}}{C}}-\overset{\overset{H}{|}}{\underset{|}{C}}=\overset{H}{\underset{H}{C}} + H-Cl \longrightarrow H-\overset{\overset{H}{|}}{\underset{\underset{H}{|}}{C}}-\overset{\overset{H}{|}}{\underset{\underset{Cl}{|}}{C}}-\overset{H}{\underset{H}{C}}-H$$

2−chloropropane

Note that the chlorine atom goes to the *central* carbon atom; the hydrogen atom goes to a terminal one:

The opposite orientation is (normally) *not* observed. The chlorine atom is always added to the more highly substituted end of the double bond, i.e. the carbon atom poorest in hydrogen. In the case of propene, the central carbon atom has the fewest hydrogens.

To which carbon atom in 1-butene do you expect the chlorine atom (from HCl) to be added?

$$\begin{array}{cccc} H & H & H & H \\ | & | & | & | \\ H-C & -C & -C & =C \\ | & | & & | \\ H & H & & H \end{array}$$    1-butene

    4  3  2  1

**10**

● To carbon number 1    →    | 10 |

● To carbon number 2    →    | 28 |

● To carbon number 3    →    | 45 |

● To carbon number 4    →    | 26 |

---

| 38 |    You do not understand the addition of hydrogen halides to acetylene.

Repeat from   | 28 | .

---

| 39 |

a) Vinyl chloride is obtained by the addition of *hydrogen chloride* to *acetylene*.

b) Vinyl chloride is obtained by the elimination of *hydrogen chloride* from *1,2-dichloroethane*.

Such an elimination of a hydrogen halide is not always easy to carry out. Chlorine is rather strongly bound, whereas bromine and in particular iodine are more readily eliminated – in the form of HBr and HI, respectively.

It is well known that inorganic halides like sodium chloride (NaCl), potassium bromide (KBr), or sodium iodide (NaI) dissociate into ions in aqueous solution. In contrast, the organic halogen compounds do not dissociate. They do *not give halide ions* in aqueous solution.

Which of the following compounds will give a precipitate with silver nitrate?

$CH_3I$, NaBr, $C_6H_5Cl$, KCl, $CaCl_2$, $CCl_2F_2$, $CCl_4$.

● All the compounds                                    → ⬚65

● NaBr, KCl, CaCl₂                                      → ⬚46

● CH₃I, C₆H₅Cl, CCl₂F₂, CCl₄                           → ⬚31

● NaBr, KCl, CaCl₂, CCl₂F₂, CCl₄                       → ⬚16

● I need further information                            → ⬚74

---

**40**

$$H\text{-}C\equiv C\text{-}H \quad + \quad HCl \longrightarrow \quad \underset{H}{\overset{H}{\diagdown}} C = C \underset{Cl}{\overset{H}{\diagup}}$$

vinyl chloride
or systematically: chloroethene

You should already know the vinyl radical from Program 9.

Give the structural formula for the vinyl radical.   →   ⬚48   .

---

**41**

$$H\text{-}\underset{Cl}{\overset{H}{\underset{|}{\overset{|}{C}}}}\text{-}\underset{Cl}{\overset{H}{\underset{|}{\overset{|}{C}}}}\text{-}H \longrightarrow \quad .................... \quad + HCl$$

The H and Cl atoms eliminated never originate from the same carbon atom. The elimination leads to the formation of a double bond.

Now try and complete the equation – including the names of all the compounds.   →   ⬚51

---

**42**          Chloroethane is formed in the reaction of *hydrogen chloride* with acetylene:

$$H\text{-}C\equiv C\text{-}H \quad + \quad HCl \longrightarrow \quad \underset{H}{\overset{H}{\diagdown}} C = C \underset{Cl}{\overset{H}{\diagup}}$$

vinyl chloride
or systematically: chloroethene

However, the question concerned the reaction between acetylene and *chlorine*.

Return to      .

---

| 43 | *a) Freon*
|    | *b) Teflon*

Let us now turn to the properties of the halogenated hydrocarbons. Like the hydrocarbons themselves, these compounds can be gases, liquids or solids. As always, low molecular weight molecules are usually gases under normal conditions of temperature and pressure. Increasing molecular masses lead to liquid and then solid compounds.

Examples:

| Compound | Molecular Mass | Normal state |
|----------|----------------|--------------|
| chloromethane (methyl chloride) | 50.5 | gaseous |
| dichloromethane (methylene chloride) | 85 | liquid |
| bromomethane (methyl bromide) | 95 | liquid |
| trichloromethane (chloroform) | 119.5 | liquid |
| tetrachloromethane (carbon tetrachloride) | 154 | liquid |
| tetrabromoethene (tetrabromoethylene) | 344 | solid |

Give the structures and names of one gaseous, one liquid, and one solid halogenated hydrocarbon.    →    |34|

| 44 | a) Fluorine, chlorine, bromine, iodine.
|    | b) Hal.

The abbreviation Hal can be used in the same way as the symbols for the elements in chemical formulas. We have already seen some examples of this.

Give the names and empirical formulas for the four compounds denoted by

$CH_3$-Hal.                                              →    |54|

| 45 |    Your answer is wrong. – Carbon atoms numbers 3 and 4 are out of the question since they do *not* belong to the *double bond*.

Repeat   |37|   carefully.

| 46 |    Your answer is correct.

The elimination of hydrogen halides (H-Hal) from halogenated hydrocarbons (R-Hal) was dealt with above. We now turn to the substitution of OH for Hal,

OH being the hydroxyl group. This exchange takes place by treatment of the halogenated hydrocarbon with water or bases:

$$R\text{-Hal} + H_2O \rightarrow R\text{-OH} + H\text{-Hal}$$

Example:

The boiling of benzyl chloride with an aqueous solution of a base such as NaOH causes the replacement of Cl by the OH group:

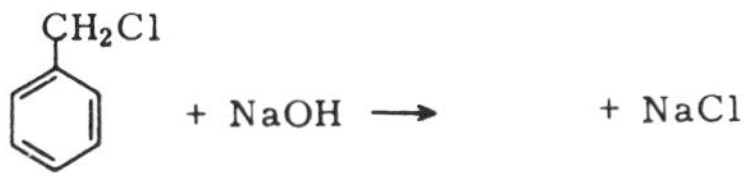

$CH_2Cl$

⬡  + NaOH ⟶  + NaCl

Write down the completed equation.     →    55

---

**47**     You do not yet understand the addition of hydrogen halides to acetylene.

Repeat from    28   .

---

**48**

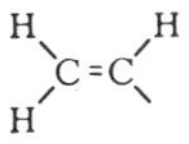

     H⟍    ⟋H          H⟍    ⟋H
       C=C              C=C
     H⟋    ⟍          H⟋    ⟍Cl

    vinyl radical 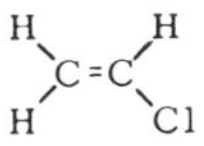            vinyl chloride

Vinyl chloride can polymerize in a manner similar to ethylene and styrene.

The polymer formed – *polyvinyl chloride* or PVC – is surely known to you.

We terminate the discussion of addition reactions with a short repetition.

Draw the structural formula for 1,2-dibromoethane and consider how it can be made.

● By addition of HBr to ethene           →    33

● By bromination (substitution) of ethane     →    62

● By addition of $Br_2$ to ethene          →    56

● By addition of $Br_2$ to acetylene       →    3

**49**     The chlorine atoms in benzyl chloride and chlorobenzene are not equally strongly bonded.

Compare the reaction conditions necessary for the replacement of Cl by OH in benzyl chloride and chlorobenzene.

This will help you answer the question correctly.

Go back to  [46] .

---

**50**     No, propane is not formed.

When HI is eliminated from the compound

$$H-\overset{\overset{\displaystyle H}{|}}{\underset{\underset{\displaystyle H}{|}}{C}}-\overset{\overset{\displaystyle H}{|}}{\underset{\underset{\displaystyle I}{|}}{C}}-\overset{\overset{\displaystyle H}{|}}{\underset{\underset{\displaystyle H}{|}}{C}}-H \qquad C_3H_7I$$

the product will be $C_3H_6$. Propane is $C_3H_8$!

Give the name of $C_3H_6$!   →   [59]

---

**51**

$$H-\overset{\overset{\displaystyle H}{|}}{\underset{\underset{\displaystyle Cl}{|}}{C}}-\overset{\overset{\displaystyle H}{|}}{\underset{\underset{\displaystyle Cl}{|}}{C}}-H \longrightarrow \underset{H}{\overset{H}{>}}C=C\underset{Cl}{\overset{H}{<}} \quad + HCl$$

1,2–dichloroethane      vinyl chloride      hydrogen chloride

You now know *two* methods for producing vinyl chloride. Summarize both methods using key-words:

a) Vinyl chloride is obtained by the addition of ..............
b) Vinyl chloride is obtained by the elimination of .............. from ..............
→  [39]

---

**52**     *a) Dichlorodifluoromethane*
      *b) Tetrafluoroethylene* (tetrafluoroethene).

Give the trade names of

a) dichlorodifluoromethane: ..............
b) the polymer of tetrafluoroethylene: ..............   →   [43]

---

$\boxed{53}$    Tetrafluoroethylene (tetrafluoroethene) is derived from ethylene (ethene) by the replacement of all four hydrogen atoms by fluorine. This should help you to write the structure.    →    $\boxed{60}$

---

$\boxed{54}$

| Fluoromethane | $CH_3\text{-}F$ |
|---|---|
| chloromethane | $CH_3\text{-}Cl$ |
| bromomethane | $CH_3\text{-}Br$ |
| iodomethane | $CH_3\text{-}I$ |
| generally: | $CH_3\text{-}Hal$ |

The general abbreviation "Hal" for halogen is very useful. Similarly, the family of hydrocarbon radicals can be denoted by a single symbol, R. Example:

| chloromethane | $CH_3\text{-}Cl$ |
|---|---|
| chloroethane | $C_2H_5\text{-}Cl$ |
| chlorobenzene | $C_6H_5\text{-}Cl$ |
| general formula: | $R\text{-}Cl$ |

What is the general formula for halogenated hydrocarbons?    →    $\boxed{61}$

**10**

---

$\boxed{55}$

benzyl chloride

Another example:

the reaction below takes place at 300 °C and increased pressure:

chlorobenzene

Deduce from the different reaction conditions which of the chlorine atoms is more strongly bonded – the one in benzyl chloride or the one in chlorobenzene?

● Both are equally strongly bonded      →    $\boxed{49}$

● The one in benzyl chloride      →    $\boxed{72}$

● The one in chlorobenzene      →    $\boxed{67}$

| 56 | Your answer is correct.

Give the product of the reaction of one mole of hydrogen chloride with one mole of acetylene.

- Chloroethane → | 38 |

- Vinyl chloride → | 63 |

- Dichloroethene → | 47 |

---

| 57 | Reaction between a halogen compound and magnesium turnings in ether.

It is very important to exclude water from this reaction. Why?

Write your answer using key-words. → | 71 |

---

| 58 | $CH_3\text{-}I + Mg \rightarrow CH_3\text{-}Mg\text{-}I$

The reaction with water exemplifies the reactivity of Grignard reagents:

$$CH_3\text{-}Mg\text{-}I + H_2O \rightarrow CH_4 + Mg(OH)I$$

Thus, Grignard reagents are *cleaved* by water, forming a *hydrocarbon* and a basic magnesium halide. What is formed in the reaction between

**Mg-Br**

and water?

Give the names of the products. → | 66 |

---

| 59 | Yes, *propene* (propylene) is formed.

$$\begin{array}{ccc} \text{H H H} & & \text{H H H} \\ \text{H-C-C-C-H} & \longrightarrow & \text{H-C-C=C} \\ \text{H I H} & & \text{H H} \end{array} + HI$$

2-iodopropane        propene

The reaction can also be written as follows:

$$CH_3\text{-}CHI\text{-}CH_3 \rightarrow CH_3\text{-}CH=CH_2 + HI$$

It takes place with a yield of 94% in alcoholic potassium hydroxide (i.e. KOH dissolved in alcohol) at 80 °C. (The alcohol only serves as solvent and does not take part in the reaction.)

The base serves to bind the hydrogen iodide in the form of potassium iodide, thereby preventing the reverse reaction from occurring.

However, the presence of a base is not always necessary. Eliminations of hydrogen halides are often achieved industrially by cleavage of the halogenated hydrocarbon vapor at high temperatures (gas-phase reaction).

Example:

The gas-phase elimination of *one* mole of HCl from *one* mole of 1,2-dichloro-ethane gives an industrially important product which you already know.

Write out the reaction equation including the names of all the compounds, and check your answer at   |51|  , or, if you need more help, go instead to   |41| .

**10**

---

|60|   Compare:

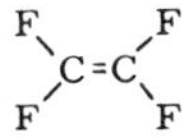

$$\underset{F}{\overset{F}{\diagdown}}C=C\underset{F}{\overset{F}{\diagup}}$$   tetrafluoroethylene (tetrafluoroethene)

Like ethylene (ethene) and other olefins (e.g. styrene and vinyl chloride) tetrafluoroethylene can also be polymerized. The polymer formed is called "teflon".

Complete the sentences:

a) The chemical name for freon is ...................
b) Teflon is polymerized ...................   →   |52|

---

|61|   R-Hal.

R is used as an abbreviation for *r*adical (or *r*esidue) e.g. hydrocarbon radicals like $CH_3-$, $C_2H_5-$, $H_2C=CH-$, $C_6H_5-$, etc.

The symbol R will be used frequently in the following pages.

The *functional group* is another term which we will use repeatedly in the next programs.

A functional group is an atom or group of atoms attached to an organic radical (R). The chemical reactions of the compound will normally take place exclusively at the functional group. Thus, this group determines the chemical "function" of the compound in question.

Examples:

R-Hal   are the halogenated hydrocarbons.
            R is the organic radical
            Hal is the functional group.

R-OH   are the alcohols
            R is the organic radical
            OH is the functional group.

R-NH$_2$  are the amines
              R is the ....................
              NH$_2$ is the ....................

Complete the last two sentences.   →   68

(Don't worry about names and formlas which may be unknown to you – we will come to these later.)

---

**62**     Your answer is wrong.

It was expressly stated at the beginning of this program that the *direct* halogenation of saturated hydrocarbons gives mixtures of difficult to separate compounds. Therefore, the direct bromination of ethane is not a viable way to 1,2-dibromoethane. There is a much better method of preparation.

Return to   48   .

---

**63**     Your answer is correct.

Before coming to the properties of the halogenated hydrocarbons, let us mention two industrially important *fluorine* compounds. You have probably already seen one of the trade names for dichlorodifluoromethane, such as

    Freon®

Write down the structural formula for freon   →   70

**64**     A correct answer would be:

"The organic compound is decomposed by heat. The halogen reacts with copper. The copper halide evaporates (in contrast to metallic copper). The copper halide vapor colors the flame."

What is the color of the flame?     →    73

---

**65**     Return to   39   where through careful reading you should find the correct answer.

(If you need to recapitulate the precipitation of halide ions with silver nitrate, go instead to   74   .)

---

**66**     *Benzene* and *basic magnesium bromide.*

How is a Grignard reagent prepared?

In what solvent is the Grignard reaction carried out?

Write your answers in the form of key-words.     →    57

---

**67**     Correct; chlorine atoms in the *side-chain* react more easily than those in the aromatic *ring.*

Write down the equation for the reaction between chlorobenzene and sodium hydroxide.     →    76

---

**68**     R is the organic group (or residue);
Hal, OH, and $NH_2$ are functional groups.

Let us now desribe some of the reactions of halogenated hydrocarbons.

These compounds can react with water or bases

      a) with elimination of hydrogen halide, *or*
      b) with substitution of OH for the halogen atom.

Whether a) or b) occurs depends not only on the starting compound but also on the reaction conditions (e.g. temperature and pressure).

Firstly, the *elimination* of a hydrogen halide gives rise to *unsaturated* reaction products.

As an exercise, complete the equation:

$$
\underset{\substack{|\ |\ | \\ H\ I\ H}}{\overset{\substack{H\ H\ H \\ |\ |\ |}}{H-C-C-C-H}} \quad \xrightarrow[\text{of HI}]{\text{elimination}} \quad + \text{ HI}
$$

What is the name of the product?

● Propane                     →    50

● Propene                      →    59

● I need further information      →    75

---

**69**     R-Hal + Mg → R-Mg-Hal

Such magnesium compounds are called

        *Grignard* reagents

after the French chemist V. Grignard, (pronounced: "grinyard").

Give the equation for the formation of the Grignard reagent from methyl iodide.    →    58

---

**70**

$$
\underset{\substack{| \\ F}}{\overset{\substack{Cl \\ |}}{Cl-C-F}} \quad \text{or} \quad \underset{\substack{| \\ F}}{\overset{\substack{F \\ |}}{Cl-C-Cl}}
$$

       dichlorodifluoromethane
       (a trade name is "Freon")

The high stability of dichlorodifluoromethane and other fluorocarbons makes them suitable for a variety of uses, including aerosol propellants or cooling liquids in refrigerators and freezers. There has been some concern that their use may cause depletion of the ozone layer.

Tetrafluoroethylene is another important fluorine compound.

Write down the structure of this compound and check at   60  , or, if you need help, go instead to   53  .

---

**71**     It is important to exclude water when preparing Grignard reagents since water destroys the reagent.

We terminate with a description of a chemical test for halogen compounds.

A piece of copper wire is immersed in the compound to be tested. The wire is then inserted into a Bunsen flame. If a halogen atom is present a distinct *green* coloration occurs.

The reaction taking place is as follows: the organic compound is decomposed by the heat. The halogen which is liberated reacts with the copper forming a copper halide, e.g. $CuCl_2$, which evaporates (the metallic copper does not evaporate). The vapor gives rise to the green color.

A clean copper wire gives no flame coloration. Explain briefly why a green flame coloration is obtained after applying a halogen compound to the wire. → $\boxed{64}$

**10**

---

$\boxed{72}$    Your answer is wrong. Compare the reaction conditions necessary for the replacement of Cl by OH in benzyl chloride and in chlorobenzene. This should help you find the correct answer.

Return to  $\boxed{46}$ .

---

$\boxed{73}$    An organic halogen compound applied to a copper wire causes a *green* coloration when held in a Bunsen flame.

This is the end of Program 10.

---

$\boxed{74}$    The salts of the hydrogen halides dissociate in aqueous solution, e.g.

$$NaCl \rightarrow Na^{\oplus} + Cl^{\ominus}$$

Addition of a silver nitrate solution causes precipitation of insoluble silver halide, e.g.

$$Na^{\oplus} + Cl^{\ominus} + Ag^{\oplus} + NO_3{}^{\ominus} \rightarrow AgCl + Na^{\oplus} + NO_3{}^{\ominus}$$

Thus, the presence of *halide ions* is a necessary condition for the formation of a precipitate on the addition of silver nitrate.

Since organic halogen compounds do not dissociate, *no halide ions* are present.

Return to  $\boxed{39}$ .

| 75 |

Examine the equation

$$\underset{\substack{| \ \ | \ \ | \\ H \ \ \ \ H}}{\overset{\substack{H \ \ H \ \ H \\ | \ \ | \ \ |}}{H-C-C=C}} + HI \longrightarrow \underset{\substack{| \ \ | \ \ | \\ H \ \ I \ \ H}}{\overset{\substack{H \ \ H \ \ H \\ | \ \ | \ \ |}}{H-C-C-C-H}}$$

2–iodopropane

which describes the addition of HI to an alkene.

The opposite reaction is the elimination of hydrogen iodide from 2-iodopropane.

Return to    | 68 |    .

**10**

| 76 |

chlorobenzene

In general, halogen atoms in aliphatic compounds or in the side-chains of aromatic compounds are more reactive than those directly bonded to *aromatic rings*.

We have already seen an example where halogen compounds were used as the starting materials for the introduction of other functional groups, e.g. the OH group.

The reaction of a halogen compound, R-Hal, with magnesium metal leads to the formation of organomagnesium compounds, i.e. organic compounds containing magnesium. Generally, such compounds are known as *organometallic* compounds.

Example:

The reaction is carried out by the dropwise addition of a solution of bromobenzene in ether to magnesium turnings.

The formula shows that the magnesium atom inserts between the organic radical and the halogen atom.

Complete the general equation:

R-Hal + Mg →                                          →    | 69 |

# Program 11

## The Hydroxyl Group

**1**

The group

$-OH$

is the *hydroxyl group*.

All compounds containing a hydroxyl group have some chemical properties in common. *Differences* in chemical behavior will be due to the position of the hydroxyl group within the molecule. We first have to learn some new concepts.

You already know that carbon chains can be arranged in different ways in organic molecules. For example, a carbon atom can be directly bonded to one, two, three, or four other carbon atoms. A carbon atom having only *one* carbon neighbor is said to be *primary*.

**11**

Give the number of primary carbon atoms in the compound

$$CH_3-\underset{\underset{\displaystyle CH_3}{|}}{CH}-CH_3$$

- One → 22
- Two → 15
- Three → 10
- Four → 12

**2**     Correct. The formulas are repeated below:

$$CH_3-CH_3 \qquad CH_3-\overset{2^\circ}{CH_2}-CH_3 \qquad CH_3-\overset{2^\circ}{CH}-\overset{2^\circ}{CH_2}-CH_3$$
$$\underset{CH_3}{|}$$

and the secondary carbon atoms are labeled $2^\circ$.
You already know:

a *primary* carbon atom is directly bonded to *one* further carbon atom;
a *secondary* carbon is directly bonded to *two* further carbon atoms.

To how many other carbons is a *tertiary* carbon atom bonded? Does one of the formulas above contain a tertiary carbon atom? If yes, write it down, and label the tertiary carbon 3°.

Compare at ⬚14 .

---

**3**

1° primary
2° secondary
3° tertiary
4° quaternary

Examine the following compound and note the numbers of primary, secondary, tertiary, and quaternary carbon atoms.

$$CH_3-\underset{\underset{CH_3}{|}}{\overset{\overset{CH_3}{|}}{C}}-CH_2-\underset{\underset{CH_3}{|}}{CH}-CH_2-CH_3$$

How many of each kind did you find?

| primary | secondary | tertiary | quaternary | | |
|---------|-----------|----------|------------|---|---|
| ● 3 | 4 | 1 | 1 | → | ⬚16 |
| ● 4 | 3 | 1 | 1 | → | ⬚19 |
| ● 5 | 2 | 1 | 1 | → | ⬚13 |
| ● 5 | 2 | 0 | 1 | → | ⬚23 |
| ● 5 | 3 | 1 | 0 | → | ⬚26 |
| ● I found other numbers | | | | → | ⬚29 |

---

**4** Compare:

secondary alcohol:

$$\underset{R}{\overset{R}{\diagdown}}\underset{OH}{\overset{H}{\diagup}}C \quad \text{or shorter} \quad \underset{R}{\overset{R}{\diagdown}}CH(OH) \quad \text{or} \quad \underset{R}{\overset{R}{\diagdown}}CHOH$$

In the short-hand notations the OH group is often enclosed in a parenthesis in order to emphasize the nature of the functional group and to avoid confusion. Although the third formula above is correct, the presence of the OH group is not as clear.

Compare again:

primary alcohol:        $R-CH_2OH$   or   $R-\overset{\displaystyle H}{\underset{\displaystyle H}{C}}-OH$

secondary alcohol:      $\overset{\displaystyle R}{\underset{\displaystyle R}{\diagdown}}CHOH$   or   $\overset{R}{\underset{R}{\diagdown}}C\overset{H}{\diagup}_{OH}$

Give the general formula for *tertiary* alcohols and compare at   |17|   .

---

| 5 |   A *secondary* carbon atom is bonded to *two* further carbon atoms. Thus the compound

$$CH_3-CH_2-CH_3$$

contains one secondary carbon atom.

Now study   |20|   in detail, and you will find the proper answer.

---

| 6 |   Compare:

$$R-C\overset{H}{\diagup}_{O-\boxed{H}}\boxed{\underset{H}{}} + [O] \longrightarrow R-C\overset{H}{\diagdown}_{O} + H_2O$$

$$\overset{R}{\underset{R}{\diagup}}C\overset{\boxed{H}}{\diagup}_{O-\boxed{H}} + [O] \longrightarrow \overset{R}{\underset{R}{\diagdown}}C=O + H_2O$$

$$\overset{R}{\underset{R}{\diagdown}}R-C-OH + [O] \longrightarrow \text{no reaction}$$

The compounds formed in these reactions will be discussed later.

Now follows the *nomenclature of alcohols.*

By international convention the names of the alcohols are derived from the corresponding hydrocarbons by changing the ending *-ane* to *-anol*.

Write down the following formulas and the names of the compounds:

$CH_3OH$            $CH_3-CH_2OH$

methanol            .........................            →   |21|

---

**7**     Your answer is wrong.

Continue at   24   , second paragraph.

---

**8**     Read the text once more, in detail, and you will find the answer.   →   22

---

**9**     A secondary carbon atom is bonded to *two* further carbon atoms. Thus the compound

$$CH_3-CH_2-CH_3$$

contains one secondary carbon atom.

Now study   20   in detail, and you will find the proper answer.

---

**10**     Correct; the compound

$$CH_3-\underset{\underset{CH_3}{|}}{CH}-CH_3$$

has *three* carbon atoms with only *one* carbon neighbor each.

Such carbon atoms are called ................... →   20

---

**11**     Your answer is wrong.

Continue at   24   , second paragraph.

---

**12**     Examine the formula more closely. The four carbon atoms are *not* all arranged in the same manner.   →   1

---

| **13** | Correct. |

Now copy the following formula and indicate the primary, secondary, tertiary, and quaternary carbon atoms by 1°, 2°, 3°, and 4°, respectively.

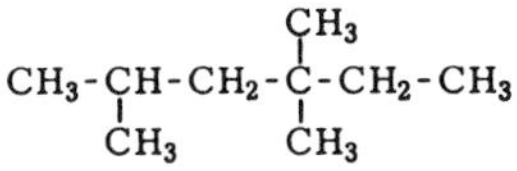

$$CH_3-\underset{\underset{CH_3}{|}}{CH}-CH_2-\underset{\underset{CH_3}{|}}{\overset{\overset{CH_3}{|}}{C}}-CH_2-CH_3$$

Compare at   24   .

---

| **14** |

$$CH_3-\underset{\underset{CH_3}{|}}{\overset{\overset{3°}{}}{CH}}-CH_2-CH_3$$

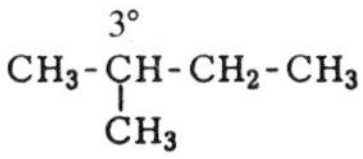

A carbon atom attached to *four* other carbon atoms is said to be *quaternary*.

Now write down the meaning of the four symbols:

1° ...............
2° ...............
3° ...............
4° ...............

and check your answer at   3   .

---

| **15** | Examine the formula more closely. There are *three* carbon atoms in identical environments.   →   1 |

---

| **16** | Your answer is wrong. |

Read the explanation at   29   .

---

| **17** |

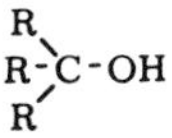

$$R-\underset{\underset{R}{|}}{\overset{\overset{R}{\diagdown}}{C}}-OH$$

In a tertiary alcohol *three* R groups are attached to the carbon atom carrying the hydroxyl group.

Consider the question:

Do quaternary alcohols exists?

● Yes            →    $\boxed{31}$

● No             →    $\boxed{27}$

● I don't know    →    $\boxed{33}$

---

$\boxed{18}$     Your answer is wrong.

Continue at   $\boxed{24}$  , second paragraph.

---

$\boxed{19}$     Your answer is wrong.

Read the explanation at   $\boxed{29}$  .

---

$\boxed{20}$     *Primary.*

A carbon atom attached to *two* other carbon atoms is said to be *secondary*.

How many of the compounds below contain a secondary carbon atom?

$$CH_3-CH_3 \qquad CH_3-CH_2-CH_3 \qquad CH_3-\underset{\underset{CH_3}{|}}{CH}-CH_2-CH_3$$

● One      →    $\boxed{5}$

● Two      →    $\boxed{2}$

● Three    →    $\boxed{9}$

---

$\boxed{21}$    $CH_3-OH$        $CH_3-CH_2-OH$
         methan*ol*          ethan*ol*

(The emphasis is on the first syllable in these words.)

A shorter notation for ethanol is $C_2H_5OH$.
The next higher alcohol is derived from *propane,* a hydrocarbon with three carbon atoms:

$$CH_3-CH_2-CH_3.$$

How many different alcohols can be derived from propane?

- One          →    35

- Two          →    32

- Three        →    42

- I don't know →    54

---

**22**    No.

There are *three* carbon atoms with only *one* carbon neighbor each in the compound:

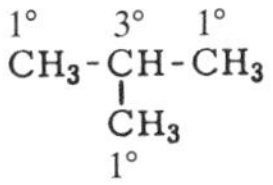

$$
\begin{array}{ccc}
1° & 3° & 1° \\
CH_3 - CH - CH_3 \\
| \\
CH_3 \\
1°
\end{array}
$$

These are labeled 1°. The fourth carbon atom (3°) has *three* neighbors.

A carbon atom with only *one* carbon neighbor is said to be *primary*.

How many primary carbon atoms are there in the formula above?

- One          →    8

- Three        →    10

- Four         →    12

---

**23**    Your answer is wrong.

Read the explanation at    29   .

---

**24**    Please check your answer carefully:

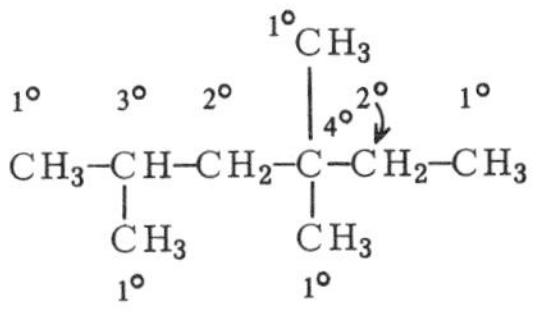

$$
\begin{array}{c}
1° CH_3 \\
1° \quad 3° \quad 2° \quad |\,4°\,2° \quad 1° \\
CH_3 - CH - CH_2 - C - CH_2 - CH_3 \\
| \qquad\qquad | \\
CH_3 \qquad CH_3 \\
1° \qquad\quad 1°
\end{array}
$$

If you had *any* mistakes, repeat $\boxed{2}$ .

Let us now return to the hydroxyl group.

When the *hydroxyl group is directly attached to a primary carbon atom,* the compound is said to be a *primary alcohol.*

Examples:

$$\begin{array}{c} \text{H H} \\ | \ | \\ \text{H-C-C-OH} \\ | \ | \\ \text{H H} \end{array} \ _{1°} \quad \text{and} \quad \begin{array}{c} \text{H H H} \\ | \ | \ | \\ \text{H-C-C-C-OH} \\ | \ | \ | \\ \text{H C H} \\ \text{H}_3 \end{array} \ _{1°}$$

Designate these primary alcohols as follows:

$CH_3\text{-}CH_2OH$    and    ..............    $\rightarrow$    $\boxed{40}$

---

$\boxed{25}$    Correct.

Three primary alcohols:

$$CH_3\text{-}CH_2OH \qquad \begin{array}{c} C_2H_5\text{-}CH\text{-}CH_3 \\ | \\ CH_2OH \end{array} \qquad \begin{array}{c} CH_3 \\ | \\ CH_3\text{-}C\text{-}CH_3 \\ | \\ CH_2OH \end{array}$$

A secondary alcohol:

$$CH_3\text{-}CHOH\text{-}C_2H_5$$

A tertiary alcohol:

$$\begin{array}{c} H_3C \\ \phantom{} \diagdown \\ H_3C\text{-}C\text{-}OH \\ \phantom{} \diagup \\ H_3C \end{array}$$

The differences between primary, secondary, and tertiary alcohols are seen not just in their formulas, but also in their chemical behavior, e.g. their reactions with oxidizing agents. Primary and secondary alcohols can be oxidized to compounds with new functional groups:

$$R\text{-}C\overset{H}{\underset{O\text{-}\boxed{H}}{\diagup\ \boxed{H}}} + [O] \longrightarrow R\text{-}C\overset{H}{\underset{O}{\diagup}} + H_2O \quad (1)$$

$$\overset{R}{\underset{R}{>}}C\overset{\boxed{H}}{\underset{O\text{-}\boxed{H}}{}} + [O] \longrightarrow \ \ldots\ldots + H_2O \quad (2)$$

Potassium permanganate or potassium dichromate can be used as oxidizing agents. The reaction equations are often abbreviated using [O] to symbolize the oxidizing agent as shown above.

In contrast, tertiary alcohols are *not* attacked by oxidizing agents under the same reaction conditions:

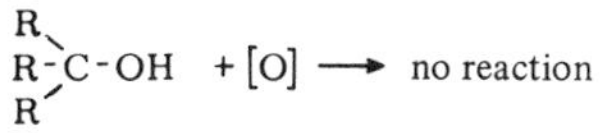

$$
\begin{array}{c}
R \\
R-C-OH \\
R
\end{array}
+ [O] \longrightarrow \text{no reaction} \qquad (3)
$$

Copy down these three equations, and complete equation (2).   →   $\boxed{6}$

---

$\boxed{26}$     Your answer is wrong.

Read the explanation at   $\boxed{29}$ .

---

$\boxed{27}$     Correct; a *quaternary* carbon atom has *four* carbon neighbors. Hence, there is no room for a functional group.

Thus, we can distinguish

primary, secondary, and tertiary alcohols.

How many of the following are *secondary* alcohols?

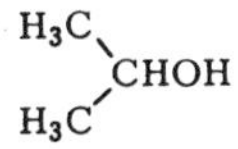

$$
\begin{array}{c}
H_3C \\
\qquad CHOH \\
H_3C
\end{array}
\qquad CH_3-CH(OH)-CH_3 \qquad R-CHOH-R
$$

● Three        →     $\boxed{37}$

● Two          →     $\boxed{39}$

● One          →     $\boxed{41}$

● None         →     $\boxed{43}$

---

$\boxed{28}$     Your answer is wrong. – Study the formulas more closely!   →   $\boxed{42}$

**29**     In the following examples the hydrogen atoms are omitted for simplicity:

primary carbon atoms (1°):

$$\overset{1°}{C}-\overset{1°}{C}$$

secondary carbon atom (2°):

$$C-\overset{2°}{C}-C$$

tertiary carbon atom (3°):

$$C-\underset{\overset{|}{C}}{\overset{3°}{C}}-C$$

quaternary carbon atom (4°):

$$C-\underset{\overset{|}{C}}{\overset{\overset{C}{|}}{C}}{}^{4°}-C$$

Repeat [3] .

---

**30**     Your answer is wrong. – Study the formulas more closely!    →   [42]

---

**31**     No.

A quaternary carbon atom has four carbon atoms as neighbors, e.g.

$$CH_3-\underset{\overset{|}{CH_3}}{\overset{\overset{CH_3}{|}}{C}}{}^{4°}CH_3$$

Hence, there is *no room* for a functional group at the quaternary carbon.
→   [27]

---

**32**     Correct; there are *two* alcohols derived from propane. They are

$$H-\underset{\overset{|}{H}}{\overset{\overset{H}{|}}{C}}-\underset{\overset{|}{H}}{\overset{\overset{H}{|}}{C}}-\underset{\overset{|}{H}}{\overset{\overset{H}{|}}{C}}-OH \qquad \text{and} \qquad H-\underset{\overset{|}{H}}{\overset{\overset{H}{|}}{C}}-\underset{\overset{|}{OH}}{\overset{\overset{H}{|}}{C}}-\underset{\overset{|}{H}}{\overset{\overset{H}{|}}{C}}-H$$

or shorter: $CH_3-CH_2-CH_2OH$ and $CH_3-CH(OH)-CH_3$

Obviously, the name *propanol* would not suffice to describe both compounds.

Do you recall the nomenclature rules well enough to name these two propanols systematically?

● Yes        →    |44|

● No         →    |47|

---

|33|        A quaternary carbon atom has four other carbon neighbors, e.g.

$$CH_3-\underset{\underset{CH_3}{|}}{\overset{\overset{CH_3}{|}}{C}}\!\!\!\!\xleftarrow{4°}CH_3$$

Hence, there is *no room* for a functional group at the quaternary carbon.    →    |27|

---

|34|

| Formula | Systematic name | Trivial name |
|---------|-----------------|--------------|
| $CH_3OH$ | methanol | methyl alcohol |
| $CH_3–CH_2OH$ | ethanol | ethyl alcohol |
| $CH_3–CH_2–CH_2OH$ | 1-propanol | propyl alcohol |
| $CH_3–CH(OH)–CH_3$ | 2-propanol | isopropyl alcohol |

The number of the carbon atom is used to indicate the position of the functional group.

We will now describe the *synthesis of alcohols*.

The *hydrolysis of alkyl halides* is a generally applicable reaction:

$$R–Hal + H_2O → R–OH + H–Hal$$

As in previous programs, –Hal stands for halogen, i.e. –Cl, –Br, or –I. R is the aliphatic residue (*alkyl* radical).

Give a feasible reaction equation for the preparation of ethanol from an alkyl chloride.

● From methyl chloride?    →    |50|

● From ethyl chloride?     →    |46|

● From propyl chloride?    →    |53|

| 35 | Your answer is wrong.

Read the explanation at    54   .

---

| 36 | Compare:

$$\begin{array}{l} H_3C \\ H_3C-C-Cl \\ H_3C \end{array} + H_2O \longrightarrow \begin{array}{l} H_3C \\ H_3C-C-OH \\ H_3C \end{array} + HCl$$

The compound formed is tertiary butyl alcohol, or more correctly tertiary butanol, abbreviated *tert*-butanol.

Tertiary alkyl halides are more easily hydrolyzed than secondary, and secondary easier than primary.

Other general procedures for the preparation of alcohols will be described in due course. First, we shall consider a few individual alcohols.

Methanol is the simplest of the alcohols. It is primary, and it represents a special case of the general formula

$$R-CH_2OH$$

where R = H.

Give the detailed structural formula for methanol.    $\longrightarrow$    48

---

| 37 | Correct.

Now examine the following formulas. – How many primary, secondary, and tertiary alcohols are there?

$$CH_3-CH_2OH \qquad \begin{array}{l} C_2H_5-CH-CH_3 \\ \qquad CH_2OH \end{array} \qquad \begin{array}{l} CH_3 \\ CH_3-C-CH_3 \\ CH_2OH \end{array}$$

$$\begin{array}{l} H_3C \\ H_3C-COH \\ H_3C \end{array} \qquad CH_3-CHOH-C_2H_5$$

| primary | secondary | tertiary | | |
|---------|-----------|----------|---|---|
| ● 2 | 2 | 1 | → | $\boxed{7}$ |
| ● 3 | 1 | 1 | → | $\boxed{25}$ |
| ● 4 | 0 | 1 | → | $\boxed{11}$ |
| ● 4 | 1 | 0 | → | $\boxed{18}$ |

---

**38**   Compare:

$$CH_3-CH=CH_2 + H-OH \rightarrow CH_3-CH(OH)-CH_3$$

propene                    2-propanol (or isopropyl alcohol)

The addition of water to an olefin is *reversible:* at higher temperature *water can be eliminated* from the alcohol, thereby regenerating the olefin.

Write down the equations for the elimination of water from

　　a) ethanol
　　b) isopropyl alcohol

Compare your answer at   $\boxed{49}$   .

---

**39**   Your answer is wrong.

Re-examine the formula for secondary alcohols at   $\boxed{4}$   .

---

**40**   $CH_3-CH_2OH$        and        $CH_3-\underset{\underset{CH_3}{|}}{C}H-CH_2OH$

Both of these are primary alcohols. In a primary alcohol you always have the group

$$-CH_2OH \qquad or \qquad -\overset{\overset{H}{|}}{\underset{\underset{H}{|}}{C}}-OH$$

Hence, the general formula for primary alcohols is

$$R-CH_2OH \qquad \text{or} \qquad R-\overset{\overset{\displaystyle H}{|}}{\underset{\underset{\displaystyle H}{|}}{C}}-OH$$

The hydroxyl group is attached to a *primary* carbon atom.

In *secondary* alcohols the OH-group is attached to a *secondary* carbon.

Write down the general formula for secondary alcohols and compare at   $\boxed{4}$ .

---

**41**     Your answer is wrong.

Re-examine the formula for secondary alcohols at   $\boxed{4}$ .

---

**42**     You have made a mistake. Although there are three carbon atoms in propane

$$CH_3-CH_2-CH_3$$

two of them are in *identical* environments. If you replace in turn one hydrogen atom at each carbon by OH, you obtain

$$CH_2OH-CH_2-CH_3 \qquad CH_3-CHOH-CH_3 \qquad CH_3-CH_2-CH_2OH$$

           (a)                      (b)                      (c)

However, *two* of these formulas are identical. Which ones?

- (a) and (b)      →    $\boxed{28}$
- (a) and (c)      →    $\boxed{32}$
- (b) and (c)      →    $\boxed{30}$

---

**43**     Your answer is wrong.
Re-examine the formula for secondary alcohols at   $\boxed{4}$ .

---

**44**     $CH_3-CH_2-CH_2OH$        $CH_3-CHOH-CH_3$

    1-propanol                2-propanol

A "trivial" name for 1-propanol is "propyl alcohol", and 2-propanol is *"isopropyl alcohol"*.

In other words, the name of the alcohol can be derived from that of the alkyl radical, adding "alcohol". The two nomenclatures are compared in the following table:

| Formula | Systematic name | Trivial name |
|---|---|---|
| $CH_3OH$ | methanol | methyl alcohol |
| $CH_3-CH_2OH$ | ......................... | ......................... |
| $CH_3-CH_2-CH_2OH$ | ......................... | propyl ................... |
| $CH_3-CH(OH)CH_3$ | ......................... | isopropyl ............... |

Complete this table and check your result at ⏍34⏍ .

---

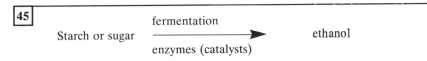

**45**

Starch or sugar $\xrightarrow[\text{enzymes (catalysts)}]{\text{fermentation}}$ ethanol

The major industrial route to ethanol is the hydration of *ethylene.*

Are you able to formulate the equation for the addition of water (H–OH) to the double bond in ethylene?

● Yes        →    ⏍55⏍

● No         →    ⏍65⏍

---

**46**    $CH_3-CH_2Cl + H_2O \rightarrow CH_3-CH_2OH + HCl$

ethyl chloride              ethanol

The reaction is carried out by heating the alkyl chloride with water or a base. Using sodium hydroxide as the base, NaCl will form in place of HCl:

$$CH_3-CH_2Cl + NaOH \rightarrow CH_3-CH_2OH + NaCl$$

ethyl chloride              ethanol

As we saw in Program 10, elimination reactions are also possible in these systems:

$$CH_3-CH_2Cl + NaOH \rightarrow CH_2{=}CH_2 + H_2O + NaCl$$

ethyl chloride              ethylene

Therefore, a proper choice of reaction conditions is necessary in order to obtain a good yield of alcohol in hydrolysis reactions.

Tertiary alkyl chlorides hydrolyze with particular ease.

Write out the equation for the hydrolysis of the *simplest tertiary alkyl chloride.*
Compare your answer at $\boxed{36}$
or, if you need more information, go instead to $\boxed{56}$ .

---

$\boxed{47}$ When more than one position of a functional group is possible, the position must be specified. Therefore, the carbon atoms in the hydrocarbon chain are numbered, e.g. in propane:

**11**

$$\overset{3}{C}H_3 - \overset{2}{C}H_2 - \overset{1}{C}H_3$$

The *position* of a functional group is specified by the *number of the carbon atom.*

Now write down the formulas and names of the two propanols:

$CH_3-CH_2-CH_2OH$ $CH_3-CHOH-CH_3$

$\rightarrow$ $\boxed{44}$

---

$\boxed{48}$ Compare:

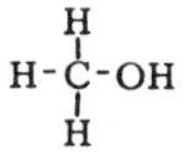

$$H-\overset{\overset{\displaystyle H}{|}}{\underset{\underset{\displaystyle H}{|}}{C}}-OH$$

Methanol is manufactured industrially from "synthesis gas". Synthesis gas is a mixture of carbon monoxide and hydrogen:

$$CO + \ldots\ldots\ldots \rightarrow CH_3OH$$

How many moles of hydrogen are required?

● One $\rightarrow$ $\boxed{60}$

● Two $\rightarrow$ $\boxed{58}$

● Three $\rightarrow$ $\boxed{62}$

---

**49**

$$CH_3-CH_2OH \quad \leftrightarrows \quad CH_2=CH_2 + H-O-H$$

　ethanol　　　　　　　　ethylene

$$CH_3-CHOH-CH_3 \quad \leftrightarrows \quad CH_3-CH=CH_2 + H-O-H$$

　isopropyl alcohol　　　　　propene

The double arrows indicate that the reactions can proceed in both directions. The chemist has to find the right conditions of temperature, pressure, and catalyst which favor reaction in the desired direction.

Program 6 deals in detail with chemical equilibria.

The solubility of isopropyl alcohol in water is less than for ethanol and methanol. The higher alcohols are even less soluble. Already the pentanols – i.e. the alcohols with five carbon atoms – are practically insoluble in water.

Methanol, ethanol, and isopropyl alcohol are used in the synthesis of other organic chemicals, i.e. they are synthetic *intermediates*. Ethanol is not only used for making alcoholic beverages, but also many cosmetics (perfumes; after-shave lotion).

*Benzyl alcohol* is another important alcohol. It is a primary alcohol possessing a phenyl group.

Give the formula for benzyl alcohol on the basis of this information.　→　72

**11**

---

**50**　　Look up the formula for ethanol in the table at　44　. You should then be able to answer the question. Repeat　44　.

---

**51**　　Fermentation is not a general procedure for the preparation of alcohols. Fermentation is a specific procedure for the preparation of ethanol.

Repeat　72　.

---

**52**　　There are three types of alcohols:

$R-CH_2OH$ 　　　　　$\underset{R}{\overset{R}{\diagdown}}CHOH$ 　　　　　$\underset{R}{\overset{R}{\diagdown}}C-OH$

　primary alcohol　　　　secondary alcohol　　　tertiary alcohol

Examine the formula for glycerol once more:

$$CH_2OH$$
$$CHOH$$
$$CH_2OH$$

Which types of hydroxyl groups are there in glycerol?

- ● Three primary                    → 79

- ● Three secondary                  → 81

- ● Two primary and one secondary    → 87

- ● One primary and two secondary    → 83

---

**11**

53    Look up the formula for ethanol in the table at 44 . You should then be able to answer the question. Repeat 44 .

---

54    Propane has the formula

$$CH_3-CH_2-CH_3$$

Since there are *two* kinds of carbon atoms, there is more than one way of attaching a hydroxyl group.

How many different alcohols can be derived from propane? Give the formulas.

- ● Two alcohols      → 32

- ● Three alcohols    → 42

---

55    Compare:

$$CH_2{=}CH_2 + H{-}O{-}H \rightarrow CH_3{-}CH_2OH$$

  ethylene                    ethanol

This reaction is carried out over a catalyst at elevated pressure.

Similarly, water can be added to propene, giving a *secondary* alcohol.

Give the equation for this reaction and compare at 38 , or, if you need more information, go instead to 66 .

---

**56** The simplest tertiary alkyl chloride is

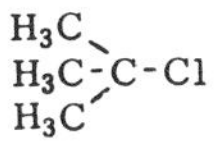

$$\begin{array}{l} H_3C \\ H_3C-C-Cl \\ H_3C \end{array}$$

Now give the equation for the hydrolysis of this compound.   →   36

---

**57** The hydrogenation of carbon monoxide is *not* a general procedure for the preparation of alcohols. *This reaction only gives methanol.*

Repeat   72   .

---

**58** The equation is

$$CO + 2 H_2 \rightarrow CH_3OH$$

and it takes place at ca. 400 °C at elevated pressure and in the presence of catalysts.

Methanol is very soluble in water.

**Methanol tastes like ethanol but is exceedingly poisonous. Even the consumption of small quantities leads to blindness or death.**

There is a *homologous series* of alcohols just as is the case for hydrocarbons and alkyl halides. The first two members of the series are *methanol* and *ethanol*. Ethanol is also poisonous when taken in large doses. In smaller doses it is intoxicating.

One of the oldest chemical processes known to mankind is the formation of ethanol by *fermentation.*

The fermentation process converts starch or sugar to ethanol under the action of certain catalysts present in yeast. These catalysts are called *enzymes.*

Complete the following scheme:

```
                ............... (process)
starch of sugar  ──────────────────────→  ethanol.
                ............... (catalysts)
```

Compare at   45   .

**59** Like all alkoxides, sodium methoxide and sodium ethoxide are *unstable toward water*. The alkoxides hydrolyze, re-forming the alcohol and a metal hydroxide, e.g.

$$R-ONa + H_2O \rightarrow \text{...............} + \text{...............}$$

Complete the equation and check at  75  .

---

**60** Note that there is a total of four hydrogen atoms on the right of the equation. You must also have four on the left.

Repeat  48  .

**11**

---

**61** The addition of water to olefins allows the preparation of a wide variety of alcohols. Ethanol and isopropyl alcohol are examples. Nevertheless, this process is *not a general one*.

Repeat  72  .

---

**62** Note that there is a total of four hydrogen atoms on the right of the equation. You must also have four on the left.

Repeat  48  .

---

**63** $CH_3-ONa$ and $C_2H_5-ONa$

What are these two alkoxides called?

Compare your answer at  73  or, if you need help, go instead to  75  .

---

**64**

$$\text{(benzene ring)}-CH_2Cl + H_2O \longrightarrow \text{(benzene ring)}-CH_2OH + HCl$$

Give the names of the two organic compounds.   →   74

---

**65**     Ethylene is

$$CH_2 = CH_2$$

When water (H–OH) is added to the double bond, ethanol is obtained.

Write down the complete equation.     →     55

---

**66**     Propene has the formula

$$CH_3-CH = CH_2$$

Add water (H–OH) in such a way that a secondary alcohol is formed.

Write down the equation.     →     38

**11**

---

**67**     
$$\begin{array}{c} CH_2OH \\ | \\ CH_2OH \end{array}$$     or          $HOCH_2-CH_2OH$

This *dihydric alcohol* is called *glycol* (strictly speaking: ethylene glycol).

*Glycerol* is a *trihydric alcohol:*

$$\begin{array}{c} CH_2OH \\ | \\ CHOH \\ | \\ CH_2OH \end{array}$$

What kind of hydroxyl groups do you see in glycerol?

● Three primary                              →     79

● Three secondary                          →     81

● Two primary and one secondary      →     77

● One primary and two secondary      →     83

● I cannot answer this question          →     52

---

**68**     We now turn to compounds in which the hydroxyl group is *directly* attached to an *aromatic* ring.

These hydroxy compounds are called *phenols*. The name is derived from *phenyl* and the ending *-ol.*

*Phenol* itself is the simplest such compound. It is the alcohol composed of a phenyl radical and a hydroxyl group.

Give the formula for phenol.   →   78

---

69     Compare:

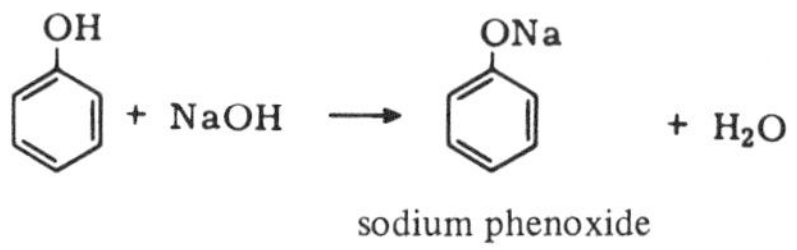

sodium phenoxide

*Thus, phenols are weak acids.* The salts are called *phenoxides.*

Aliphatic alcohol can also form metal salts. However, in this case it is necessary to react the alcohol with a free metal. The products are hydrogen gas and a metal *alkoxide.*

The reaction between ethanol and sodium is formulated below:

$$2\ C_2H_5\!-\!OH + 2\ Na \rightarrow 2\ C_2H_5\!-\!ONa + H_2$$

Now formulate the *general* equation for the reaction of an alcohol (R–OH) with sodium.   →   85

---

70     *Hydroxy-toluenes*

There are three hydroxy-toluenes, the *o-*, *m-*, and *p-*compounds:

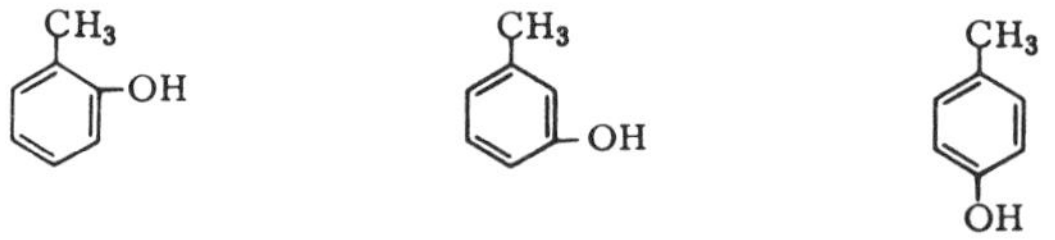

The correct names of these three phenols are *o-*, *m-*, *and p-cresol* (or *o-*, *m-*, and *p-*methylphenol).

Similarly, there are three different *chlorophenols.*

Give the formulas and names of the chlorophenols.   →   80

**71**    CH₃–ONa          + H₂O →    CH₃–OH + NaOH

sodium methoxide                methanol

This reaction is a *hydrolysis.*
*The aliphatic alcoholates differ from the phenolates in their instability toward water.*

This is the end of Program 11.

---

**72**    Benzyl alcohol:

Benzyl alcohol can be prepared using the general synthetic method for alcohols. What is this *general* method?

● The fermentation process          →    51

● Hydrogenation of carbon monoxide    →    57

● Addition of water to olefins        →    61

● Hydrolysis of alkyl halides         →    90

---

**73**    CH₃–ONa                    C₂H₅–ONa

sodium methoxide              sodium ethoxide

Are these two compounds stable toward water?

● Yes          →    82

● No           →    86

● I don't know →    59

**11**

| **74** | Benzyl chloride, benzyl alcohol. |

Till now, we have described alcohols containing only *one* hydroxyl group.

Such alcohols are said to be *monohydric*.

Molecules possessing *more than one hydroxyl group* are called diols, triols, etc., or generally: polyols or polyhydric alcohols.

Normally, the hydroxyl group in polyhydric alcohols must be joined to *different* carbon atoms.

Compounds having more than one hydroxyl group at one carbon atom are unstable since they easily eliminate water, e.g.:

$$R-C\overset{H}{\underset{O-H}{\overset{O-H}{|}}} \longrightarrow R-C\overset{H}{\underset{O}{\diagdown}} + H_2O$$

$$R-C\overset{O-H}{\underset{O-H}{\overset{O-H}{|}}} \longrightarrow R-C + H_2O$$

Copy these equations and complete the last one.   →   [84]

---

| **75** | Hydrolysis of alkoxides: |

$$R-ONa + H_2O \rightarrow R-OH + NaOH$$

Alkoxides are frequently used in laboratory reactions. The most important alkoxides are the sodium salts of methanol and ethanol. Their names are

> sodium methoxide and
> sodium ethoxide

Give the formulas for these two compounds.   →   [63]

---

| **76** | Compare: |

$$CH_3-CH=CH_2 + Cl_2 \rightarrow CH_2Cl-CH=CH_2 + HCl$$

We now add hypochlorous acid to the reaction product:

$$CH_2Cl-CH=CH_2 + HO-Cl \rightarrow CH_2Cl-CH(OH)-CH_2Cl$$

Finally, a hydrolysis replaces the two chlorine atoms by OH groups.

Formulate this latter reaction and compare at $\boxed{88}$ .

---

$\boxed{77}$     Correct.

Glycerol is used in cosmetic creams and ointments. Glycerol is also an important component in the manufacture of resins (see Program 22).

Write down the formulas for glycol and glycerol, and indicate the primary (1°) and secondary (2°) hydroxyl groups.   →   $\boxed{87}$

---

$\boxed{78}$     Phenol:

Phenol is present in *coal tar* from which it can be isolated. However, this route is not sufficient to satisfy the demand for phenol, so several synthetic procedures have been developed. One of these is based on the reaction of *chlorobenzene* with water.

Formulate the equation for this reaction.   →   $\boxed{89}$

---

$\boxed{79}$     No, the three carbon atoms in glycerol are not identical!

Repeat   $\boxed{67}$

---

$\boxed{80}$

| OH—Cl | OH—Cl | OH Cl |
|---|---|---|
| *o*−chlorophenol | *m*−chlorophenol | *p*−chlorophenol |

Many phenols are partially soluble in water. The aqueous solutions are *weakly acidic*. Phenol can be neutralized by the addition of sodium hydroxide or other strong bases, whereby a hydrogen atom from the hydroxyl group is replaced by metal.

Complete the equation:

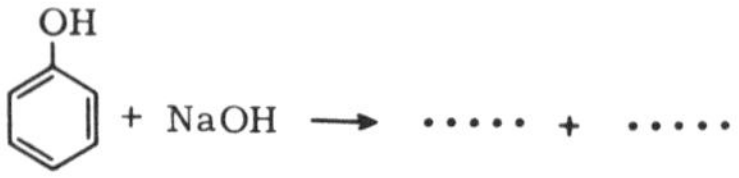

OH

$\bigcirc$ + NaOH $\longrightarrow$ ····· + ·····

$\rightarrow$   69

---

**81**    No, the three carbon atoms in glycerol are not identical!

Repeat   67   .

---

**82**    Your answer is wrong.

Repeat from   85   , last paragraph.

---

**83**    Your answer is wrong.

Read the explanation at   52   .

---

**84**

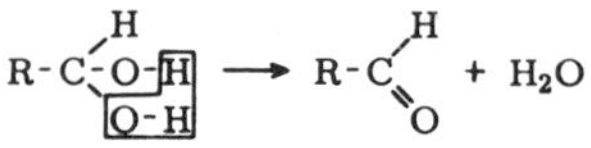

$$R - C\overset{H}{\underset{\boxed{O-H}}{-O-\boxed{H}}} \longrightarrow R - C\overset{H}{\underset{O}{\diagdown}} + H_2O$$

aldehyde

$$R - C\overset{O-H}{\underset{\boxed{O-H}}{-O-\boxed{H}}} \longrightarrow R - C\overset{OH}{\underset{O}{\diagdown}} + H_2O$$

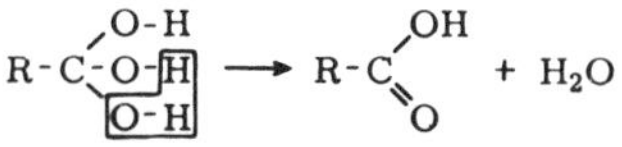

carboxylic acid

The products of these reactions will be described in subsequent programs.

The *simplest polyhydric alcohol (polyol)* is derived from ethane,

$$CH_3CH_3$$

by the replacement of one hydrogen atom at each methyl group by OH.

Write the formula for the diol, and compare at   67   .

**85**
$$2 \text{ R–OH} + 2 \text{ Na} \rightarrow 2 \text{ R–ONa} + H_2$$

This reaction is completely analogous to the reaction between sodium and water:

$$2 \text{ H–OH} + 2 \text{ Na} \rightarrow 2 \text{ H–ONa} + H_2$$

but sodium does not react as violently with alcohols as with water.

The alcoholates are *not stable* in the presence of water: they *hydrolyze*. Hydrolysis gives the original alcohol and a metal hydroxide:

$$\text{R–ONa} + H_2O \rightarrow \ldots\ldots\ldots + \ldots\ldots\ldots$$

Complete this equation.    $\rightarrow$    75

**11**

---

**86**      Correct; sodium methoxide and ethoxide are *unstable toward water*.

Formulate the reaction between sodium methoxide and water. What is such a reaction called?    $\rightarrow$    71

---

**87**

| | |
|---|---|
| CH$_2$OH 1° | CH$_2$OH 1° |
| CH$_2$OH 1° | CHOH 2° |
| | CH$_2$OH 1° |
| glycol | glycerol |

Glycerol is a constituent of fats and vegetable oils. These will be described in Program 23.

Glycerol is also manufactured industrially. Much interesting chemistry can be learned from the glycerol synthesis, which will now be discussed. If, however, you wish to omit this part, continue at   68   .

One synthesis of glycerol starts from propene (= propylene) which is treated with chlorine at elevated temperature. Under these conditions, chlorine does *not* add to the double bond. Instead, a methyl group undergoes *substitution* by chlorine.

$$CH_3\text{–}CH{=}CH_2 + Cl_2 \rightarrow \ldots\ldots\ldots + HCl$$

Try to complete this equation.    $\rightarrow$    76

|88|

Compare:

$$CH_2Cl-CH(OH)-CH_2Cl + 2\ H_2O \rightarrow$$

$$CH_2OH-CH(OH)-CH_2OH + 2\ HCl$$

You will notice that the product is glycerol.

This ends our discussion of glycerol for the time being. Thank you for working through these sections.

We shall now consider compounds in which the hydroxyl group is *directly* attached to an *aromatic ring*.

**11**

Such compounds are called *phenols*. The name is derived from the radical *phen*yl and the suffix *ol*.

The simplest of these compounds is phenol itself. It is the alcohol composed of a phenyl radical and a hydroxyl group.

Give the formula for phenol.    $\rightarrow$    |78|

|89|

Compare:

chlorobenzene              phenol

Since the halogen atom in chlorobenzene is not very reactive, this reaction must be carried out under pressure and at a temperature of ca. 400 °C.

Other methods for preparing phenol will be discussed later on.

Phenol could also be described as *"hydroxy-benzene"*. This is not a systematic name, however. Using the "hydroxy-benzene" name, what would be the names of the compounds derived from toluene by joining an OH group to the aromatic ring?

Compare your answer at    |70|   .

90   The *hydrolysis of alkyl halides* is a general method for the preparation of alcohols. Benzyl alcohol is made from benzyl chloride in this way:

$$\text{C}_6\text{H}_5\text{CH}_2\text{Cl} + \text{H}_2\text{O} \longrightarrow$$

Complete the equation and compare your answer at   64  .

11

# Program 12

## Ethers, Thioethers, and Thiols

---

**1** Ethers are compounds of the type

R–O–R

in which the hydrocarbon groups R can be either *identical or different*. The name of the ether is composed of the names of the two groups (in alphabetical order) followed by the word "ether".

What, then, is the name of the following compound

$CH_3-O-CH_2-CH_3$?

Compare your answer at [10] or, if you need help, go instead to [20] .

**12**

---

**2** Methyl phenyl ether:

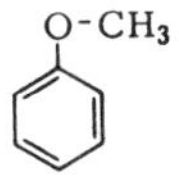

*Ethers can be prepared* by the reaction of alkoxides or phenoxides with alkyl halides:

R–ONa + R–Hal

Complete this equation. → [13]

---

**3** $CH_3-CH_2-ONa + CH_3I$ → $CH_3-CH_2-O-CH_3 + NaI$

    sodium ethoxide    methyl     ethyl methyl ether
                           iodide

Alkyl chlorides can also be used. Since not all compounds are equally reactive, the starting materials should be chosen carefully.

How would you prepare methyl phenyl ether?

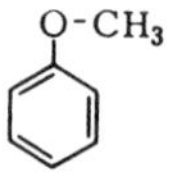

O⁻CH₃ (on benzene ring)

- From sodium phenoxide and methyl chloride          →          17
- From chlorobenzene and sodium methoxide          →          19
- The two methods just mentioned are equally good          →          23
- I don't know          →          28

---

**12**

**4**     The chlorine atom in methyl chloride is more *reactive* (or *labile*) than that in chlorobenzene. Therefore, the reaction with methyl chloride proceeds more *smoothly*.

Ethers can also be made from alcohols by the *elimination of water*. For example, under certain reaction conditions, diethyl ether is formed from ethanol:

$$CH_3-CH_2-O\boxed{-H + H-O}-CH_2-CH_3 \rightarrow \text{...............}$$

Complete the equation.    →    14

---

**5**     You made a mistake.

Remember the names of the groups:

           CH₂⁻

benzyl: (ring with CH₂⁻)    phenyl: (ring)

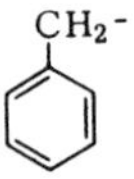

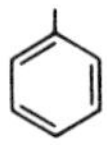

Now give the formula for benzyl methyl ether.    →    15

---

**6**     Correct.

Write the equations for the following reactions and then answer the question:

Which of these reactions are *intramolecular*?

a) isopropyl alcohol $\xrightarrow[\text{(dehydration)}]{-H_2O}$ propene

b) isopropyl alcohol   $\xrightarrow[\text{(dehydrogenation)}]{-H_2}$   acetone

c) isopropyl alcohol   $\xrightarrow[\text{(dehydration)}]{-H_2O}$   di-isopropyl ether

● None                                    →   $\boxed{18}$

● a) and b)                               →   $\boxed{16}$

● a) and c)                               →   $\boxed{39}$

● b) and c)                               →   $\boxed{41}$

● All three                               →   $\boxed{26}$

● Please give me further information      →   $\boxed{29}$

**12**

---

$\boxed{7}$   The chlorine atom in methyl chloride is more reactive (or labile) than that in chlorobenzene. Hence, the reaction of sodium phenoxide with methyl chloride proceeds more smoothly than that between sodium methoxide and chlorobenzene.

Give the equation for the sodium phenoxide / methyl chloride reaction.   →   $\boxed{17}$

---

$\boxed{8}$   Your answer is wrong.

Study   $\boxed{25}$   attentively and be sure you understand the meaning of:

   *inter* = between (two or more molecules)
   *intra* = within (a single molecule).

---

$\boxed{9}$   Your answer is wrong.

Study   $\boxed{25}$   attentively and be sure you understand the meaning of

   *inter* = between (two or more molecules)
   *intra* = within (a single molecule).

---

| 10 |
Ethyl methyl ether.

Write formulas for dimethyl ether and diethyl ether, and compare at    | 21 |    .

| 11 |
Your answer is wrong.

You will find an explanation at    | 7 |    .

| 12 |
Your answer is wrong.

Study    | 25 |    attentively and be sure you understand the meaning of

   *inter* = between (two or more molecules)
   *intra* = within (a single molecule).

**12**

| 13 |
R–O–Na + R–Hal → R–O–R + NaHal

Alkyl *bromides* or *iodides* are often used in this reaction; the alkyl chlorides are less reactive.

Ethyl methyl ether,

$$CH_3–CH_2–O–CH_3$$

can be prepared from sodium ethoxide and methyl iodide. Give the equation for this reaction. (The equation at the beginning of this section will be of help.)   →   | 3 |

| 14 |
$$CH_3–CH_2–OH + HO–CH_2–CH_3 → CH_3–CH_2–O–CH_2–CH_3 + H_2O$$

The elimination of water (dehydration) from an alcohol can, however, also take a different course.

We have already seen this in the preceding program.

What is the product of the elimination of wateer from *one* molecule of ethanol?

$$H-\overset{\overset{\displaystyle H}{|}}{\underset{\underset{\displaystyle H}{|}}{C}}-\overset{\overset{\displaystyle H}{|}}{\underset{\underset{\displaystyle H}{|}}{C}}-OH \longrightarrow$$

Complete the equation.    →    | 25 |

---

**15**    Benzyl methyl ether:

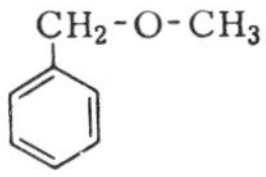

$$\text{CH}_2\text{-O-CH}_3$$

Now give the formula for methyl phenyl ether.    →    $\boxed{2}$

---

**16**    Correct.

To recapitulate the two terms, complete the following:

a) Reactions between two or more molecules are called ................................

b) Reactions taking place within one single molecule are
   called ...................................
                                   →   $\boxed{27}$

**12**

---

**17**    The reaction between sodium phenoxide and methyl chloride is the best method:

a) 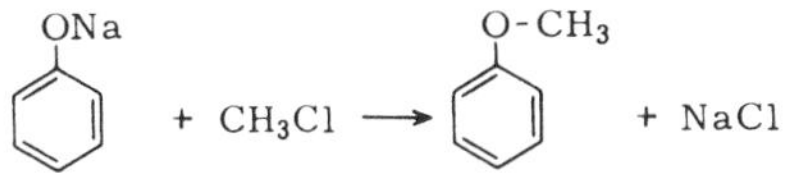    ONa    + CH$_3$Cl →    O-CH$_3$    + NaCl

Route b), the reaction between chlorobenzene and sodium methoxide,

b) 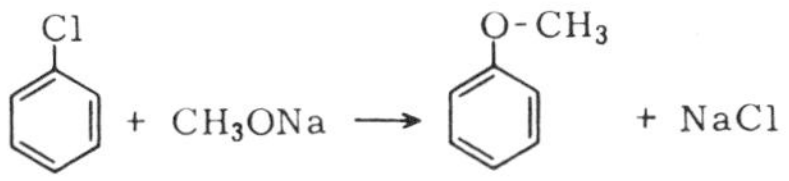    Cl    + CH$_3$ONa →    O-CH$_3$    + NaCl

gives the same product, but the two methods are *not equally good.*

Why is method a) the better one?    →   $\boxed{4}$

---

**18**    Your answer is wrong.

A reaction is intramolecular if only one molecule of starting material is involved in the conversion.

Make sure your equations are balanced.    →   $\boxed{6}$

---

---

**19**    Your answer is wrong. The question was in fact a difficult one; you will find an explanation at    28 .

---

**20**    The compound

$$CH_3-O-CH_2-CH_3$$

contains a methyl and an ethyl group, joined by an oxygen atom. Thus, the compound is an ether. The name is obtained simply by giving the names of the two groups, in alphabetical order, followed by the word "ether".

What is this compound called?    →    10

---

**21**    $CH_3-O-CH_3$                    $CH_3-CH_2-O-CH_2-CH_3$

dimethyl ether                    diethyl ether

The formula for diethyl ether can be written in a simpler form as:

$$C_2H_5-O-C_2H_5.$$

Now give the name of the compound

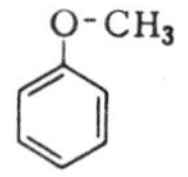

● Benzyl methyl ether?    →    5

● Methyl phenyl ether?    →    2

---

**22**    Your answer is wrong.

Repeat    40    in detail, and you will find the correct answer.

---

**23**    Your answer is partly correct; although the two reactions do give the *same product*, the two methods are *not equally suitable*.

You will see why at    28 .

**24**     Correct.

The hydrogen bonds between alcohol molecules also manifest themselves in the boiling points, which are different from those of the corresponding alkyl chlorides.

Do you expect the alcohols to have higher or lower boiling points than the alkyl chlorides with the same chain-length? (Compare for example methanol or ethanol with chloromethane or chloroethane, respectively.)

- Lower.     $\rightarrow$    44

- Higher.    $\rightarrow$    34

---

**25**

a)

$$H-\overset{\overset{\displaystyle H}{|}}{\underset{\underset{\displaystyle H}{|}}{C}}-\overset{\overset{\displaystyle H}{|}}{\underset{\underset{\displaystyle H}{|}}{C}}-OH \longrightarrow \overset{H}{\underset{H}{\diagup}}C=C\overset{H}{\underset{H}{\diagdown}} + H_2O$$

ethylene

or simpler:

$$CH_3-CH_2-OH \rightarrow CH_2=CH_2 + H_2O$$

*Contrast:*

b)     $CH_3-CH_2-OH + HO-CH_2-CH_3 \rightarrow$
       $CH_3-CH_2-O-CH_2-CH_3 + H_2O$

Reactions like the ones shown in a) and b) can be distinguished in terms of two important concepts:

I:    *intramolecular reactions* are those taking place within a *single molecule.*

II:   *intermolecular reactions* are those taking place between *two or more molecules* (inter = between).

Are the reactions shown in a) and b) above *intra-* or *inter*molecular?

- Both are intramolecular          $\rightarrow$    8

- both are intermolecular          $\rightarrow$    9

- a) is intramolecular, b) intermolecular    $\rightarrow$    6

- a) is intermolecular, b) intramolecular    $\rightarrow$    12

| 26 |
|---|

Your answer is wrong.

A reaction is intramolecular if only one molecule of starting material is involved in the conversion.

Make sure your equations are balanced.   →   | 6 |

---

| 27 |
|---|

a) Reaction *between* two or more molecules are called *intermolecular* reactions.

b) Reaction taking place *within* a single molecule are called *intramolecular* reactions.

Let us turn now to another important reaction:

*Condensation reactions* are reactions in which *two or more molecules combine* to form larger entities by *elimination of water.*

Thus, the formation of diethyl ether from two molecules of ethanol is a condensation reaction:

$$2 \; C_2H_5-OH \rightarrow \; .............. \; + H_2O$$

Complete this equation.   →   | 43 |

---

| 28 |
|---|

The two possible routes to methyl phenyl ether were:

a)      + $CH_3Cl$   →     + NaCl

b)      + $CH_3$-ONa   →     + NaCl

Although the two methods give the same product, they are not equally suitable.

Remember that the chlorine atoms in methyl chloride and chlorobenzene are not equally reactive. The reactants should be chosen in order that the reaction which takes place is that which occurs most readily.

Now can you tell which method is the best?

- a)                    →    17

- b)                    →    11

- I don't know    →    7

---

29       The three equations are

a)       $CH_3-CH(OH)-CH_3 \longrightarrow CH_3-CH=CH_2 + H_2O$

b)       $CH_3-CH(OH)-CH_3 \longrightarrow CH_3-CO-CH_3 + H_2$

c)       $CH_3-CH(OH)-CH_3 + CH_3-CH(OH)-CH_3 \longrightarrow$
$$\begin{array}{c} CH_3-CH-CH_3 \\ | \\ O \\ | \\ CH_3-CH-CH_3 \end{array} + H_2O$$

**12**

The difference between these reactions which concerns us here is that some of them take place *within* a single molecule; one of them is a reaction *between* two molecules.

Point out the *intramolecular* reactions:

- a) and b)      →    16

- a) and c)      →    39

- b) and c)      →    41

---

30       *Condensation reaction*

The formation of diethyl ether from two molecules of ethanol through the elimination of water is a condensation reaction.

Do you expect diethyl ether to have a higher or a lower boiling point than ethanol?

- Lower                  →    48

- Higher                 →    40

- About the same    →    42

| 31 | Correct. |

The hydrogen bonds are *significantly weaker* than ordinary chemical bonds.

The formation of hydrogen bonds between molecules results in

- association?          →     49

- condensation?          →     62

- I don't know          →     52

---

| 32 | Hazards of working with ether: |

**12**

1. Flammability

2. Vapor forms explosive mixtures with air.

3. Explosive peroxides are formed.

4. The vapor is an anesthetic.

For these reasons, the utmost care should be taken when working with ether. No open flames or installations which may cause ignition should be used. A water bath is used for heating ether. The ether should be freed from peroxides. This is done by shaking with an aqueous solution of iron(II) sulfate.

Ether has a sweetish smell and has been widely used as an anesthetic.

What is the *chemical reactivity* of diethyl ether?

- It is very reactive          →     53

- It is quite unreactive          →     47

- I don't know.          →     56

---

| 33 | No; the hydrogen bond is significantly weaker than ordinary chemical |

bonds.     →     31

**34**     Correct; alcohols have higher boiling points than the corresponding chloroalkanes:

| chloromethane: | bp 24 °C; | methanol: | bp 65 °C |
|---|---|---|---|
| chloroethane: | bp 12 °C | ethanol: | bp 78 °C |

The alcohol molecules are associated with each other through ................, e.g.:

Such ................ bonds cannot be formed by alkyl chlorides.

Write down the missing word.   →   |57|

**12**

---

**35**     Sodium thiolates are compounds of the type

    R–SNa

in which a thiol hydrogen atom is replaced by a sodium atom.

Dialkyl sulfides are obtained by the reaction of thiolates with alkyl halides:

    R–SNa + R–Hal → ................ + NaHal

Complete this equation (the two R groups may be different).   →   |55|

---

**36**     No; the hydrogen bond is *significantly weaker* than ordinary chemical bonds.   →   |31|

---

**37**     R-Hal     + NaSH ⟶ R–SH     + NaHal

| alkyl | sodium | thiol |
| halide | hydrogen | |
| | sulfide | |

The analogous reaction of an alkyl halide with *sodium sulfide* ($Na_2S$) in place of sodium *hydrogen sulfide* leads to a *thioether* (= *dialkyl sulfide*).

How many moles of alkyl halide react with one mole of sodium sulfide to give a thioether?

- One                                    →  58

- Two                                    →  54

- Three                                  →  63

- I need further information             →  66

---

**38**  R–SH
thiol

The *thiols* are thio-alcohols, i.e. alcohols in which oxygen is replaced by sulfur. Thus, the compound

$$CH_3-SH$$

is methanethiol.

Write down the name and formula for the corresponding *ethyl* compound.   →  51

---

**39**    Your answer is wrong.

A reaction is intramolecular if only one molecule of starting material is involved in the conversion.

Make sure your equations are balanced.   →  6

---

**40**    Diethyl ether is formed by the condensation of two molecules of ethanol. (A molecule of water is eliminated.)

Therefore, diethyl ether has a *much larger molecular weight than ethanol,* and one could have expected a *higher* boiling point. This is not, however, the case:

    ethanol:         bp 78 °C
    diethyl ether:   bp 35 °C (bp =boiling point)

The reason is that alcohol molecules are held together by *hydrogen bonds* in the liquid phase, e.g. for ethanol:

The individual ethanol molecules are not free, but weakly bonded to neighboring molecules. Hence, the boiling point of ethanol is higher than it would have been in the absence of hydrogen bonds.

The *hydrogen bonds* are, however, much weaker than ordinary chemical bonds.

Ethers have no hydroxylic hydrogen and *cannot*, therefore, form hydrogen bonds. Hence their boiling points are normal, corresponding to their molecular masses.

Now, which substance has the higher boiling point – methanol or dimethyl ether?

**12**

- Methanol          →    24

- Dimethyl ether          →    22

---

**41**     Your answer is wrong.

A reaction is intramolecular if only one molecule of starting material is involved in the conversion.

Make sure your equations are balanced.     →     6

---

**42**     Diethyl ether is formed by the condensation of two molecules of ethanol. Therefore, the molecular mass of the ether is higher than that of ethanol. Hence, equal boiling points cannot be expected for the two compounds.

Repeat     30   .

---

**43**     $2\ C_2H_5-OH \rightarrow C_2H_5-O-C_2H_5 + H_2O$

The elimination of water is carried out with the aid of sulfuric acid at an elevated temperature.

What type of reaction is this? The formation of larger molecules from smaller ones by the elimination of water is a ............... reaction.

Complete the sentence and compare at  30  , or, if you need help, go instead to  27  .

---

**44**      Alcohol molecules are held together in larger aggregates by hydrogen bonds. This is not the case for alkyl halides. Therefore, alcohols have ...............
(higher/lower) boiling points than the corresponding alkyl chlorides.   →   34

---

**45**      The complete table is given below:

| Formula | Name |
| --- | --- |
| R-SH | thiol |
| R-S-R | thioether (dialkyl sulfide) |
| R-S-R<br>$\overset{\parallel}{O}$ | sulfoxide |
| $\overset{O}{\overset{\parallel}{R\text{-}S\text{-}R}}$<br>$\parallel$<br>$O$ | sulfone |

The R groups can be different in the above formulas. This can be expressed as

$R^1\text{-}S\text{-}R^2$      for dialkyl sulfides, and

$\overset{O}{\overset{\parallel}{R^1\text{-}S\text{-}R^2}}$      for sulfones.
$\underset{\parallel}{\phantom{x}}$
$O$

Organic compounds can be analyzed for sulfur by heating with a piece of ............... When sulfur is present, ............... is formed. This gives a ...............
precipitate with lead salts after acidification.

Complete this sentence.   →   59

---

**46**

$\overset{\phantom{O}}{\underset{\overset{\parallel}{O}}{CH_3\text{-}S\text{-}CH_3}}$          $\overset{\overset{O}{\parallel}}{\underset{\overset{\parallel}{O}}{CH_3\text{-}S\text{-}CH_3}}$

Dimethyl sulfoxide          dimethyl sulfone

The *detection of sulfur in organic compounds* is carried out by heating the compound with a piece of sodium. This leads to decomposition of the organic material and formation of sodium sulfide. Acidification gives $H_2S$ which forms a black precipitate after the addition of a solution of a lead salt.

To recapitulate the names of organic sulfur compounds, complete the following:

| Formula | Name |
| --- | --- |
| R–SH | ................... |
| R–S–R | ................... |

$\rightarrow$   64

---

**47**     Correct; *ether is chemically resistant toward most reagents.*

We will now describe the *thioethers.*

12

In most organic oxygen compounds the oxygen atom(s) can be replaced by sulfur. The sulfur compounds are often referred to as thio-organic compounds. While it is easy to replace O by S in a formula, the actual *synthesis* of the thioorganic compound may be carried out by different means.

Replacement of O by S in an ether

     R–O–R

leads to a *thioether* or *dialkyl sulfide.*

Write down the general formula and the names used for this class of compounds.
$\rightarrow$   60

---

**48**     Diethyl ether is formed by the condensation of two molecules of ethanol with the elimination of water. The molecular mass of the ether is, therefore, *higher* than that of the alcohol. Hence, a *higher* boiling point could be expected.

The reality is different. As you may know, diethyl ether has a *lower* boiling point than ethanol.

The reason for this is given at   40   .

---

**49**     You are mistaken.

Continue at   52   .

---

| 50 |

Compare:

$CH_3-S-CH_3$                    $CH_3-S-C_2H_5$

dimethyl sulfide                 methyl ethyl sulfide

Give the general formula for the *thio-organic* compounds obtained by replacing the O in an alcohol by S.   →   |38|

---

| 51 |

$C_2H_5-SH$

ethanethiol

Alkanethiols (= thiols) can be prepared by the reaction of alkyl halides with sodium hydrogen sulfide, NaHS.

Give the equation for this reaction and compare at   |37|   , or, if you need help, go instead to   |61|   .

---

| 52 |

In a *condensation reaction* two or more molecules combine under elimination of water to give a new compound.

*Association* is another phenomenon, viz. the formation of aggregates of molecules held together by *hydrogen bonds*.

Make sure you understand and can distinguish these two concepts and continue at   |62|   .

---

| 53 |

Your answer is wrong.

Read about the chemical properties of ethers at   |62|   .

---

| 54 |

*Two moles of an alkyl halide* react with one mole of sodium sulfide:

$$2 R-Hal + Na_2S \rightarrow R-S-R + 2 NaHal$$

This leads to dialkyl sulfides in which the two substituents (R) are *identical*.

In order to synthesize thioethers in which the *substituents* (R) are *different*, *thiolates* are used as the starting materials.

Thiolates are derived from thiols,

    R−SH.

Give the general formula for the sodium thiolates.     →     35

---

**55**     R−S−Na + R−Hal → R−S−R + NaHal

Most thiols and dialkyl sulfides (thioethers) have an unpleasant, penetrating odor. These compounds are chemically less important than their oxygen analogs.

In contrast to oxygen, sulfur can have valences higher than two. Hence, dialkyl sulfides can be *oxidized*. Depending on the reaction conditions, one or two oxygen atoms can be taken up, to give *sulfoxides* and *sulfones:*

12

$$R-\underset{\underset{O}{\overset{\|}{}}}{S}-R \qquad\qquad R-\underset{\underset{O}{\overset{\|}{}}}{\overset{\overset{\overset{O}{\|}}{}}{S}}-R$$

sulfoxide          sulfone

What are the valences of sulfur in these two compounds?

● Two and four          →     67

● Two and six          →     68

● Four and six          →     65

---

**56**     Read about the chemical properties of ethers at     62   .

---

**57**     *Hydrogen bonds*

Hydrogen bonding between molecules leads to the formation of larger aggregates. This phenomenon is called *association*.

How strong is the hydrogen bond compared with ordinary chemical bonds?

● The hydrogen bond is much weaker          →     31

● The hydrogen bond is stronger          →     33

● They are equally strong          →     36

---

**58**   Your answer is wrong.

Read the explanation at   66   .

---

**59**   The missing words were *sodium, sodium sulfide,* and *black.*

This is the end of Program 12.

---

**60**   R−S−R

Dialkyl sulfide (thioether)

The names of the dialkyl sulfides are formed the same way as for the ethers:

the names of the two hydrocarbon groups (R) are given in alphabetical order, followed by the word "sulfide".

Give the formulas for dimethyl sulfide and ethyl methyl sulfide.   →   50

---

**61**   Alkyl halides have the general formula

R−Hal

Complete the general equation for reaction with sodium hydrogen sulfide:

R−Hal + Na−S−H → .............. + NaHal     →   37

---

**62**   Correct, *hydrogen bonding* leads to the *association* of molecules.

The chemical reactions of ethers are quickly dealt with: they are resistant toward most reagents. In other words, *ethers are unreactive.*

For this reason, and because diethyl ether (often just called "ether") is a good solvent for many organic compounds, it is widely used in the laboratory as a solvent or diluent.

Caution has to be exercised when working with ether:

1. Ether is *highly flammable.*

12

2. Ether is very volatile. Its vapor is heavier than air and forms *explosive mixtures with air.*

3. Liquid ether forms peroxides in contact with atmospheric oxygen. Upon distillation of ether, the distillation residue will become enriched in highly *explosive* peroxides.

4. Ether is an anesthetic and may cause nausea. The vapor should not be inhaled.

Write down the four hazards of ether in key-word fashion. 　→ 　 |32|

---

|63|　　Your answer is wrong. Read the elucidation at 　|66| 　.

---

|64|

| Formula | Name |
|---|---|
| R–SH | thiol |
| R–S–R | dialkyl sulfide (thioether) |

**12**

Now complete the following table for two further classes of thio-organic compounds

| Formula | Name |
|---|---|
| R–S–R<br>‖<br>O | ........................................ |
| ............ | ........................................ |

and compare at 　|45| 　.

---

|65|　　Correct; sulfur is tetra- and hexavalent in the sulfoxides and sulfones, respectively.

Give the formulas for dimethyl sulfone and dimethyl sulfoxide, writing the name of each compound below the formula.

Check your result at 　|46| 　, or, if you need help, go instead to 　|55| 　.

---

|66|　　Dialkyl sulfides (thioethers) are compounds of the type

　　R–S–R.

They are prepared by the reaction between sodium sulfide ($Na_2S$) and alkyl halides,

R—Hal.

Therefore, with how many moles of alkyl halide must *one* mole of $Na_2S$ react?   →   $\boxed{54}$

---

$\boxed{67}$   Your answer is wrong.

Examine the formulas for sulfoxides and sulfones more closely. Count the number of bonds to sulfur, and you will surely find the correct answer.   →   $\boxed{55}$

---

$\boxed{68}$   Your answer is wrong.

Examine the formulas for sulfoxides and sulfones more closely. Count the number of bonds to sulfur, and you will surely find the correct answer.   →   $\boxed{55}$

# Program 13

## The Carbonyl Group (I)

---

**1**     The functional group to be discussed in this program, the *carbonyl group*, has the general structure

$$\overset{\diagdown}{\underset{\diagup}{C}} = O$$

in which the two unspecified valences can represent bonds to either hydrocarbon groups (R) or hydrogen.

Give the general formula for carbonyl compounds carrying two hydrocarbon groups.    →   10

---

**2**

$C_2H_5 - CO - C_2H_5$         $CH_3 - CO - C_2H_5$        ⬡$- CO - CH_3$

diethyl ketone           ethyl methyl         methyl phenyl
                       ketone                 ketone

**13**

Now give the names of the following aromatic ketones:

$C_6H_5 - CO - C_6H_5$       $C_6H_5 - CO - CH_3$     ⬡$- CH_2 - CO - CH_3$

                                                     →    19

---

**3**     Correct; ketones and aldehydes contain the carbonyl group.

Therefore, these compounds are also know as *carbonyl compounds* (or *oxo*-compounds):

$$R - \underset{\underset{O}{\|}}{C} - R \qquad R - C\overset{\diagup H}{\diagdown_{O}}$$

    ketone       aldehyde

The simplest carbonyl compound,

$$H - C\overset{\diagup H}{\diagdown_{O}}$$

is also an aldehyde.

A carbonyl group is also present in the carboxylic acids with the general formula:

$$R-C\underset{O}{\overset{OH}{\diagdown}}$$

However, these compounds are referred to as carboxylic acids rather than carbonyl compounds, and their chemistry will be dealt with in subsequent programs. Carbon dioxide ($O=C=O$) can also be thought of as a carbonyl compound, but its geometry is different from that of the normal carbonyl group. In this program, therefore, carboxylic acids and carbon dioxide are not treated as carbonyl compounds.

The typical carbonyl group is only connected to hydrocarbon radicals or hydrogen atoms. Using this definition, how many of the following compounds are carbonyl compounds?

**13**

$$R-C\underset{O}{\overset{H}{\diagdown}} \qquad H-C\underset{O}{\overset{H}{\diagdown}} \qquad CH_3-CO-CH_3$$

$$CH_3OH \qquad R-C\underset{O}{\overset{OH}{\diagdown}} \qquad CH_3-CHO$$

- Six        →    18
- Five       →    21
- Four       →    13
- Three      →    25
- Two        →    28

**4**    The three aldehydes are

$$CH_3-C\underset{O}{\overset{H}{\diagdown}} \qquad C_2H_5-C\underset{O}{\overset{H}{\diagdown}} \qquad C_3H_7-C\underset{O}{\overset{H}{\diagdown}}$$

acetaldehyde          propionaldehyde          butyraldehyde

Compare your answer, and write down the three names (they are pronounced acet-aldehyde, propion-aldehyde, and butyr-aldehyde).

The systematic names of the aldehydes are derived from those of the hydrocarbons (with the same number of carbon atoms) by changing the ending *-ane* to *-al*.

Examples:

$$H-C\underset{O}{\overset{H}{\diagdown}} \qquad\qquad CH_3-C\underset{O}{\overset{H}{\diagdown}}$$

methanal             ethanal
(formaldehyde)    (acetaldehyde)

Formaldehyde (trivial name) is derived from methane, so the systematic name is methanal.

Acetaldehyde has the same number of carbon atoms as ethane, hence its name is ethanal.

Give the systematic names of the following two compounds.

$$C_2H_5-C\underset{O}{\overset{H}{\diagdown}} \quad \text{and} \quad C_3H_7-C\underset{O}{\overset{H}{\diagdown}} \qquad\qquad \rightarrow \boxed{15}$$

---

**5**

No, *the oxidation of secondary alcohols gives ketones.*

$$\rightarrow \boxed{37}$$

---

**6**

$$R-CH_2OH \quad \longrightarrow \quad R-C\underset{O}{\overset{H}{\diagdown}}$$

primary alcohol        aldehyde

$$R-CH(OH)-R \quad \longrightarrow \quad R-CO-R$$

secondary alcohol     ketone

Write down the completed equations for the *dehydrogenation* of the following two alcohols:

$$CH_3-CH_2OH \quad \longrightarrow$$
ethanol

$$CH_3-CH(OH)-CH_3 \quad \longrightarrow$$
isopropyl alcohol

$$\rightarrow \boxed{11}$$

---

**7** | *1. Oxidation:*

$$R-CH_2OH + [O] \rightarrow R-CHO + H_2O$$

*2. Dehydrogenation:*

$$R-CH_2OH \xrightarrow[\text{catalyst}]{\text{ca. } 300\,^{\circ}C} R-CHO + H_2$$

In order to obtain ketones, one has to start from *secondary alcohols:*

R, H
\C
R' OH

Formulate the general equation for the *oxidation* of secondary alcohols and check your result at  |17| , or, if you need help, go instead to  |37| .

**13**

---

**8** |   You must look more closely at the *structural formulas.* It is of the utmost importance that you are able to recognize the functional groups in the formulas in order to have a complete understanding of organic chemistry.   →  |1|

---

**9**

H-C⟨H / O      formaldehyde or methanal          CH$_3$-C⟨H / O      acetaldehyde or ethanal

*Ketones* are named by giving the names of the two hydrocarbon residues in alphabetical order, followed by the word "ketone", e.g.

CH$_3$-C-C$_3$H$_7$   methyl propyl ketone
      ‖
      O

CH$_3$-C-CH$_3$   dimethyl ketone
      ‖
      O

Write down the following formulas, and give the name of each compound.

C$_2$H$_5$-CO-C$_2$H$_5$,   CH$_3$-CO-C$_2$H$_5$,   ⬡-CO-CH$_3$

→  |2|

**10**    The correct formula is

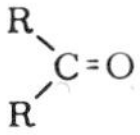

$$
\begin{array}{c}
R \\
\diagdown \\
\phantom{R}C = O \\
\diagup \\
R
\end{array}
$$

which may also be written

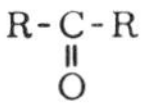

R-C-R    or    R-CO-R
   ‖
   O

Compounds with this general formula are called *ketones*. What is the name of the functional group in a ketone?

● Hydroxyl group        →    20

● Carbonyl group        →    23

● I am not sure          →    1

**13**

**11**    $CH_3-CH_2OH \rightarrow CH_3-CHO + H_2$
        ethanol

    $CH_3-CH(OH)-CH_3 \rightarrow CH_3-CO-CH_3 + H_2$
        isopropyl alcohol

Give the names of the compounds formed:

● Acetaldehyde and ethanal              →    33

● Acetaldehyde and dimethyl ketone      →    22

● Acetone and dimethylketone            →    36

● I don't know                          →    40

**12**    R–C–R    or    R–CO–R;   ketone
     ‖
     O

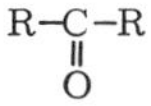

The carbonyl group need carry only *one hydrocarbon residue*. The other substituent is then a hydrogen atom:

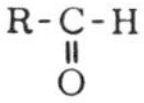

R-C-H
   ‖
   O

What is the name of the functional group?    →    24

| **13** | Correct. |

The four carbonyl compounds are

$$R-C\overset{H}{\underset{O}{\diagdown}} \qquad H-C\overset{H}{\underset{O}{\diagdown}} \qquad CH_3-CO-CH_3 \qquad CH_3-CHO$$

How many of these are *aldehydes?*

● Four      →   | 30 |

● Three     →   | 27 |

● Two       →   | 35 |

● One       →   | 39 |

**13**

| **14** | You must look more closely at the *structural formulas*. It is of the utmost importance that you are able to recognize the functional groups in the formulas in order to have a complete understanding of organic chemistry.   →   | 1 | |

| **15** | *Propanal, butanal.* |

Now give the structural formulas for formaldehyde (= methanal) and acetalde-hyde (ethanal). Compare your results at   | 29 |  , or, if you have difficulties, repeat   | 27 |   instead.

| **16** | Correct; *water* is the by-product in the preparation of formaldehyde from methanol. |

The complete equation is repeated below:

$$2\ CH_3-OH + O_2 \rightarrow 2\ H-CHO + 2\ H_2O$$

Having described the *general* methods of preparing ketones and aldehydes, let us now consider some syntheses.

Acetone is manufactured industrially from acetic acid at ca. 400 °C in the presence of catalysts:

$$2\ CH_2-COOH \xrightarrow[\text{catalysts}]{400\,°C} CH_3-CO-CH_3 + \text{............} + \text{..........}$$

What are the other products formed?

- $CO + H_2O$                    →    47

- $CO_2 + H_2O$                  →    42

- $CO_2 + 2 H_2O$                →    50

- I need further information      →    53

---

**17**

$$\underset{\substack{\text{secondary} \\ \text{alcohol}}}{\underset{R}{\overset{R}{>}}C\underset{OH}{\overset{H}{<}}} + [O] \longrightarrow \underset{\text{ketone}}{\underset{R}{\overset{R}{>}}C=O} + H_2O$$

or simpler:

$$R-CH(OH)-R + [O] \rightarrow \qquad R-CO-R + H_2O$$
secondary alcohol                    ketone

The *dehydrogenation* of secondary alcohols also affords ketones.

Give the general equation for the dehydrogenation of secondary alcohols of the type $R-CH(OH)-R$. If possible, specify also the reaction conditions.   →   32

---

**18**       Not every organic compound containing oxygen is a carbonyl compound!

Repeat   3  , writing *structural* formulas for each compound in order to recognize the functional group more readily.

---

**19**

$C_6H_5-CO-C_6H_5$              $C_6H_5-CO-CH_3$              ⬡–$CH_2-CO-CH_3$

diphenyl ketone                methyl phenyl                benzyl methyl
                               ketone                       ketone

You will have noticed that $C_6H_5$ is just an abbreviation for phenyl,   ⬡–

The preparation of carbonyl compounds will be treated next.

We have already presented a very *versatile synthesis* in the program on alcohols.

From which alcohols can *aldehydes* be prepared by oxidation?

● Primary alcohols          →    31

● Secondary alcohols        →    5

● Tertiary alcohols         →    34

● I need further information →    57

---

**20**    You are mistaken:

−OH is the *hydroxyl group*
$\geq$C=O is the *carbonyl group*

When two hydrocarbon groups are attached to the carbonyl group,
we obtain ................

**13**

Give the name of the class of compounds.    →    23

---

**21**    Not every compound containing the C=O grouping is considered a
carbonyl compound in the context of this program.

See    3  .

---

**22**    Correct; acetaldehyde and dimethyl ketone are formed.

*Dimethyl ketone,* $CH_3-CO-CH_3$, is better known under the trivial name of
*acetone.*

Acetone is prepared industrially by dehydrogenation of isopropyl alcohol.

The simplest aldehyde, formaldehyde, is also prepared from an alcohol. Which
one?

● Methanol           →    38

● Ethanol            →    66

● Isopropyl alcohol  →    45

● I don't know       →    48

---

**23**

*Ketones* contain the *carbonyl group,*

$$>C=O$$

Give the general formula and the generic name of the carbonyl compounds carrying two hydrocarbon residues.    →    12

---

**24**

*Carbonyl group.* Compounds of the type

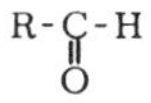

$$R-\underset{O}{\overset{\|}{C}}-H$$

are *aldehydes* (emphasis on 'al).

The formula for an aldehyde is often written:

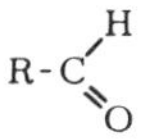

$$R-C\overset{H}{\underset{O}{\diagdown}}$$

13

or shorter: R–CHO

The functional group in these compounds,

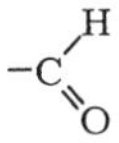

$$-C\overset{H}{\underset{O}{\diagdown}} \quad or \quad -CHO$$

is the *aldehyde group.*

What is the group *common* to aldehydes and ketones?

$$>C=O \qquad → \qquad \boxed{3}$$

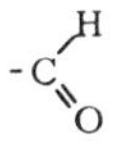

$$-C\overset{H}{\underset{O}{\diagdown}} \qquad → \qquad \boxed{14}$$

$$-CHO \qquad → \qquad \boxed{8}$$

---

**25**    Repeat   3   , studying the formulas more attentively!

---

| 26 | Cumene; cumene-phenol process. |
|---|---|

The "cumene-phenol process" is industrially important for the preparation of phenol and acetone. Here is the equation again:

benzene     propene     isopropylbenzene     phenol     acetone
                                    (cumene)

The *oxo process* is another technical procedure used for the preparation of *higher aldehydes,* e.g.

$$CH_3-CH=CH_2 + CO + H_2 \rightarrow CH_3-CH_2-CH_2-CHO$$

Write down this equation and also give the names of all the compounds.    →    |44|

**13**

| 27 | Correct. |
|---|---|

Let us now describe some individual carbonyl compounds.

The simplest aldehyde is formaldehyde:

Remembering the general formula for aldehydes,

you see that formaldehyde is a unique case where R = H.

Give the structural formulas for the next three aldehydes (i.e. R = $CH_3$, $C_2H_5$, and $C_3H_7$).    →    |4|

| 28 | Your answer is wrong. |
|---|---|

Repeat   |3| , writing *structural* formulas for each compound, and you will easily recognize the carbonyl compounds.

29

Formaldehyde: $H-C\overset{H}{\underset{O}{\big\langle}}$

acetaldehyde: $CH_3-C\overset{H}{\underset{O}{\big\langle}}$

Now give the *systematic* names of these two compounds.

Check your answer at  9  , or, if you need assistance, go instead to  4  .

30    Not every compound containing the C=O grouping is considered a carbonyl compound in the context of this program. See  3  .

13

31    *Aldehydes* are formed by oxidation of *primary* alcohols:

$$R-CH_2OH + [O] \rightarrow R-CHO + H_2O$$

*In practice,* this reaction can be carried out in two different ways:

*1. Oxidation*

Oxidizing agents like potassium permanganate, potassium dichromate, or even atmospheric oxygen may be used (the latter particularly in industrial preparations). The oxidizing agents are symbolized by [O] for simplicity:

$$R-CH_2OH + [O] \rightarrow R-CHO + \ldots\ldots\ldots\ldots$$

*2. Dehydrogenation*

Aldehydes are also formed by the catalytic dehydrogenation of primary alcohols at elevated temperature:

$$R-CH_2OH \xrightarrow[\text{catalyst}]{\text{ca. 300 °C}} \ldots\ldots\ldots\ldots + H_2$$

Write down the two completed equations as well as the names of the procedures.   →   7

**32**

$$R-CH(OH)-R \xrightarrow[\text{catalyst}]{\text{ca. } 300\,°C} R-CO-R + H_2$$

Summarize the oxidation reactions of alcohols by writing the following equations, including the names of the products:

$$R-CH_2OH \longrightarrow R-C\overset{H}{\underset{O}{\diagdown}}$$

primary alcohol   · · · · · · · ·

$$R-CH(OH)-R \longrightarrow R-CO-R$$

secondary alcohol   · · · · · · · ·

$$\overset{R}{\underset{R}{\diagdown}} R-C-OH \nrightarrow \text{(no reaction)}$$

tertiary alcohol                 →   **6**

**13**

---

**33**    "Acetaldehyde" and "ethanal" are two names for the *same* compound. The systematic nomenclature for aldehydes is based on the names of the corresponding hydrocarbons, changing the ending *-ane* to *-al*, e.g.

$CH_3-CHO$: two carbon atoms as in ethane; hence *ethanal*.

$CH_3-CH_2-CHO$: three carbon atoms as in propane; hence *propanal*.

Repeat   **11**   .

---

**34**    No; *tertiary alcohols are not attacked by oxidizing agents.* They are un-reactive.    →   **37**

---

**35**    Repeat   **13**   , writing *structural* formulas for the compounds given. This will help you recognize the aldehydes.

---

**36**    "Acetone" and "dimethyl ketone" are two different names for the *same compound.* The names of the ketones are formed by naming the two hydrocarbon residues (in alphabetical order), followed by the word "ketone", e.g.

$CH_3-CO-CH_3$ dimethyl ketone

$CH_3-CO-C_2H_5$ ethyl methyl ketone

Repeat  $\boxed{11}$ .

---

$\boxed{37}$    Let us recapitulate the reactions of alcohols with mild oxidizing agents. ([O] symbolizes an oxidizing agent such as potassium permanganate or potassium dichromate.)

*Primary alcohols:*

$$R-CH_2OH + [O] \rightarrow R-CHO + H_2O$$

*Secondary alcohols:*

$$R^1-CH(OH)-R^2 + [O] \rightarrow R^1-CO-R^2 + H_2O$$

*Tertiary alcohols:*

$$\begin{matrix} R^1 \\ R^2 \\ R^3 \end{matrix}\!\!\!\diagdown\!\!\!\text{C—OH} \quad + \quad [O] \quad \xrightarrow{\;\;/\!/\;\;} \text{(no reaction)}$$

Complete the sentence:

The oxidation of primary alcohols gives ................... $\rightarrow$ $\boxed{31}$

---

$\boxed{38}$    Correct!

Methanol is the starting material for the preparation of formaldehyde. The reaction is carried out by passing methanol vapor over a heated copper catalyst together with an excess of air.

What is formed apart from formaldehyde? Complete the equation:

$$2\ CH_3OH + O_2 \rightarrow$$

- $H_2$             $\rightarrow$  $\boxed{49}$

- $H_2O$           $\rightarrow$  $\boxed{16}$

- I don't know    $\rightarrow$  $\boxed{51}$

---

**39**    Your answer is wrong. You should examine the *structural* formulas for the compounds.

Continue at   24   .

**40**    The two products were

$$CH_3-CHO \text{ and } CH_3-CO-CH_3$$

The first is an aldehyde, the second a ketone. You should now be able to answer the question at   11   .

**41**

Chloral:   $Cl_3C-C \overset{H}{\underset{O}{\diagdown}}$

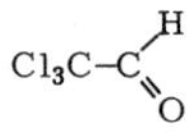

Which of the following names for chloral is correct?

● Chloroacetaldehyde          →   59

● Chloroaldehyde              →   61

● Trichloroacetaldehyde        →   56

**13**

**42**    Correct; one mole of $CO_2$ and one mole of $H_2O$ are formed per mole of acetone.

The following scheme illustrates from which parts of acetic acid the products are derived:

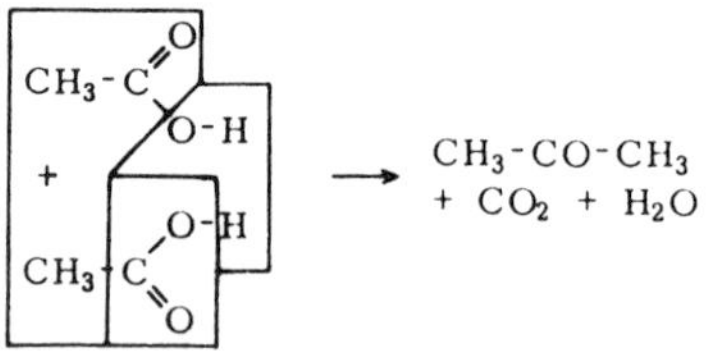

$$CH_3-CO-CH_3$$
$$+ CO_2 + H_2O$$

Do you know the reaction conditions? Write them down and compare at   52   .

**43**    *Phenol* and *acetone* are formed.

The reaction is important for the industrial preparation of these two compounds.

Isopropylbenzene is also known under the trivial name *cumene,* and the reaction is, therefore, also known as the *cumene-phenol process.*

The preparation of isopropylbenzene is indicated in the following equation. Insert the formulas for isopropylbenzene as well as the two end products and their names.

| benzene | propene | isopropyl benzene | .... | .... |

Check your result at ⟨63⟩ , or, if you need help, go instead to ⟨52⟩ .

---

**44**

$$CH_3–CH=CH_2 \;+\; CO \;+\; H_2 \longrightarrow CH_3–CH_2–CH_2–CHO$$

| propene | carbon monoxide | hydrogen | butyraldehyde = butanal |

The following olefin

$$CH_3–CH_2–CH_2–CH_2–CH_2–CH=CH_2$$

is used in another technical application of the oxo process.

Derive the formula for the aldehyde formed.   →   ⟨75⟩

**13**

---

**45**      You made two mistakes:

1. Isopropyl alcohol has a *three*-carbon chain. Hence, a compound possessing *three* carbon atoms must be formed in the oxidation reaction. Formaldehyde has but one carbon atom.

2. Isopropyl alcohol is a *secondary* alcohol. *Ketones are formed by oxidation of secondary alcohols.*

Write down the last sentence.

Which alcohol can be used in the preparation of formaldehyde?

● Methanol          →      ⟨38⟩

● Ethanol            →      ⟨66⟩

● I don't know        →      ⟨48⟩

---

46      Benzene is a typical aromatic compound. It has three conjugated double bonds in a six-membered ring.

Quinone is

Is this an aromatic compound?

● Yes      →      84

● No      →      55

47      Your answer is wrong. Admittedly, the question was a difficult one. Read the explanation at  53  .

48      Aldehydes can be prepared from primary alcohols:

$$R-CH_2OH + [O] \rightarrow R-CHO + H_2O$$

The aldehyde formed has the same number of carbon atoms as the alcohol.

Formaldehyde, H–CHO, has only one carbon atom.

Therefore, a *primary* alcohol possessing only *one* carbon atom must be used as starting material.

Which alcohol is this?

● Methanol      →      38

● Ethanol      →      66

49      Methanol vapor is passed with an excess of air over heated copper. The hydrogen formed combines immediately with oxygen from the air to give
....................

Continue at   16  .

**50**     Your answer is wrong. Admittedly, the question was a difficult one.
Read the explanation at    53  .

**51**     The two methods of preparation of aldehydes from alcohols are:

1. Oxidation:

$$R–CH_2OH + [O] \rightarrow R–CHO + H_2O$$

2. Dehydrogenation:

$$R–CH_2OH \rightarrow R–CHO + H_2$$

Formaldehyde is prepared by oxidation of methanol by passing a mixture of methanol vapor and *excess air* over a heated copper catalyst. Give the equation for this reaction.

What is the other reaction product?

**13**

● Hydrogen       →     49

● Water          →     16

**52**     The preparation of acetone from acetic acid is carried out at

ca. 400 °C in the presence of catalysts.

Another preparation of acetone is outlined below. It is a little complicated so carefully study the formulas.

First, *isopropylbenzene* is synthesized from *benzene* and *propene:*

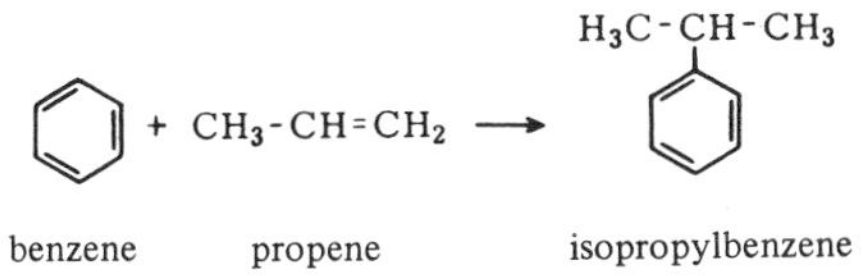

benzene        propene        isopropylbenzene

The oxidation of isopropylbenzene gives a so-called hydroperoxide:

H₃C-CH-CH₃ (benzene ring) + O₂ ⟶ "hydroperoxide" structure with $H_3C-C-CH_3$ and O-O-H group on benzene ring

isopropylbenzene          "hydroperoxide"

This hydroperoxide decomposes with migration of the hydroxyl group onto the benzene ring.

The final products are phenol and acetone.

Write down the equation for the formation of the final products from the hydroperoxide.   →   82

**13**

---

53    The formation of acetone from acetic acid can be visualized as follows

$CH_3-C$ with O and O-H
+
$CH_3-C$ with O-H and O

$\xrightarrow[\text{catalyst}]{400\ °C}$

$CH_3-CO-CH_3$
+ ...................
+ ...................

Give the formulas (and numbers of moles) of the other two products formed.   →   42

---

54    Correct; quinone is a diketone.

Now try to formulate the structure of quinone. It is a diketone possessing six carbon atoms in a ring. If your structure appears reasonable, check it at   79   .
If you are uncertain, please repeat from   78   .

---

55    Correct; quinone is *not* an aromatic compound. This is reflected in its chemical behavior:

Quinone is unstable and easily reduced.

The reduction of quinone gives hydroquinone, a compound which we have already mentioned. The hydroquinone can be oxidized back to quinone:

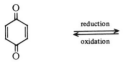

reduction
oxidation

quinone                          hydroquinone

Write down the completed equation.   →   68

---

**56**    *Trichloroacetaldehyde* (chloral) was used previously as a soporific (sleep-inducing drug).

In the aldehydes described so far,

R was always an aliphatic hydrocarbon (alkyl) residue. Aromatic aldehydes are also known, i.e. compounds with R = aryl (aromatic hydrocarbon residue). The simplest of these is *benzaldehyde*.

What is the structure of benzaldehyde?   →   69

---

**57**    Inspect the three formulas:

phenol            benzaldehyde              benzyl alcohol

Obviously, benzaldehyde cannot be produced by the oxidation of phenol – there is one carbon atom too few!

Benzaldehyde is prepared from ...............   →   76

**58** Correct; *compounds with several hydroxyl groups attached to one C-atom are unstable.*

In the present case, elimination of water gives benzaldehyde:

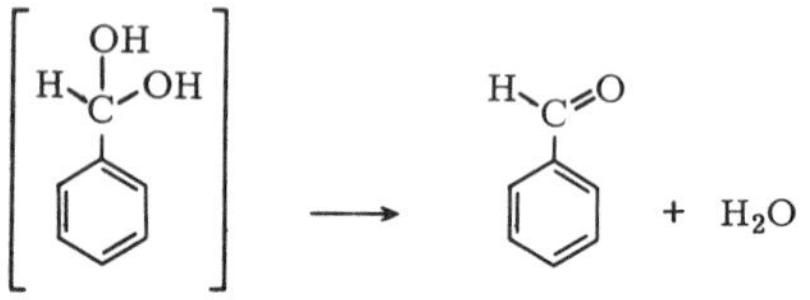

Benzaldehyde smells like bitter almonds. Like other aromatic aldehydes, it is used as a fragrance.

Let us consider two aromatic ketones:

acetophenone              benzophenone

These names are trivial ones. What are the correct, systematic names of these two ketones? (Remember: acetone, $CH_3-CO-CH_3$, is dimethyl ketone.)

Write down the two formulas and the systematic names.   →   $\boxed{78}$

---

**59** The compound $CCl_3-CHO$ is *a* chloroacetaldehyde, but this name is incomplete. There are *three* chlorine atoms. Therefore, the compound is
..................................

Continue at   $\boxed{56}$   .

---

**60** Your answer is wrong.

We had learned in connection with alcohols that *compounds with more than one hydroxyl group attached to a single carbon atom are unstable:*

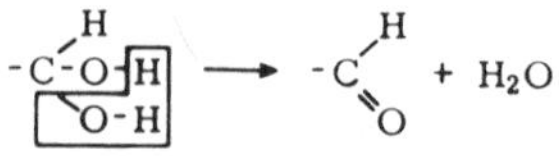

Write down the italicized phrase.   →   $\boxed{58}$

**61**
     The name "chloroaldehyde" only denotes an aldehyde containing chlorine. It is not a specific name.

Which is the correct name?

- Chloroacetaldehyde      →   59

- Trichloroacetaldehyde      →   56

**62**

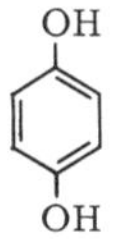

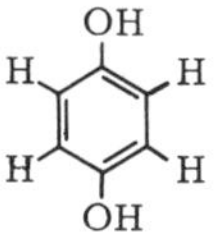

A common name for this compound is *hydroquinone*.

Oxidation of hydroquinone gives quinone:

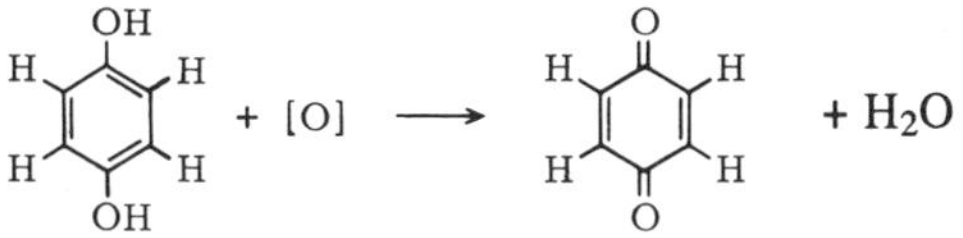

Examine the formula for quinone carefully in order to identify the functional groups. Then answer the question:

What kind of compound is quinone?

- An aldehyde      →   70

- A ketone      →   71

- A dialdehyde      →   72

- A diketone      →   54

- I cannot answer      →   77

**63**     Compare:

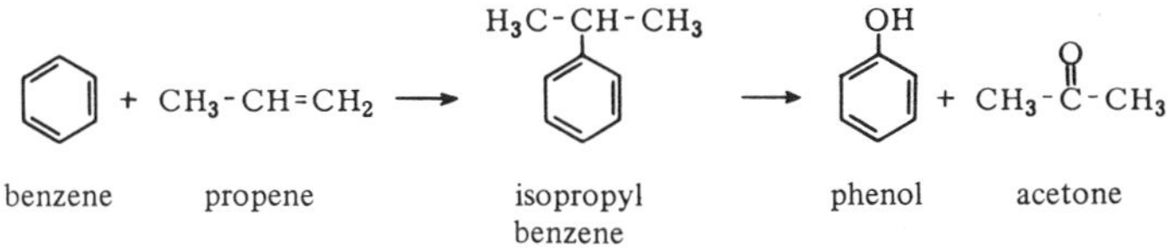

benzene     propene        isopropyl benzene       phenol     acetone

The trivial name for isopropylbenzene is ............... Therefore, this preparation of phenol (and acetone) is also called the ............... process.

Complete the two sentences. → ☐26

---

**64**    Oxo process.

Give the three starting materials used in the oxo process:

a) ...............
b) ...............
c) ...............

What is the difference in number of carbon atoms between starting material and product?

Check your answers at ☐74 , or, if you are uncertain, continue instead at ☐26 , second paragraph.

---

**65**    Formaldehyde:      gaseous
acetaldeyde:       gaseous
propionaldehyde:   liquid
butyraldehyde:     liquid

The aqueous solution of formaldehyde is called ...............   →   ☐73

---

**66**    There are *two* carbon atoms in ethanol. The aldehyde formed by oxidation, viz. acetaldehyde, also has *two* carbon atoms.

$$CH_3\text{-}CH_2OH + [O] \longrightarrow CH_3\text{-}C\overset{H}{\underset{O}{\diagdown}} + H_2O$$

There is only *one* carbon atom in formaldehyde, HCHO.

Which alcohol can be oxidized to formaldehyde?

● Methanol              →   ☐38

● Isopropyl alcohol     →   ☐45

● I don't know          →   ☐48

**67** *Compounds with more than one hydroxyl group per carbon atom are unstable:*

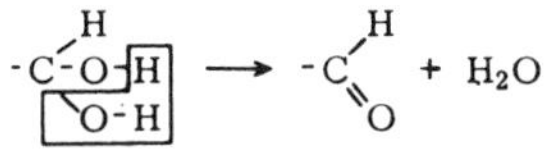

$$-\overset{H}{\underset{O-H}{\underset{|}{C-O{-}H}}} \longrightarrow -\overset{H}{\underset{O}{C}} + H_2O$$

Please memorize this fact.   →   58

**68**

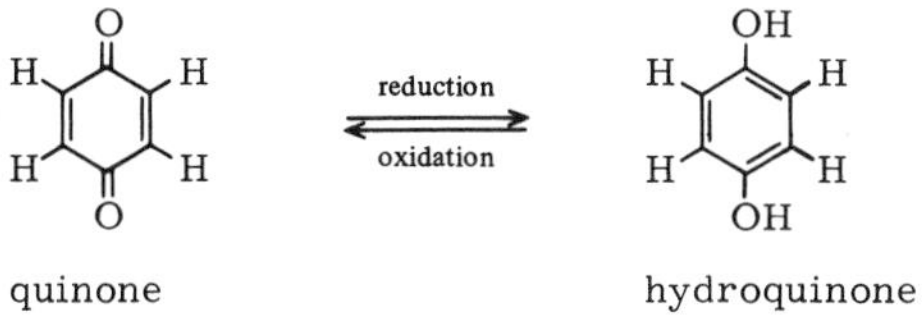

quinone          hydroquinone

In contrast to quinone, hydroquinone is an aromatic compound.

This is the end of Program 13.

**13**

**69** *Benzaldehyde:*   $H{\diagdown}_C{\diagup}^O$

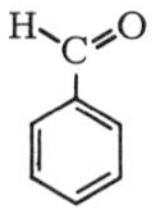

How can benzaldehyde be prepared?

● From phenol          →   80

● From benzyl alcohol          →   76

● I don't know          →   83

**70** Aldehydes are compounds possessing the group

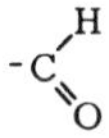

$$-\overset{H}{\underset{O}{C}}$$

Look at the quinone structure more carefully, and you will realize that there is no aldehyde group.

Repeat   62   .

**71**     Your answer is partially correct. Ketones are compounds possessing the $\gtrsim$C=O group. There are *two* such groups in quinone:

Therefore, quinone is better described as

● A ketone?          →     81

● A diketone?          →     54

---

**72**     It is correct that there are two $\gtrsim$C=O groups in quinone. However, these are not *aldehyde* groups, aldehydes being compounds of the type

Examine the structure of quinone more carefully.     →     62

---

**73**     *Formalin*

Formalin is used, for example, as a preservative.

*Chloral* is another important aldehyde. You can develop its formula by replacement of the three hydrogen atoms in the methyl group of acetaldehyde by chlorine atoms.

Check your result at     41     .

---

**74**     The starting materials for the oxo process are:

    a) an olefin
    b) carbon monoxide
    c) hydrogen

The end product has *one* carbon atom *more* than the olefin used.

We will consider the *physical properties* of the aldehydes very briefly: the first two members of the series are gases at room temperature. The next two are liquids.

Formaldehyde, being a gas, is often used as a ca. 40% aqueous solution, known as *formalin*.

Give the names and physical states of the first four aldehydes.   →   65

---

75   $CH_3-CH_2-CH_2-CH_2-CH_2-CH_2-CH_2-CHO$

You will notice that the reaction product has *one carbon atom more* than the starting material.

Do you remember the name of the synthesis of aldehydes from olefins, CO, and $H_2$?

If yes, write it down, and compare at   64   .

If no, please repeat from   44   .

---

13

76   Correct!

Preparation of *benzaldehyde from benzyl alcohol:*

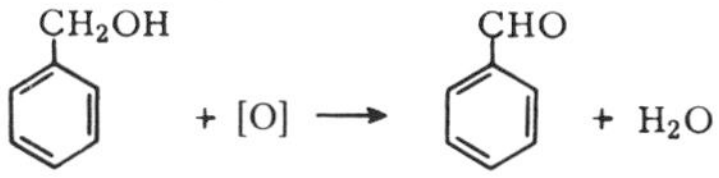

Benzaldehyde can also be synthesized from toluene: in the first step benzal chloride ist obtained by chlorination:

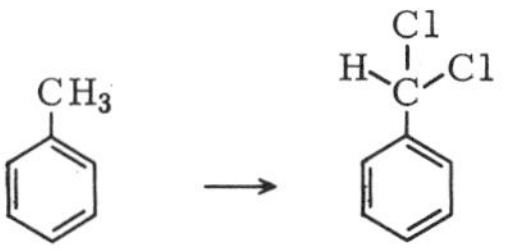

    toluene          benzal chloride

This is hydrolyzed by boiling with water or a base:

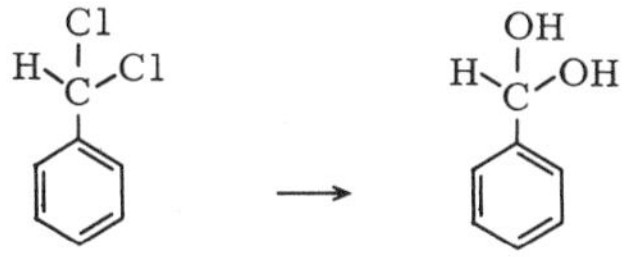

As you will notice, the product of the second step carries *two* hydroxyl groups at *one* carbon atom.

Are such compunds stable?

- Yes          →     60
- No          →     58
- I don't know.    →     67

---

**77**    Aldehydes are compounds possessing the group

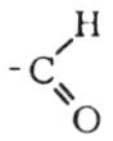

$$-C\underset{O}{\overset{H}{\lessgtr}}$$

whereas ketones are

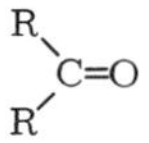

$$\underset{R'}{\overset{R}{\diagdown}}C=O \quad (R \neq H)$$

**13**

Which of these groups is present in quinone? Repeat    62   .

---

**78**

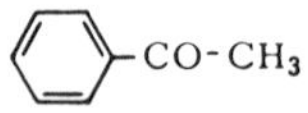

methyl phenyl ketone 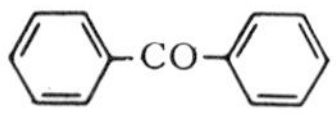          diphenyl ketone

A method of preparation of aromatic ketones will be discribed later.

*The quinones* represent a special class of carbonyl compounds.

The formula for the simplest quinone is derived from benzene by first introducing two hydroxyl groups in positions 1 and 4 (*para*-relationship).

Construct his formula.    →    62

---

**79**    Quinone:

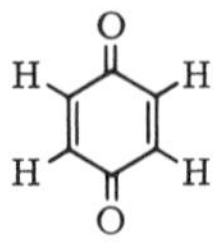

Is quinone an aromatic compound?

(Consider your answer well!)

● Yes          →    84

● No           →    55

● I don't know    →    46

---

**80**    Study the two formulas,

OH                 CH₂OH

        and

phenol            benzyl alcohol

and compare them with the formula for benzaldehyde. This will enable you to answer the question:

How can benzaldehyde be prepared?

● From phenol        →    57

● From benzyl alcohol     →    76

---

**81**    Your answer is wrong.

Since there are *two* C=O groups in quinone, it is a ................... ketone.
                                      →    54

---

**82**    Check your result carefully:

CH₃-C-CH₃      OH        O

                 →      + CH₃-C-CH₃

"hydroperoxide"

Thus, the compounds formed are .............. and ............... .
Give the names. →    43

---

**83**     Remember that aldehydes are obtained by oxidation of primary alcohols. The formulas for phenol and benzyl alcohol are given below:

OH

phenol

CH₂OH

benzyl alcohol

How can benzaldehyde be prepared by oxidation?

● From phenol          →     57

● From benzyl alcohol  →     76

**84**     No, quinone is *not* an aromatic compound. An explanation is given at     46    .

# Program 14

## The Carbonyl Group (II)

**1** We will now consider the *chemical properties* of carbonyl compounds.

First a question:

Can aldehydes and ketones be reduced?

- Yes $\rightarrow$ 10
- No $\rightarrow$ 20
- I don't know $\rightarrow$ 13

---

**2** Compare:

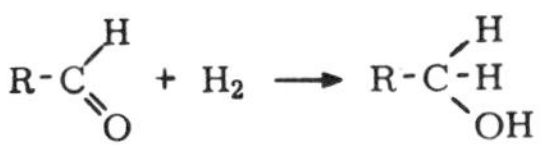

$$R-C\begin{smallmatrix}H\\ \\ \diagdown O\end{smallmatrix} + H_2 \longrightarrow R-C\begin{smallmatrix}H\\ |\\ \diagdown OH\end{smallmatrix}H$$

or shorter:

$$R-CHO + H_2 \rightarrow R-CH_2OH$$

Ketones can be reduced in the same manner. Complete the equation:

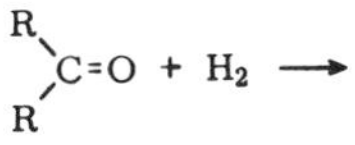

$$\begin{smallmatrix}R\\ \diagdown\\ \diagup\\ R\end{smallmatrix}C{=}O + H_2 \longrightarrow$$

or in shortened form:

$$R-CO-R + H_2 \longrightarrow$$

Check your result at 12 .

---

**3** *Noble metals* or *finely divided nickel* (Raney nickel) are used as hydrogenation catalysts.

The process is known as *catalytic hydrogenation,* and one speaks of *catalytically activated hydrogen.*

14

In another hydrogenation procedure the hydrogen is generated in *immediate contact with the compound to be hydrogenated.*

As an example, hydrochloric acid is added to a mixture of an aldehyde or ketone with zinc dust. Hydrogen is formed due to reaction between zinc and the acid. The *freshly formed* hydrogen is particularly reactive. One speaks of *nascent* hydrogen, or hydrogen *in statu nascendi* (Latin, "in the state of birth").

The reason why hydrogen *in statu nascendi* is so reactive is that hydrogen *atoms* are formed initially. These can react with the compound to be hydrogenated before combination to inactive hydrogen *molecules* takes place.

Thus, hydrogenation can be carried out in two different ways. Complete the sentences:

a) by using a ...................
b) by generating the hydrogen *in* ...................  →  14

---

**14**

**4** Correct; the *catalytically activated* hydrogen is more reactive.

Carbonyl compounds can also be oxidized. The oxidation of aldehydes gives carboxylic acids:

$$R-C{\overset{H}{\underset{O}{}}} + [O] \longrightarrow R-C{\overset{OH}{\underset{O}{}}}$$

With the aid of this general equation formulate the oxidations of acetaldehyde, propionaldehyde, and butyraldehyde:

$$CH_3-C{\overset{H}{\underset{O}{}}} + [O] \longrightarrow$$

acetaldehyde

.................... + [O] →
propionaldehyde

.................... + [O] →
butyraldehyde

→  16

| **5** |
|---|

Aldehyde + [O] → carboxylic acid.
Ketone + [O] → no reaction.

We will need the general formulas for aldehydes and ketones in the next type of reaction to be discussed.

Please give these formulas. → |18|

---

| **6** |
|---|

When a *double bond* is present in a molecule, this is usually the *most reactive site*. For example, addition reactions would take place at the double bond.

Repeat |18| .

---

| **7** |
|---|

No, alkaline catalysts do not effect hydrogenation. *Noble metals or finely divided nickel (Raney nickel) are hydrogenation catalysts.*

Write down the last sentence. → |3|

---

| **8** |
|---|

*Addition* or *addition reaction.*

The product of such a reaction is called an *addition product.*

When a compound H–X is added to a carbonyl group, C=O, to which end of the group will the hydrogen atom become bonded?

● To the oxygen atom → |19|

● To the carbon atom → |26|

● I don't know → |29|

---

| **9** |
|---|

When a *double bond* is present in a molecule, this is usually the *most reactive site*. For example, addition reactions would take place at the double bond.

Repeat |18| .

---

**10**          Correct; aldehydes and ketones can be reduced to alcohols. *Which compounds are formed by reduction of aldehydes?*

- Primary alcohols          →          21

- Secondary alcohols          →          25

- Tertiary alcohols          →          28

- I don't know          →          30

---

**11**          *Vanadium* compounds are used as *oxidation* catalysts! The *hydrogenation* catalysts are the *noble metals* and finely divided *nickel* (Raney nickel).

Copy down the last sentence and continue then at          3          .

---

**12**          R–CO–R + $H_2$ → R–CH(OH)–R

Give the name of the product:

- A primary alcohol          →          24

- A secondary alcohol          →          22

- A tertiary alcohol          →          27

- I don't know          →          30

---

**13**          *Aldehydes and ketones can,* in fact, *be reduced.*

Remember that aldehydes are *formed* by oxidation of primary alcohols. Similarly, ketones are obtained from secondary alcohols. Since aldehydes and ketones can be *prepared* by *oxidation,* it must be possible to transform them back into alcohols by *reduction.*

Continue at          10          .

---

**14**          Hydrogenations can be carried out

a) by using a catalyst;
b) by generating the hydrogen *in statu nascendi.*

It is not unimportant which method is used for a given reaction. Thus, the *nascent* hydrogen is not capable of reducing *all* double bonds: the reaction takes place with the carbonyl group, but $C=C$ double bonds are *not* reduced.

Hence, the more reactive hydrogen is present in:

- Catalytic hydrogenation?    →    $\boxed{4}$

- Nascent hydrogen?    →    $\boxed{23}$

---

$\boxed{15}$    Compare:

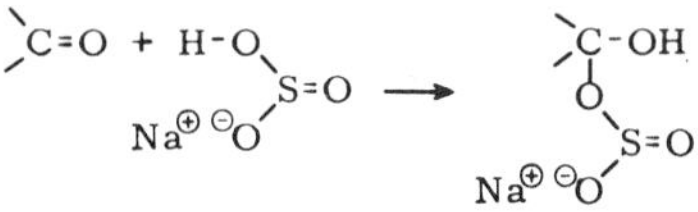

$$\begin{array}{c} \diagdown \\ \diagup \end{array} C{=}O \ + \ H{-}O \diagdown \underset{Na^{\oplus}\ ^{\ominus}O}{\diagup} S{=}O \ \longrightarrow \ \begin{array}{c} \diagdown \\ \diagup \end{array} \underset{\underset{\underset{Na^{\oplus}\ ^{\ominus}O}{\diagup}}{\overset{O}{|}}}{\underset{|}{C}}{-}OH \ \ S{=}O$$

Now give the equation for the reaction of acetone (= dimethyl ketone) with sodium hydrogen sulfite, and check your result at   $\boxed{36}$   .

**14**

---

$\boxed{16}$

$$CH_3{-}C\overset{H}{\underset{O}{\diagup}} \ + \ [O] \ \longrightarrow \ CH_3{-}C\overset{OH}{\underset{O}{\diagup}}$$

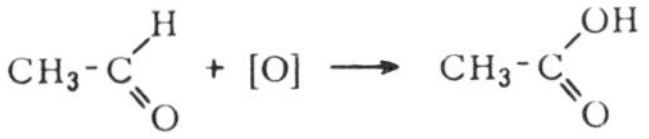

acetaldehyde    acetic acid

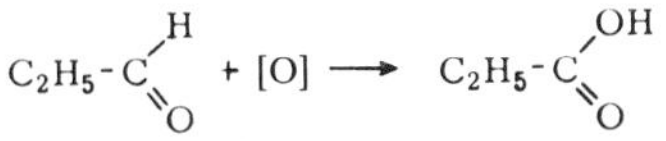

$$C_2H_5{-}C\overset{H}{\underset{O}{\diagup}} \ + \ [O] \ \longrightarrow \ C_2H_5{-}C\overset{OH}{\underset{O}{\diagup}}$$

propionaldehyde    propionic acid

$$C_3H_7{-}C\overset{H}{\underset{O}{\diagup}} \ + \ [O] \ \longrightarrow \ C_3H_7{-}C\overset{OH}{\underset{O}{\diagup}}$$

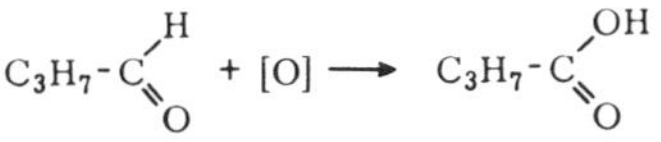

butyraldehyde    butyric acid

You will realize that the names of the aldehydes are derived from those of the acids formed upon oxidation:
*acetic* acid is formed from *acet*aldehyde;
*propionic* acid is formed from *propion*aldehyde, and
*butyric* acid from *butyr*aldehyde.

Ketones are *not oxidizable* under mild conditions. Under more forceful conditions, the ketone molecules will be cleaved into smaller fragments (e.g. during combustion). Such oxidation processes are, however, of little interest.

Complete the following scheme:

Aldehyde + [O] →
Ketone  + [O] →

→  [5]

---

[17]

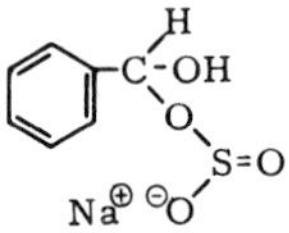

The addition compounds formed from aldehydes or ketones and sodium hydrogen sulfite are easily *converted back to the starting materials*. Usually, this already takes place on heating an aqueous solution. The decomposition of the addition product of acetaldehyde and sodium hydrogen sulfite is an example:

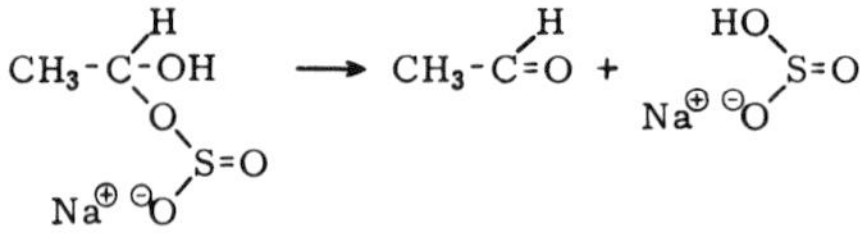

Give the products of decomposition of the following addition compounds:

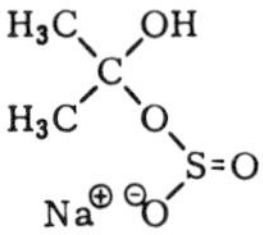

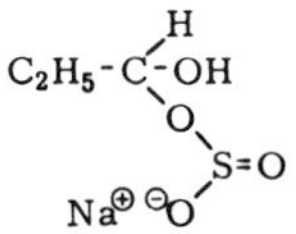

- Acetone and propionaldehyde      →   [37]

- Acetone and acetaldehyde      →   [42]

- Acetone and butyraldehyde      →   [45]

- I need more information      →   [54]

**14**

**18**

$$R-C\overset{H}{\underset{O}{\diagup}} \qquad R-\overset{O}{\underset{\|}{C}}-R$$

     aldehyde          ketone

Aldehydes and ketones undergo *addition reactions* with numerous reagents. Deduce from the formulas at which point the addition reactions will occur.

● Between R and C          →     $\boxed{6}$

● Between C and H          →     $\boxed{9}$

● At the C=O double bond          →     $\boxed{29}$

● I don't understand the question    →     $\boxed{31}$

---

**19**      Correct; the hydrogen atom in H–X is added to the oxygen atom of the carbonyl group.

Formulate the equation.    →    $\boxed{32}$

**14**

---

**20**      *Aldehydes and ketones can, in fact, be reduced.*

Remember that aldehydes are formed by oxidation of primary alcohols. Similarly, ketones are obtained from secondary alcohols. Since aldehydes and ketones can be *prepared* by *oxidation,* it must be possible to transform them back into alcohols by *reduction.*

Continue at    $\boxed{10}$   .

---

**21**      Correct.

*Primary alcohols are formed by the reduction of aldehydes.*

Try now and formulate the general equation for this reaction, using hydrogen as the reducing agent:

$$R-C\overset{H}{\underset{O}{\diagup}} + H_2 \longrightarrow$$

                              →    $\boxed{2}$

| **22** |
|---|

*Secondary alcohols are formed by the reduction of ketones.* When hydrogen is the reducing agent, the reaction is a *hydrogenation.*

You already know that catalysts are required in hydrogenation reactions. Do you remember which ones?

- Alkaline catalysts          →     ⬜ 7

- Noble metals                →     ⬜ 3

- Vanadium compounds     →     ⬜ 11

| **23** |
|---|

Nascent hydrogen will reduce the C=O double bond in carbonyl compounds, but *not* the C=C double bond.

In contrast, catalytically activated hydrogen *also reduces the C=C double bond.*

Which is the more active form of hydrogen?     →     ⬜ 4

**14**

| **24** |
|---|

Your answer is wrong.

Read the explanation at     ⬜ 30   .

| **25** |
|---|

Your answer is wrong.

Read the explanation at     ⬜ 30   .

| **26** |
|---|

Your answer is wrong.

Please pay more attention; the correct answer is given at     ⬜ 29   .

| **27** |
|---|

Your answer is wrong.

Read the explanation at     ⬜ 30   .

| **28** |
|---|

Your answer is wrong.

Read the explanation at     ⬜ 30   .

**29** Addition takes place at the C=O double bond.

Aldehydes and ketones undergo the same types of addition reactions. Therefore, we will use the general formula for carbonyl compounds:

$$\ce{>C=O}$$

Compounds of the type

H-X

are frequent reaction partners in addition reactions. The compounds H–X always add in such a way that the hydrogen atom becomes bonded to the oxygen end of the carbonyl group; the group X then forms a bond to carbon:

$$\ce{>C=O + H-X -> >C-OH}$$
$$\ \ \ \ \ \ \ \ \ \ \ \ \ \ \ \ \ \ \ \ \ \ \ | \atop X$$

This reaction is called ........................ → 8

**14**

**30** Let us summarize the relationship between primary and secondary alcohols, aldehydes, and ketones:

$$R-CH_2OH \underset{\text{reduction}}{\overset{\text{oxidation}}{\rightleftharpoons}} R-C{\overset{H}{\underset{O}{\diagdown}}}$$

primary          aldehyde
alcohol

$$R-CH(OH)-R \underset{\text{reduction}}{\overset{\text{oxidation}}{\rightleftharpoons}} R-CO-R$$

secondary      ketone
alcohol

Write down these equations.

Which kind of alcohol is formed on reduction of an aldehyde? → 21

**31**     *Addition reactions* have already been mentioned. The reaction between ethylene and chlorine is an example; chlorine adds to the C=C double bond:

$$\underset{H}{\overset{H}{>}}C=C\underset{H}{\overset{H}{<}} + Cl_2 \longrightarrow H-\underset{Cl}{\overset{H}{C}}-\underset{Cl}{\overset{H}{C}}-H$$

The addition reactions of carbonyl compounds are quite similar.

Repeat   18  .

---

**32**

$$>C=O + H-X \longrightarrow >\underset{X}{C}-OH \longrightarrow$$

Consider now some examples of compounds H–X used in addition reactions. Sodium hydrogen sulfite is one such compound:

$$>C=O + \underset{Na^{\oplus \ominus}O}{\overset{H-O}{>}}S=O \longrightarrow$$

This equation may look a little complicated. Try and complete it on the basis of the general equation above. Check your answer at   15  , or, if you need help, go instead to   33  .

---

**33**     We repeat the general equation:

$$>C=O + H-X \longrightarrow >\underset{X}{C}-OH$$

Sodium hydrogen sulfite is just one example of a compound H–X in which X stands for the group

$$\underset{Na^{\oplus}\,\ominus O}{\overset{-O}{>}}S=O$$

Surely, you can now complete the following equation:

$$>C=O + \underset{Na^{\oplus}\,\ominus O}{\overset{H-O}{>}}S=O \longrightarrow$$

$$\longrightarrow \quad 15$$

**34**

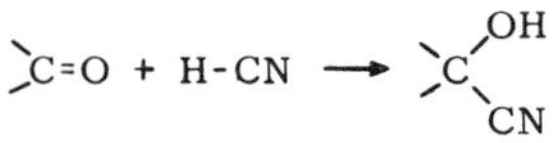

$$\text{>C=O} + \text{H-CN} \longrightarrow \text{>C} \begin{smallmatrix} \text{OH} \\ \text{CN} \end{smallmatrix}$$

On the basis of this general equation complete the following:

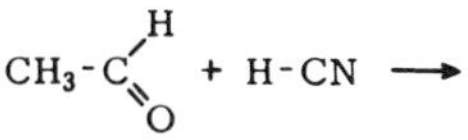

$$\text{CH}_3\text{-C} \begin{smallmatrix} \text{H} \\ \text{\\} \\ \text{O} \end{smallmatrix} + \text{H-CN} \longrightarrow$$

acetaldehyde

$$\begin{smallmatrix} \text{H}_3\text{C} \\ \text{H}_3\text{C} \end{smallmatrix} \text{C=O} + \text{H-CN} \longrightarrow$$

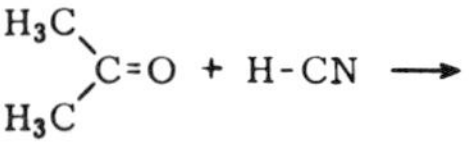

acetone

Check your result at ⬚44 .

---

**35**

Compare:

propionaldehyde cyanohydrin:    $\text{C}_2\text{H}_5\text{-C} \begin{smallmatrix} \text{H} \\ \text{OH} \\ \text{CN} \end{smallmatrix}$

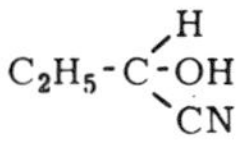

ethyl methyl ketone cyanohydrin:    $\begin{smallmatrix} \text{H}_3\text{C} \\ \text{C}_2\text{H}_5 \end{smallmatrix} \text{C} \begin{smallmatrix} \text{OH} \\ \text{CN} \end{smallmatrix}$

**14**

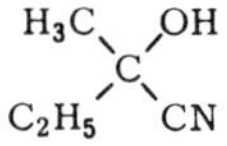

*Ammonia,* too, readily adds to the carbonyl group, particularly in aldehydes. Our general formula for the addition reagent was H–X. Thus, in ammonia (NH$_3$), X stands for the NH$_2$ group:

H–NH$_2$ reacts as H–X

You can now formulate the reaction between *ammonia* and *acetaldehyde.*    → ⬚46

---

**36**

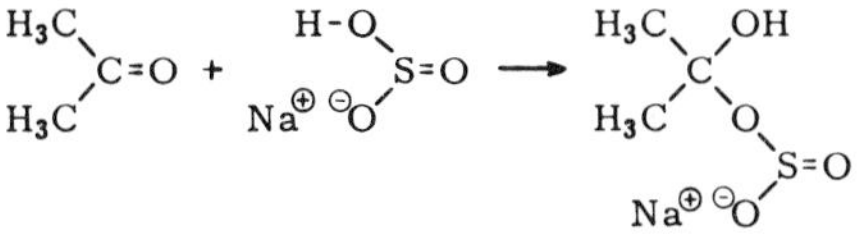

$$\begin{smallmatrix} \text{H}_3\text{C} \\ \text{H}_3\text{C} \end{smallmatrix} \text{C=O} + \begin{smallmatrix} \text{H-O} \\ \text{Na}^{\oplus} \, {}^{\ominus}\text{O} \end{smallmatrix} \text{S=O} \longrightarrow \begin{smallmatrix} \text{H}_3\text{C} \\ \text{H}_3\text{C} \end{smallmatrix} \text{C} \begin{smallmatrix} \text{OH} \\ \text{O} \\ \phantom{x}\text{S=O} \\ \text{Na}^{\oplus} \, {}^{\ominus}\text{O} \end{smallmatrix}$$

The addition product can be more simply written:

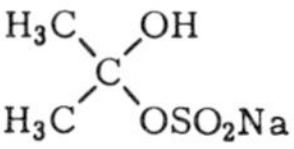

$$\begin{smallmatrix} \text{H}_3\text{C} \\ \text{H}_3\text{C} \end{smallmatrix} \text{C} \begin{smallmatrix} \text{OH} \\ \text{OSO}_2\text{Na} \end{smallmatrix}$$

This type of addition reaction takes place very easily. It suffices to shake a concentrated aqueous solution of sodium hydrogen sulfite with the aldehyde or ketone. The addition products are mostly crystalline compounds which separate from the concentrated solutions. They are soluble in water.

The addition products can be named according to the compounds from which they are formed, adding "bisulfite addition product" to the name of the carbonyl compound (bisulfite = hydrogen sulfite; it is tacitly understood that it is a sodium salt), e.g.

"acetone bisulfite addition product" and
"acetaldehyde bisulfite addition product".

Give the formula for the benzaldehyde bisulfite addition product, remembering that benzaldehyde is the simplest of the aromatic aldehydes.   →   |17|

---

|37|

     Correct; acetone and propionaldehyde are the products.

**14**

The reaction of carbonyl compounds with sodium hydrogen sulfite is often used for the purpose of purification:

Treatment of the aldehyde or ketone with a concentrated aqueous solution of sodium hydrogen sulfite gives a crystalline addition compound. Filtration removes undesirable soluble by-products. Heating an aqueous solution of the isolated bisulfite addition product causes decomposition into the starting materials, i.e., the aldehyde or ketone and sodium hydrogen sulfite.

*Hydrogen cyanide* is another example of a molecule which readily adds to the C=O bond. The formula for hydrogen cyanide is

$$H-C\equiv N.$$

Consider the general orientation of addition of compounds of the type H−X to carbonyl groups. To which atom will the CN group in HCN become bonded?

● Oxygen       →   |53|

● Carbon       →   |51|

● I don't know    →   |57|

**38**

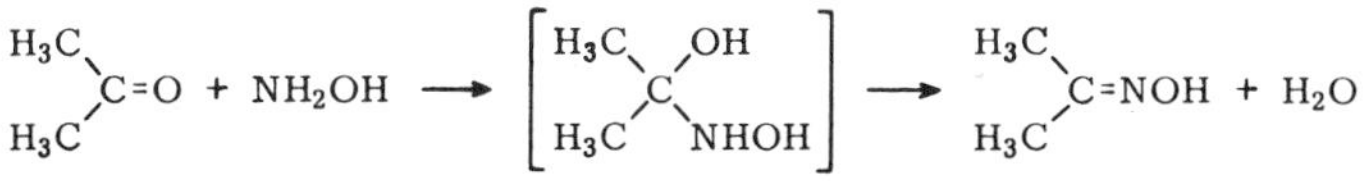

$$\underset{H_3C}{\overset{H_3C}{\diagdown}}C{=}O \;+\; NH_2OH \;\longrightarrow\; \left[ \underset{H_3C}{\overset{H_3C}{\diagdown}}\underset{NHOH}{\overset{OH}{\diagup}}C \right] \;\longrightarrow\; \underset{H_3C}{\overset{H_3C}{\diagdown}}C{=}NOH \;+\; H_2O$$

The reaction product is an *oxime*. This particular product is *acetone oxime*.
*Oximes are derived from carbonyl compounds by replacement of O by =NOH.*

Give the name of the compound

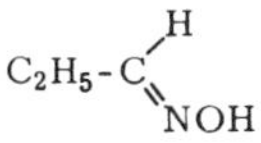

$$C_2H_5{-}\underset{NOH}{\overset{H}{C}}$$

● Aldehyde oxime                         →   50

● Propionaldehyde hydroxylamine          →   60

● Propionaldehyde oxime                  →   49

---

**14**

**39**          Compare:

cyclohexane:   $\underset{H_2C}{\overset{H_2C}{}}\underset{CH_2}{\overset{CH_2}{}}\cdots$ CH₂ ring

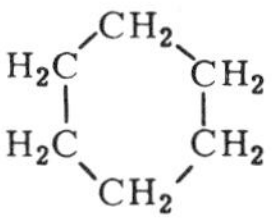

Cyclohexanone is the corresponding ketone:

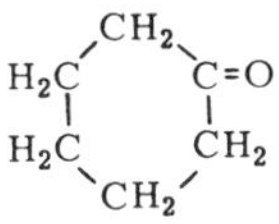

Reaction of this ketone with hydroxylamine gives *cyclohexanone oxime*. Give
the equation for this latter reaction, omitting intermediate products.   →   52

---

**40**          Correct.

A *ring expansion* has taken place in this reaction: cyclohexanone is a
six-membered ring, the product a seven-membered one. This product
is called *caprolactam* and is a starting material for the manufacture
of nylon. (See also Program 22.)

Complete the equation below, showing the formation of caprolactam from cyclohexanone oxime:

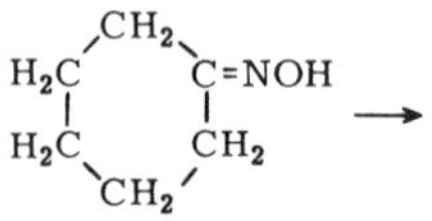

$$
\begin{array}{c}
\phantom{H_2C}\overset{\displaystyle CH_2}{\diagup\phantom{xx}\diagdown} \\
H_2C\phantom{xxxxx}C{=}NOH \\
\phantom{H_2}|\phantom{xxxxx}| \\
H_2C\phantom{xxxx}CH_2 \\
\phantom{H}\diagdown_{\displaystyle CH_2}\diagup
\end{array}
\quad\longrightarrow
$$

Check your formulas carefully before controlling at   56   .

---

**41**     Cyclohexanone oxime has a six-membered ring composed of carbon atoms only. In caprolactam there is a further ring atom.

Repeat   52   .

---

**42**     Acetaldehyde is

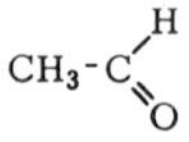

$$
CH_3{-}C\diagup^{\displaystyle H}_{\diagdown\! O}
$$

Repeat   17   .

---

**43**     Your answer is wrong.

Study the last equation at   52   more carefully.

---

**44**

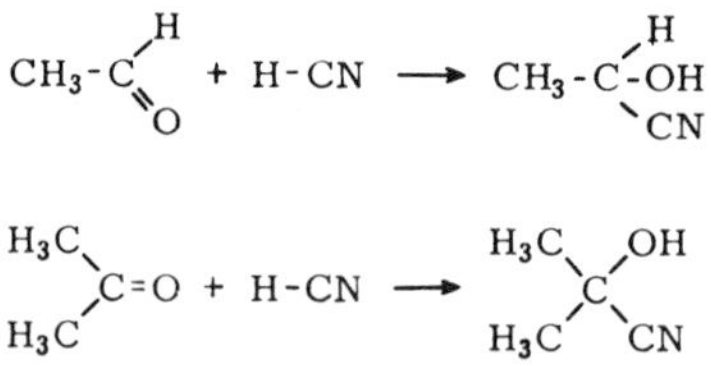

$$
CH_3{-}C\diagup^{\displaystyle H}_{\diagdown\! O} + H{-}CN \longrightarrow CH_3{-}C\overset{\displaystyle H}{\underset{\displaystyle CN}{\big|}}{-}OH
$$

$$
\begin{array}{c} H_3C \\ \diagdown \\ \phantom{x}C{=}O \\ \diagup \\ H_3C \end{array} + H{-}CN \longrightarrow
\begin{array}{c} H_3C\phantom{x}OH \\ \diagdown\phantom{x}\diagup \\ C \\ \diagup\phantom{x}\diagdown \\ H_3C\phantom{x}CN \end{array}
$$

The addition products of HCN (hydrogen cyanide) and an aldehyde or ketone are called *cyanohydrins*. The names of individual compounds are formed by adding "cyanohydrin" to the name of the carbonyl compound.

Give the names of the two cyanohydrins shown above.    →    55

---

**45**    Butyraldehyde is

$$C_3H_7-C\underset{O}{\overset{H}{\diagdown}}$$

Repeat **17** .

---

**46**    Compare:

$$CH_3-C\underset{O}{\overset{H}{\diagup}} + NH_3 \longrightarrow CH_3-\underset{NH_2}{\overset{H}{C}}-OH$$

Reaction products of this type are called "aldehyde ammonia addition products".

Now give the formulas for the ammonia addition products of propionaldehyde and butyraldehyde.    →    **59**

**14**

---

**47**    *Caprolactam.* This is used in the manufacture of synthetic fibers (nylon).

This is the end of Program 14.

---

**48**    Compare:

Hydroxylamine:    $\overset{H}{\underset{H}{\diagdown}}N-OH$    (or simpler: $NH_2OH$).

Hydroxylamine gives addition products with carbonyl compounds:

$$\diagup\!\!\!\!C=O + \overset{H}{\underset{H}{\diagdown}}N-OH \longrightarrow \left[ \diagup\!\!\!\!\overset{OH}{\underset{\underset{OH}{N}}{C}}_H \right] \quad \text{or simpler:}$$

$$\diagup\!\!\!\!C=O + NH_2OH \longrightarrow \left[ \diagup\!\!\!\!\overset{OH}{\underset{NHOH}{C}} \right]$$

Thus, $NH_2OH$ reacts like other compounds of the type H–X, –NHOH adding to the carbon end of the carbonyl group.

However, the *addition products* shown in square brackets above are *unstable* and eliminate water to give the final products:

$$\left[ \begin{array}{c} \diagup OH \\ >C \diagdown \\ NHOH \end{array} \right] \longrightarrow \ >C=NOH \ + \ H_2O$$

Formulate the two steps in the reaction between acetone and hydroxylamine. → [38]

---

[49]    Correct.

*The oxime of cyclohexanone* is of particular importance industrially. We will develop the formula for this compound step by step.

*Cyclohexanone* is a *ketone* – a derivative of *cyclohexane*.

Give the structure of cyclohexane and check it at [39]  , or, if in doubt, go to [58]  for more information.

**14**

---

[50]    This name is not precise enough. The oxime is a derivative of a particular aldehyde and should be named accordingly.

Repeat [38]  .

---

[51]    Correct – the CN group adds to the carbon end of the carbonyl group; the hydrogen atom goes to oxygen:

$$>C=O + H-CN \rightarrow$$

Complete the equation.  → [34]

---

[52]    Compare:

$$\begin{array}{c} H_2 \\ H_2C \diagup \overset{C}{\phantom{.}} \diagdown C=O \\ H_2C \diagdown \underset{C}{\phantom{.}} \diagup CH_2 \\ H_2 \end{array} + NH_2OH \longrightarrow \begin{array}{c} H_2 \\ H_2C \diagup \overset{C}{\phantom{.}} \diagdown C=NOH \\ H_2C \diagdown \underset{C}{\phantom{.}} \diagup CH_2 \\ H_2 \end{array} + H_2O$$

cyclohexanone            cyclohexanone oxime

The above *oxime rearranges* in concentrated sulfuric acid solution.

A rearrangement is an intramolecular reaction in which the empirical formula is preserved, but a structural change takes place.

The rearrangement of cyclohexanone oxime is shown below:

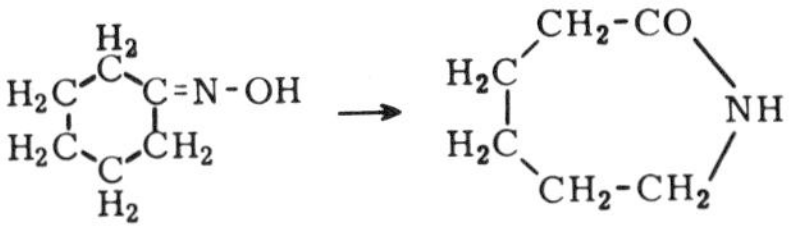

cyclohexanone oxime         caprolactam

The reaction is quite complicated: a C–C bond is broken, a new C–N bond is formed, and oxygen migrates from nitrogen to carbon.

How many ring atoms are there in these two compounds?

| Cyclohexanone oxime | Caprolactam | | |
|---|---|---|---|
| 6 | 6 | → | 41 |
| 6 | 7 | → | 40 |
| 6 | 8 | → | 43 |

**14**

---

**53**    You do not understand the general principle of addition to the C=O group.

Read   57   attentively.

---

**54**     Let us consider another example of decomposition of sodium hydrogen sulfite addition products.

The acetaldehyde "bisulfite addition product",

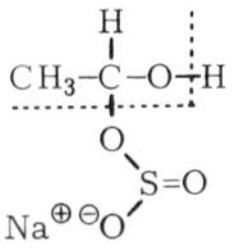

is cleaved into its constituents (cf. dashed line) by heating an aqueous solution. The constituents are *acetaldehyde* and *sodium hydrogen sulfite*.

Which carbonyl compounds are formed by decomposition of the following two addition products?

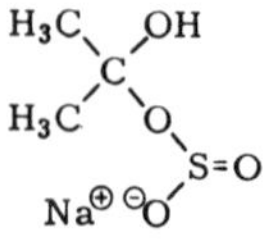

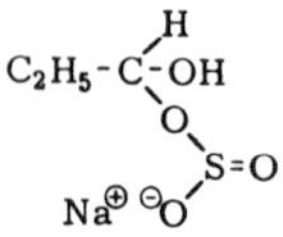

- Acetone and propionaldehyde　　　　→　　$\boxed{37}$

- Acetone and acetaldehyde　　　　→　　$\boxed{42}$

- Acetone and butyraldehyde　　　　→　　$\boxed{45}$

---

$\boxed{55}$

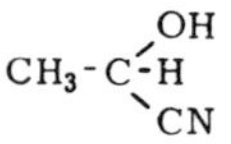

acetaldehyde cyanohydrin 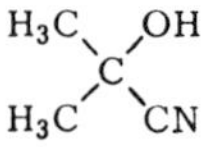　　　　　　acetone cyanohydrin

Now give the formulas for the cyanohydrins of

propionaldehyde and
ethyl methyl ketone.　→　$\boxed{35}$

---

$\boxed{56}$

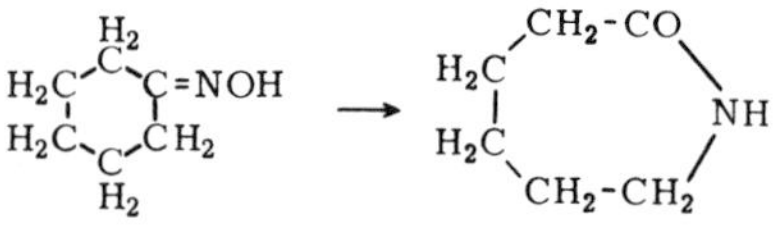

The new compound is called .................... and is used in the manufacture of
....................　→　$\boxed{47}$

---

$\boxed{57}$　It has already been stated that compounds of the type H–X *always* add to the carbonyl group in such a way that *H becomes bonded to the oxygen atom,* X to the carbon atom.

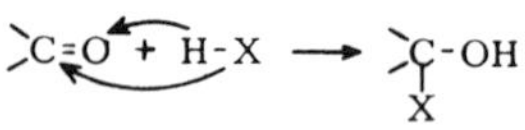

If X = CN, then H–X = H–CN, i.e. hydrogen cyanide. How does HCN add to the carbonyl group?

- CN goes to O　　　　→　　$\boxed{29}$

- CN goes to C　　　　→　　$\boxed{51}$

**58**     Cyclohexane is a *cyclic* hydrocarbon composed of six (*"hexa"*) carbon atoms in a ring. The compound is saturated, i.e. it is an *alkane,* hence the suffix *ane.*

Surely, you can now give the formula for this compound.   →   |39|


**59**     The formulas for the ammonia addition products of propionaldehyde and butyraldehyde are:

$$C_2H_5-\underset{\displaystyle \overset{\displaystyle |}{NH_2}}{\overset{\displaystyle \overset{\displaystyle H}{\diagup}}{C}}-OH$$

$$C_3H_7-\underset{\displaystyle \overset{\displaystyle |}{NH_2}}{\overset{\displaystyle \overset{\displaystyle H}{\diagup}}{C}}-OH$$

Ammonia addition products are only stable in aqueous solution.

The reaction between aldehydes or ketones and *hydroxylamine* is important.

Hydroxylamine is a derivative of ammonia in which a hydrogen atom is replaced by a hydroxyl group.

Give, on the basis of this information, the formula for hydroxylamine.   →   |48|

**14**


**60**     This name is wrong.

The compound is an *oxime* formed from propionaldehyde and hydroxylamine.

What is the name of the oxime

$$C_2H_5-C\underset{\displaystyle NOH}{\overset{\displaystyle H}{\diagup}}   ?$$

● Aldehyde oxime          →   |50|

● Propionaldehyde oxime          →   |49|

# Program 15

## The Carbonyl Group (III)

**1**    In the last program on the carbonyl group we will discuss polymerization reactions, the aldol condensation, and the Cannizzaro reaction.

Let us start with *polymerization* reactions.

A polymerization is the *combination of several unsaturated molecules of the same kind to form a larger molecule which may be either open-chain or cyclic.* Catalysts are often required.

Apart from formaldehyde and acetaldehyde, most carbonyl compounds are stable toward polymerization.

To illustrate the polymerization of formaldehyde we first "open up" the C=O double bond:

$$
\begin{array}{c} H \\ \diagdown \\ \diagup \\ H \end{array} C{=}O \quad \longrightarrow \quad 
\begin{array}{c} H \\ | \\ -C-O- \\ | \\ H \end{array}
$$

The first step of the polymerization corresponds to the combination of two such purely hypothetical "opened" carbonyl groups:

$$
\begin{array}{c} H \\ | \\ -C-O- \\ | \\ H \end{array} \;+\; 
\begin{array}{c} H \\ | \\ -C-O- \\ | \\ H \end{array} \;\longrightarrow
$$

Complete the equation.    →    |10|

---

**2**

$$
\left[ \begin{array}{c} H \\ | \\ -C-O- \\ | \\ H \end{array} \right]_x \quad \textit{polymer}
$$

Polyformaldehyde (= paraformaldehyde) consists of long chains containing e.g. tens, hundreds, or even thousands of monomer units.

If paraformaldehyde consists of one thousand monomer units, and the molecular mass of $CH_2O$ is 30, what is the molecular mass of the polymer?

300 000   →   24

30 000   →   13

3 000   →   22

---

| 3 | | |
|---|---|---|
| **Number of molecules reacting** | | **Name of the reaction** |
| two | | dimerization |
| three | | trimerization |
| several | | polymerization |

The polymerizable molecules may be described as *monomers*. The polymerized product is a *polymer*.

Complete the sentence:

.............. $\xrightarrow{\text{polymerization}}$ ..............   →   14

---

**15**

| 4 |
|---|

$$\begin{array}{c} CH_3 \\ | \\ -C-O- \\ | \\ H \end{array}$$

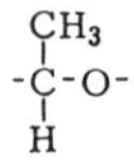

Give the product of combination of three molecules of acetaldehyde.   →   16

---

| 5 |
|---|

Correct; it is a six-membered ring:

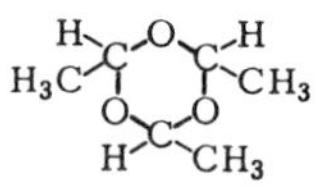

What is the correct description of this compound?

● Monomeric acetaldehyde     →   33

● Dimeric acetaldehyde        →   35

● Trimeric acetaldehyde       →   15

● Polymeric acetaldehyde      →   38

---

**6**          Correct; the hydrogen atom adds to the oxygen end of the carbonyl group, and a new carbon-carbon bond is formed:

$$CH_3-C\overset{O}{\underset{H}{\diagup}} \ + \ H-\overset{H}{\underset{H}{C}}-C\overset{H}{\underset{O}{\diagup}}$$

Give the formula for the reaction product.     →     18

---

**7**

$$CH_3-CH(OH)-CH_2-C\overset{H}{\underset{O}{\diagup}}$$
$$\underset{\text{alcohol}}{\uparrow} \qquad \underset{\text{aldehyde}}{\uparrow}$$

"*Ald*ehyde alcoh*ol*" = *aldol*

In the aldol condensation we have seen for the first time how *a functional group can influence the reactivity of other groups or atoms within the molecule.* In the present case, the *carbonyl group increases the reactivity of a neighboring hydrogen atom.*

**15**

Write the formulas for:

formaldehyde, acetaldehyde, propionaldehyde, butyraldehyde, and benzaldehyde.

Consider the structural requirements for the aldol condensation given above and then answer the question:

How many of these five aldehydes will undergo the aldol condensation?

- All five        →     32

- Four           →     36

- Three          →     17

- Two            →     39

- One            →     41

- I don't know   →     45

| **8** | *The carbonyl group activates the hydrogen atoms in the α-position.* |

Before coming to the last reaction undergone by the aldehydes, we need to recapitulate their *reduction* reaction. Write out the general equation for the reduction of an aldehyde. What is the product?

- A primary alcohol      →    19
- A secondary alcohol      →    42
- A tertiary alcohol      →    47
- A carboxylic acid      →    49

---

| **9** | Your answer is wrong. |

Read the explanation at   12   .

---

| **10** |

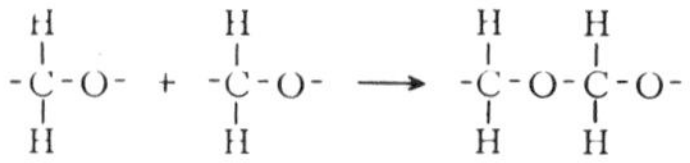

The polymerization continues when a third molecule of formaldehyde joins the chain.

Give the formula for the product of this second step.    →    21

---

| **11** | Your answer is wrong. |

You have misunderstood the requirements for the aldol condensation.

Read   45   .

---

| **12** | We have already learned that compounds of the type H–X add to the C=O group in such a way that *H goes to oxygen*, X goes to carbon. |

Complete the equation,

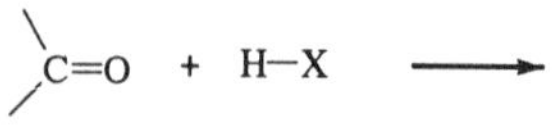

$$\diagdown\!\!\!\!\diagup C{=}O \ + \ H{-}X \ \longrightarrow \qquad\qquad \rightarrow \quad 27$$

15°

---

**13**    The molecular mass will be 30 000.

Such a polymer molecule may also be termed a *macromolecule*. Polymers can be unstable, decomposing to monomer units, especially on heating. This process is termed *depolymerization*.

The formation of a polymer from monomers is a ............... reaction.
The formation of monomers from a polymer is a ............... reaction.

Complete these sentences.    →    23

---

**14**    *Monomer* $\xrightarrow{\text{polymerization}}$ *polymer*

A polymer molecule may also be termed a *macromolecule*.

The two ends of the paraformaldehyde chain have free valences,

$$-\underset{\underset{H}{|}}{\overset{\overset{H}{|}}{C}}-O-\underset{\underset{H}{|}}{\overset{\overset{H}{|}}{C}}-O-\underset{\underset{H}{|}}{\overset{\overset{H}{|}}{C}}-O-\underset{\underset{H}{|}}{\overset{\overset{H}{|}}{C}}-O-\underset{\underset{H}{|}}{\overset{\overset{H}{|}}{C}}-O-$$

which can be saturated on reaction with small amounts of water to give:

$$HO-\underset{\underset{H}{|}}{\overset{\overset{H}{|}}{C}}-O-\underset{\underset{H}{|}}{\overset{\overset{H}{|}}{C}}-O-\underset{\underset{H}{|}}{\overset{\overset{H}{|}}{C}}-O-\underset{\underset{H}{|}}{\overset{\overset{H}{|}}{C}}-O-\underset{\underset{H}{|}}{\overset{\overset{H}{|}}{C}}-O-H$$

These *end groups* are often ignored when giving the formula for a polymer. Thus, paraformaldehyde is designated as

$$\left[-\underset{\underset{H}{|}}{\overset{\overset{H}{|}}{C}}-O-\right]_x$$

This formula indicates that a large number (x) of formaldehyde molecules have come together to form a ............... molecule.

Give the general name of such molecules.    →    2

---

**15**    *Trimeric* acetaldehyde.

This compound is also called *paraldehyde*. The trimerization of acetaldehyde is catalyzed by concentrated hydrochloric acid.

15

The formulas for paraformaldehyde and paraldehyde are repeated below:

$$\left[\begin{array}{c} H \\ | \\ -C-O- \\ | \\ H \end{array}\right]_x \qquad \begin{array}{c} H \diagdown \diagup O \diagdown \diagup H \\ H_3C \diagup C \quad C \diagdown CH_3 \\ O \diagdown \diagup O \\ H \diagup C \diagdown CH_3 \end{array}$$

Can paraformaldehyde and paraldehyde undergo addition reactions?

● Yes                  →    |29|

● No                  →    |25|

● Only paraformaldehyde    →    |34|

● Only paraldehyde        →    |37|

---

|16|

$$\begin{array}{c} CH_3 \quad CH_3 \quad CH_3 \\ | \quad\quad | \quad\quad | \\ C-O-C-O-C-O \\ | \quad\quad | \quad\quad | \\ H \quad\quad H \quad\quad H \end{array}$$

**15**

The free valences are saturated when the two ends of the chain join to form a ring.

What is the size of the ring formed?

● A three-membered ring    →    |26|

● A six-membered ring      →    |5|

● A nine-membered ring    →    |28|

● I don't know           →    |31|

---

|17|     Correct. – Of the five aldehydes mentioned, only three undergo the aldol condensation.

The five aldehydes were: formaldehyde, acetaldehyde, propionaldehyde, butyraldehyde, and benzaldehyde. Write the formulas and names of the two aldehydes which do *not* undergo the aldol condensation, and briefly state the reasons.   →    |40|

---

**18**    The completed equation is given below. Check your formula for the reaction product:

$$CH_3-C{\overset{H}{\underset{\diagdown O}{\lessgtr}}} + H-\overset{\overset{\displaystyle H}{|}}{\underset{\underset{\displaystyle H}{|}}{C}}-C{\overset{\displaystyle H}{\underset{\diagdown O}{\diagup}}} \longrightarrow CH_3-\overset{\overset{\displaystyle H}{|}}{\underset{\underset{\displaystyle OH}{|}}{C}}-\overset{\overset{\displaystyle H}{|}}{\underset{\underset{\displaystyle H}{|}}{C}}-C{\overset{\displaystyle H}{\underset{\diagdown O}{\diagup}}}$$

or simpler: $CH_3-CH(OH)-CH_2-CHO$

This reaction is called the *aldol condensation.* It is not really a condensation reaction because no molecule is eliminated. Strictly speaking, it could be termed a *dimerization,* i.e. a special case of a polymerization. However, the term "condensation" is commonly used.

You may wish to recapitulate the concepts of "condensation" and "polymerization" at   30  .

The aldol condensation takes place in the presence of a base. The name *aldol* stems from the fact that the reaction product is an *ald*ehyde as well as an alcoh*ol*.

Indicate the aldehyde and alcohol functions in the condensation product.   →    7

**15**

**19**    Correct. – The *reduction* of an aldehyde takes place according to the equation:

$$R-C{\overset{H}{\underset{\diagdown O}{\lessgtr}}} + H_2 \longrightarrow R-CH_2OH$$

As you know, aldehydes can also be *oxidized:*

$$R-C{\overset{H}{\underset{\diagdown O}{\lessgtr}}} + [O] \longrightarrow$$

Complete the equation. What is the nature of the reaction product?

● A carboxylic acid     →    43

● A primary alcohol     →    53

● A secondary alcohol     →    56

20

Correct. – The aldehyde

$$CH_3-\underset{\underset{CH_3}{|}}{\overset{\overset{CH_3}{|}}{C}}-C\overset{\nearrow H}{\underset{\searrow O}{}}$$

has *no α-hydrogen atom* (i.e. no hydrogen atom on the carbon neighboring the carbonyl group). Therefore, an aldol condensation is impossible. The requirement for the occurrence of an aldol condensation is the availability of an α-hydrogen atom. The success of the reaction is due to the fact that the carbonyl group activates the ............... atoms in the ............... position.

Write the completed sentence and check your result at ⟦8⟧.

21

$$-\underset{\underset{H}{|}}{\overset{\overset{H}{|}}{C}}-O-\underset{\underset{H}{|}}{\overset{\overset{H}{|}}{C}}-O-\underset{\underset{H}{|}}{\overset{\overset{H}{|}}{C}}-O-$$

The chain length is increased when further molecules of formaldehyde are added:

$$-\underset{\underset{H}{|}}{\overset{\overset{H}{|}}{C}}-O-\underset{\underset{H}{|}}{\overset{\overset{H}{|}}{C}}-O-\underset{\underset{H}{|}}{\overset{\overset{H}{|}}{C}}-O-\underset{\underset{H}{|}}{\overset{\overset{H}{|}}{C}}-O-\underset{\underset{H}{|}}{\overset{\overset{H}{|}}{C}}-O-$$

For the sake of clarity, the units from which the polymer is formed are indicated by vertical dashed lines in the above formula.

The polymeric formaldehyde is called *paraformaldehyde.* When only a *few* molecules condense, the term polymerization is often not used. Instead, the combination of only *two* molecules is called a *dimerization,* and of *three,* a *trimerization.* A true polymer may be formed from hundreds or thousands of molecules.

Complete the table:

| Number of molecules combining | Name of the reaction |
| --- | --- |
| two | |
| three | |
| several | |

→ ⟦3⟧

| 22 | You are wrong. Formaldehyde has a molecular mass of 30. The polymerization of one thousand such molecules gives a molecular mass of ............... $\rightarrow$ |13| |

| 23 |

$$\text{Polymer} \xrightleftharpoons[\text{polymerization}]{\text{depolymerization}} \text{monomers}$$

The depolymerization of paraformaldehyde takes place upon warming:

$$\left[\begin{array}{c} \text{H} \\ | \\ -\text{C}-\text{O}- \\ | \\ \text{H} \end{array}\right]_x \longrightarrow x\ CH_2O$$

The complete decomposition of the polymer results in the formation of x molecules of monomer.

*Acetaldehyde,* too, can polymerize.

As a prelude to polymerization, write the "opened" form of the C=O double bond:

$$\begin{array}{c} CH_3 \\ | \\ C = O \\ | \\ H \end{array} \rightarrow$$

Continue then at |4| , or, if you need assistance, go instead to |50| .

**15**

| 24 | You are wrong. Formaldehyde has a molecular mass of 30. The polymerization of one thousand such molecules gives a molecular mass of ............... $\rightarrow$ |13| |

| 25 | Correct. – Addition reactions take place at double bonds. These are no longer present in the aldehyde polymers.

Hence, *paraformaldehyde and paraldehyde do not undergo addition reactions.*

This terminates our presentation of the polymerization of carbonyl compounds.

The most important reactions of carbonyl compounds are due to the fact that *the C=O group activates α-hydrogen atoms;* that is, the carbonyl group facilitates the dissociation of a C–H bond at the neighboring carbon atom.

In other words, a carbonyl compound possessing an α-hydrogen atom can react in the same way as a compound of the type H–X, e.g.

$$H-\overset{\overset{\displaystyle H}{|}}{\underset{\underset{\displaystyle H}{|}}{C}}-C\overset{H}{\underset{O}{}} \quad = \quad H-X, \text{ where}$$

$$X \quad = \quad -\overset{\overset{\displaystyle H}{|}}{\underset{\underset{\displaystyle H}{|}}{C}}-C\overset{H}{\underset{O}{}}$$

The addition of H–X to the carbonyl group was discussed in Program 15 (e.g., H–CN). In like manner, a molecule of acetaldehyde can add to the C=O double bond of another molecule of acetaldehyde.

To which end of the carbonyl group does the hydrogen atom add in such addition reactions?

- To oxygen → $\boxed{6}$
- To carbon → $\boxed{9}$
- I don't know → $\boxed{12}$

**15**

---

**26** Your answer is wrong.

Read the explanation at $\boxed{31}$ .

---

**27** $\overset{}{\underset{}{}}C{=}O + H{-}X \longrightarrow \overset{}{\underset{\underset{X}{|}}{}}C{-}OH$

So, to which end of the C=O group does the hydrogen atom add? → $\boxed{6}$

---

**28** Your answer is wrong.

Read the explanation at $\boxed{31}$ .

---

**29** No. – Addition reactions take place at the C=O double bond in carbonyl compounds. No such double bond is present in the polymers. Hence, *paraformaldehyde and paraldehyde do not undergo addition reactions.* → $\boxed{25}$

---

**30**       A *condensation* is the combination of two or more molecules with elimination of water.

A *polymerization* is the combination of several *unsaturated* molecules.   →   |18|

---

**31**       Joining the two ends of the molecule gives a ring compound as shown below:

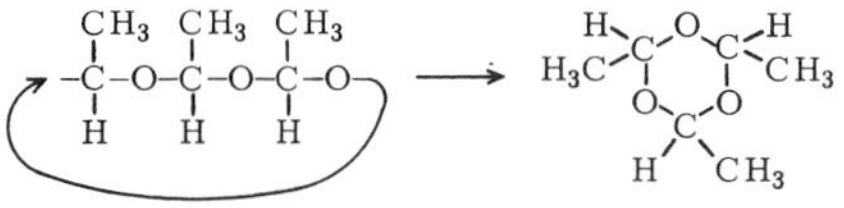

What is the size of the ring formed?   →   |5|

---

**32**       Your answer is wrong.

Read the explanation at   |45|  .

---

**33**       Unsaturated monomeric compounds are the starting materials for polymers. The *monomers polymerize,* forming a *polymer.*

Complete the scheme:

        depolymerization

..............  ⇌  ..............

        polymerization                                    →   |23|

---

**34**       No. – Addition reactions take place at the C=O double bond in carbonyl compounds. No such double bond is present in the polymers. Hence, *paraformaldehyde and paraldehyde do not undergo addition reactions.*   →   |25|

---

**35**       The combination of *two* molecules gives a *dimer.* The reaction is thus a *dimerization.*

Similarly, three molecules *trimerize* to give a *trimer.*

Repeat  |5|  .

---

---

**36**    Your answer is wrong.

Read the explanation at    45 .

---

**37**    No. – Addition reactions take place at the $C=O$ double bond in carbonyl compounds. No such double bond is present in the polymers. Hence, *paraformaldehyde and paraldehyde do not undergo addition reactions.*    →    25

---

**38**    The compound is a particular kind of polymeric acetaldehyde, namely one in which ontly *three* molecules have polymerized. Such a reaction is called a *trimerization.*

Repeat    5 .

---

**39**    Your answer is wrong.

Read the explanation at    45 .

---

**15**

**40**

| | | |
|---|---|---|
| Formaldehyde: | H-C with H and O | there is no neighboring carbon atom. |
| Benzaldehyde: | $H-C=O$ on benzene ring | there is no hydrogen atom on the neighboring carbon. |

The requirement for the aldol condensation was the availability of neighboring hydrogen atoms, which are activated by the carbonyl group.

A carbon atom next to a functional group is also referred to as the α-carbon (alpha-carbon). Therefore, we may also say that *the carbonyl group activates the* α-*hydrogen atoms* – or the hydrogen atoms in the α-position.

Does the following aldehyde undergo the aldol condensation?

$$CH_3-\underset{\underset{CH_3}{|}}{\overset{\overset{CH_3}{|}}{C}}-C\underset{O}{\overset{H}{<}}$$

● Yes     →     11

● No     →     20

---

---

**41**    Your answer is wrong.

Read the explanation at    45 .

---

**42**    No, a *primary* alcohol is formed.

Complete the equation, adding one hydrogen atom to the carbon, and one to the oxygen atom of the carbonyl group:

$$R-C\overset{H}{\underset{O}{\big<}} + H_2 \longrightarrow$$

$$\rightarrow \boxed{19}$$

---

**43**    The oxidation of an aldehyde takes place according to the equation:

$$R-C\overset{H}{\underset{O}{\big<}} + [O] \longrightarrow R-C\overset{OH}{\underset{O}{\big<}}$$

The product is a carboxylic acid.

As we have seen, aldehydes can be both *oxidized* and *reduced*. Aldehydes can also undergo a special reaction involving two molecules, one of which is oxidized, the other reduced. This takes place when the aldehyde is treated with an aqueous solution of NaOH:

**15**

$$2 R-C\overset{H}{\underset{O}{\big<}} + NaOH \longrightarrow R-CH_2OH + R-COONa$$

This reaction is named after its discoverer, the Italian chemist *Cannizzaro*.

Complete the verbal description of the Cannizzaro Reaction:

Two moles of an aldehyde react with ............... to form one mole of the corresponding ............... and one mole of the sodium salt of the corresponding carboxylic acid.    →    58

---

**44**    Correct.

Please give the names and formulas for the two aldehydes which undergo the Cannizzaro reaction with particular ease. (The four aldehydes mentioned were formaldehyde, acetaldehyde, propionaldehyde, and benzaldehyde.)

Check your answer at    65 .

---

**45**     Let us repeat the necessary requirements for the aldol condensation:

*Hydrogen atoms must be present at the neighboring carbon.* The corbonyl group *activates* these neighboring hydrogen atoms.

Inspect the five aldehydes given below, and indicate which ones *cannot* undergo the aldol condensation:

$H-C\overset{H}{\underset{O}{\diagup\!\!\!\diagdown}}$     $CH_3-C\overset{H}{\underset{O}{\diagup\!\!\!\diagdown}}$     $CH_3-CH_2-C\overset{H}{\underset{O}{\diagup\!\!\!\diagdown}}$

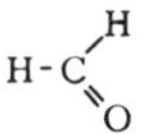

formaldehyde 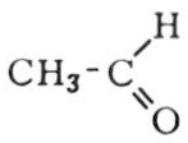     acetaldehyde 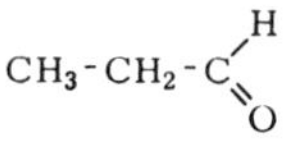     propionaldehyde

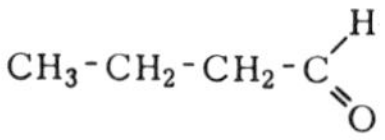

$CH_3-CH_2-CH_2-C\overset{H}{\underset{O}{\diagup\!\!\!\diagdown}}$     $H\diagdown\!\!\underset{}{C}\!\overset{O}{\diagup}$

butyraldehyde 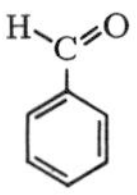     benzaldehyde

● Formaldehyde and acetaldehyde     →     60

● Formaldehyde and propionaldehyde     →     63

● Butyraldehyde and benzaldehyde     →     68

● Formaldehyde and benzaldehyde     →     72

**46**

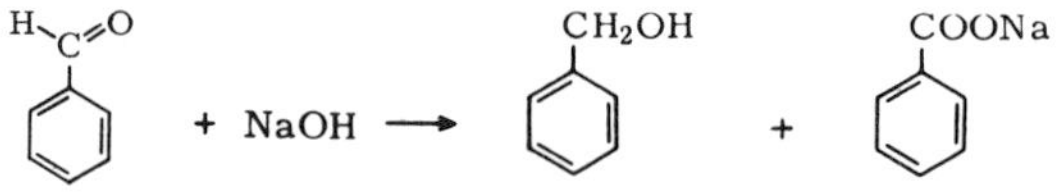

$H\diagdown\!\!\underset{}{C}\!\overset{O}{\diagup}$     + NaOH ⟶     $CH_2OH$     +     $COONa$

benzaldehyde     benzyl alcohol     benzoic acid (as the Na salt)

In the program on redox reactions we learned of a concept which can also be applied to the Cannizzaro reaction. Which concept?

● Oxidation     →     66

● Reduction     →     69

● Simultaneous oxidation and reduction     →     70

● Disproportionation     →     61

15

**47**     No, a *primary* alcohol is formed.

Complete the equation, adding one hydrogen atom to the carbon, and one to the oxygen atom of the carbonyl group:

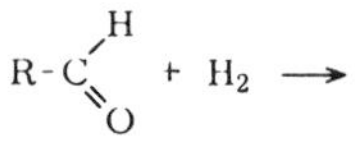

$$R-C{\overset{\displaystyle H}{\underset{\displaystyle O}{}}} \; + \; H_2 \; \longrightarrow$$

→ [19]

**48**     Correct. – At least half a mole of NaOH per mole of aldehyde must be used in order to neutralize the carboxylic acid formed.

Let us now return to the aldol condensation. Does this take place in the presence of acids or bases?

● Bases          →   [62]

● Acids          →   [64]

● I don't know          →   [67]

**49**     No, carboxylic acids are formed by the *oxidation* of aldehydes. *Reduction gives alcohols.*

Repeat   [8]  .

**50**     We have written the "opened" form of formaldehyde as follows:

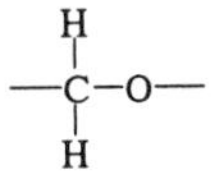

$$-\overset{\displaystyle H}{\underset{\displaystyle H}{C}}-O-$$

Can you write the corresponding form of acetaldehyde?

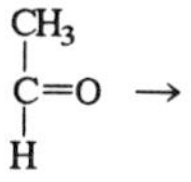

$$\underset{\displaystyle H}{\overset{\displaystyle CH_3}{C}}=O \; \longrightarrow$$

→ [4]

**51**     Your answer is wrong.

Read the explanation at   [59]  .

**52**     The aldol condensation reaction is *faster* than the Cannizzaro reaction. Because of this, all aldehydes which can undergo the aldol condensation will do so, and the Cannizzarro reaction will *not* be observed. Since acetaldehyde can undergo the aldol condensation, this is the reaction which takes place on the addition of sodium hydroxide.  →   74

**53**     No, carboxylic acids are formed by the oxidation of aldehydes.

Examine the general equation at   43  .

**54**     Your answer is wrong.

Read the explanation at   59  .

**55**     The aldol condensation reaction is *faster* than the Cannizzaro reaction. Because of this, all aldehydes which can undergo the aldol condensation will do so, and the Cannizzarro reaction will *not* be observed. Since acetaldehyde can undergo the aldol condensation, this is the reaction which takes place on the addition of sodium hydroxide.  →   74

**56**     No, carboxylic acids are formed by the oxidation of aldehydes.

Examine the general equation at   43  .

**57**     Your answer is wrong.

Read the explanation at   59  .

**58**     Two moles of an aldehyde react with *NaOH* to form one mole of the corresponding *alcohol* and one mole of the sodium salt of the corresponding carboxylic acid.

Other alkali hydroxides can be used in place of NaOH.

The Cannizzaro reaction of benzaldehyde (two moles) gives one mole of benzyl alcohol and one mole of the sodium salt of benzoic acid.

Give the equation for this reaction.   →   46

---

**59**     The aldol condensation reaction is *faster* than the Cannizzaro reaction. Using this information we can decide which of the four aldehydes,

formaldehyde, acetaldehyde, propionaldehyde, and benzaldehyde,

will undergo the Cannizzaro reaction.

Two aldehydes will undergo the *aldol* condensation because they possess α-hydrogen atoms.

Therefore, a Cannizzaro reaction will *not* occur.

The other two aldehydes *cannot* undergo aldol condensations. Accordingly, the addition of sodium hydroxide leads to the Cannizzaro reaction.

Give the names and formulas for the two aldehydes undergoing the Cannizzaro reaction.   →   65

---

**60**     Your answer is partially correct. Formaldehyde does not undergo the aldol condensation.

However, acetaldehyde does possess hydrogen atoms next to the carbonyl group. These are activated by the carbonyl group and, as a result, acetaldehyde undergoes the aldol condensation.

Repeat   45  .

15

---

**61**     The Cannizzaro reaction is a *disproportionation*.

The reaction takes place in the presence of strong bases only, e.g. NaOH.

Consider the following question carefully before answering:

How much NaOH is needed in order to complete the Cannizzaro reaction of one mole of an aldehyde?

● A catalytic amount     →   71

● Half a mole            →   48

● One mole               →   73

---

| 62 |

Correct. – The aldol condensation takes place in the presence of strong bases. This is also true of the Cannizzaro reaction.

How does one predict which of the two reactions will take place on the addition of a strong base to an aldehyde? This is a question of whether one reaction occurs faster than the other. If they were equally fast, then both would take place.

In fact, *the aldol condensation is the faster* of the two. Hence, the Cannizzaro reaction is not observed for those aldehydes which undergo the aldol condensation.

How does acetaldehyde react with sodium hydroxide?

● It undergoes the aldol condensation          →     | 74 |

● It undergoes the Cannizzaro reaction          →     | 52 |

● Aldol condensation and Cannizzaro
  reaction take place simultaneously          →     | 55 |

**15**

| 63 |

It is true that formaldehyde does not undergo the aldol condensation. However, propionaldehyde does. Propionaldehyde has hydrogen atoms at the carbon neighboring the carbonyl group. These C–H bonds are activated by the carbonyl group.

Repeat   | 45 |   .

| 64 |

No, the aldol condensation takes place in the presence of bases.    →    | 62 |

| 65 |

Formaldehyde:    H-C $\overset{H}{\underset{O}{\lesseqgtr}}$

benzaldehyde:    $\overset{H}{\underset{}{\searrow}}$C$\nearrow$O

Formulate the Cannizzaro reaction of formaldehyde.    →    | 75 |

---

**66**     Your answer is incomplete. Two moles of an aldehyde react with each other in the Cannizzaro reaction. One is *oxidized,* the other *reduced.*

Continue at     |70|   .

---

**67**     The aldol condensation takes place in the presence of bases.     →     |62|

---

**68**     It is true that benzaldehyde does not undergo the aldol condensation. However, butyraldehyde does.

Butyraldehyde has hydrogen atoms at the carbon neighboring the carbonyl group. These C–H bonds are activated by the carbonyl group.

Repeat     |45|   .

---

**69**     Your answer is incomplete. Two moles of an aldehyde react with each other in the Cannizzaro reaction. One is *oxidized,* the other *reduced.*

Continue at     |70|   .

---

**70**     In fact, both oxidation and reduction take place in the Cannizzaro reaction:

Two moles react with each other; one is being oxidized, the other reduced.

Something similar can happen in a redox reaction: one atom goes to a higher oxidation number – another atom of the same element goes to a lower one.

Such a reaction is called a

*disproportionation.*     →     |61|

---

**71**     A small amount of NaOH will *not* suffice. The Cannizzaro reaction gives rise to one mole of a primary alcohol per mole of carboxylic acid. The acid would immediately neutralize the NaOH, forming the salt of the acid. Hence, the Cannizzaro reaction would not proceed any further. The minimum amount of NaOH necessary is the amount which will just suffice to neutralize all of the carboxylic acid formed.

For one mole of aldehyde, one needs, therefore, at least ................... mol of NaOH.     →     |48|

---

**72**     Correct. – Formaldehyde and benzaldehyde do *not* undergo the aldol condensation.

Write the formulas for these two aldehydes, and give reasons *why* no aldol condensation occurs.   →   [40]

---

**73**     No, it is *not* necessary to use a full mole of NaOH per mole of aldehyde in the Cannizzaro reaction. Two moles of aldehyde give one mole of the corresponding alcohol and one mole of the carboxylic acid. Hence, *one* mole of aldehyde will furnish *half a mole* of the carboxylic acid. This neutralizes the NaOH present. In order to have sufficient NaOH for the completion of the reaction, we need ................... mole of NaOH per mole of aldehyde. → [48]

---

**74**     Correct. – Acetaldehyde undergoes the aldol condensation. Since this is faster than the Cannizzaro reaction, the latter will not take place.

How many of the following aldehydes will undergo the Cannizzaro reaction upon treatment with sodium hydroxide?

**15**

Formaldehyde, acetaldehyde, propionaldehyde, benzaldehyde.

● All four          →   [51]

● Three of them     →   [54]

● Two of them       →   [44]

● One of them       →   [57]

● I don't know      →   [59]

---

**75**

$$2\ H\text{-}C\!\!\underset{\diagdown O}{\overset{\diagup H}{}} + NaOH \longrightarrow H\text{-}\underset{\overset{|}{H}}{\overset{\overset{H}{|}}{C}}\text{-}OH + H\text{-}COONa$$

formaldehyde          methanol     sodium salt of formic acid

This completes Program 15.

---

# Program 16

## The Carboxyl Group

---

**1**

Carboxylic acids are organic compounds of the type

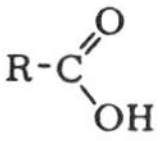

$$R-C\underset{OH}{\overset{O}{\diagup}} \quad \text{or shorter:} \quad R\text{-COOH}.$$

The functional group is the carboxyl group:

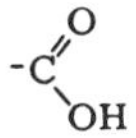

$$-C\underset{OH}{\overset{O}{\diagup}} \quad \text{or} \quad -COOH.$$

Write the structural formulas for the two simplest carboxylic acids, using

a) R = H
b) R = CH₃  →  8

---

**2**

You gave wrong formulas for acetic acid and butyric acid.

Repeat in detail from  8 .

**16**

---

**3**

No, formic acid is the product of oxidation of *methanol*. Formaldehyde is an intermediate product in this reaction:

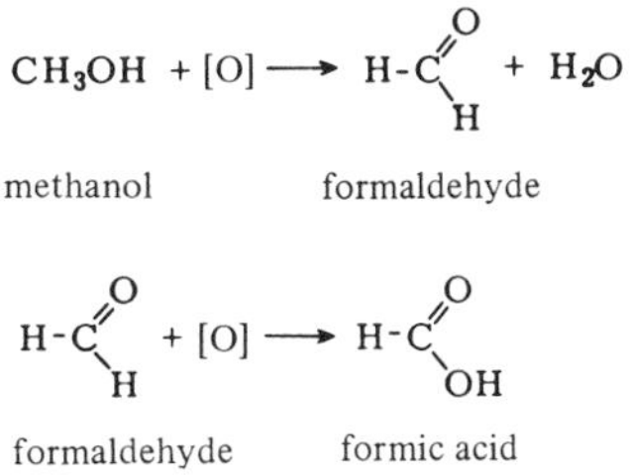

$$CH_3OH + [O] \longrightarrow H-C\underset{H}{\overset{O}{\diagup}} + H_2O$$

methanol            formaldehyde

$$H-C\underset{H}{\overset{O}{\diagup}} + [O] \longrightarrow H-C\underset{OH}{\overset{O}{\diagup}}$$

formaldehyde       formic acid

Now formulate the equation for the oxidation of ethanol, and give the name of the reaction product.   →  20

---

**4**

No; repeat  35  carefully.

---

**5**    You only gave the correct name of oxalic acid.

Repeat from   50   .

---

**6**    You did not name any of the three acids correctly. Repeat from   8  
and take care to memorize the names and formulas.

---

**7**    You gave an erroneous formula for propionic acid. Repeat carefully
from   8   .

---

**8**

a)   $H-C\overset{O}{\underset{OH}{}}$      b)   $CH_3-C\overset{O}{\underset{OH}{}}$

     formic acid           acetic acid

Surely you already know *acetic acid*. Vinegar is a 3–5% solution of acetic acid.
Pure acetic acid solidifies already at 16.6 °C to an ice-like material. It is, therefore,
known as *glacial* acetic acid.

Formic acid is the stinging substance found in ants and nettles.

Let us continue the series of aliphatic carboxylic acids. Write the formulas for
the next two members, using R = ethyl and R = propyl, respectively.

                                            →   16

---

**9**    Only your name for acetic acid was correct. Repeat from   8   and be
sure to memorize the names and formulas.

---

**10**    Remember the two isomeric alcohols:

$CH_3-CH_2-CH_2-OH$      and      $\underset{CH_3}{\overset{CH_3}{}}CH-OH$

     1–propanol                     2–propanol
     (propylalcohol)              (isopropylalcohol)

There is a similar relationship between butyric and isobutyric acids. Give the
formula for isobutyric acid.    →   24

16

**11**    Carbon monoxide (CO) and sodium hydroxide (NaOH) are the starting materials.

The product is HCOONa

Now complete the reaction equation.    →    31

---

**12**    No, propionic acid is formed by the oxidation of *1-propanol*. Propionaldehyde is an intermediate in the reaction:

$$CH_3\text{-}CH_2\text{-}CH_2\text{-}OH + [O] \longrightarrow CH_3\text{-}CH_2\text{-}C{\overset{O}{\underset{H}{\big\backslash}}} + H_2O$$

     1–propanol          propionaldehyde

$$CH_3\text{-}CH_2\text{-}C{\overset{O}{\underset{H}{\big\backslash}}} + [O] \longrightarrow CH_3\text{-}CH_2\text{-}C{\overset{O}{\underset{OH}{\big\backslash}}}$$

     propionaldehyde        propionic acid

Now write the equation for the oxidation of ethanol and give the name of the reaction product.    →    20

---

**13**    You gave the right formula for chloroacetic acid, but a wrong one for maleic acid. Repeat from   50   .

**16**

---

**14**    By now you ought to know the formula for propionic acid. Try not to make such mistakes again!

Repeat from   34   .

---

**15**    Your answer is correct.

Now assign the correct formulas to the following acids.

| Acetic acid | Butyric acid | Propionic acid | | |
|---|---|---|---|---|
| ● HCOOH | $CH_3COOH$ | $CH_3CH_2COOH$ | → | 2 |
| ● $CH_3COOH$ | $CH_3CH_2COOH$ | $CH_3CH_2CH_2COOH$ | → | 23 |
| ● $CH_3COOH$ | $CH_3CH_2CH_2COOH$ | $CH_3CH_2COOH$ | → | 30 |
| ● $CH_3COOH$ | $CH_3CH_2CH_2COOH$ | HCOOH | → | 7 |

16

$$CH_3-CH_2-C{\overset{O}{\underset{OH}{}}} \qquad CH_3-CH_2-CH_2-C{\overset{O}{\underset{OH}{}}}$$

propionic acid           butyric acid

The name "propionic" is derived from propane. Both have three carbon atoms.

Butyric acid is present in rancid butter and is also reponsible for unpleasant body odor.

There is another carboxylic acid isomeric with butyric acid. Give the formula and check your result at 24 , or, if you need help, go instead to 10 .

17   No; repeat 35 carefully.

18   Your answer is wrong. Continue at 28 .

19   You confused the three xylenes:

o–xylene           m–xylene           p–xylene

The oxidation of o-xylene gives an *ortho*-dicarboxylic acid.

Similarly, p-xylene gives a *para*-dicarboxylic acid.

Repeat 73 .

20

$$CH_3CH_2OH + [O] \longrightarrow CH_3-C{\overset{O}{\underset{H}{}}} + H_2O$$

ethanol           acetaldehyde

$$CH_3-C{\overset{O}{\underset{H}{}}} + [O] \longrightarrow CH_3-C{\overset{O}{\underset{OH}{}}}$$

acetaldehyde           acetic acid

Acetic acid is manufactured *industrially* in this way. Atmospheric oxygen is used as the oxidant.

We will now consider two other methods used for the production of formic and acetic acids.

The sodium salt of formic acid is obtained by reacting sodium hydroxide and carbon monoxide at 150 °C and 7 bar.

Give the equation for this reaction and check your result at [31] , or, if you need help, go instead to [11] .

---

**21**     Your answer is only partially correct. The reaction between NaOH and CO gives *formic acid only.*

Repeat [56] .

---

**22**     The general formula for a carboxylic acid is:

R–COOH,

where R is the organic residue (or H in the case of formic acid);

–COOH is the functional group, the carboxyl group.

The carboxyl group contains an acidic hydrogen atom. The acidic character is due to a dissociation of the O—H bond. The hydrogen atoms in the organic group R do *not* dissociate.

Thus, the general equation for the dissociation of a carboxylic acid is:

$$R–COOH \rightarrow R–COO^{\ominus} + H^{\oplus}$$

Write the corresponding equation for acetic acid.     →     [37]

---

**23**     You confused the formulas for propionic and butyric acids. Repeat from [16] .

---

24

$$H_3C$$
$$\phantom{H_3}\big\backslash CH-C \overset{\displaystyle O}{\underset{OH}{\big/\big/}}$$
$$H_3C\big/$$

isobutyric acid

Assign the correct names to the following acids:

HCOOH             CH₃COOH             CH₃–CH₂–COOH

● acetic          propionic          butyric acid        →   6

● propionic       acetic             butyric acid        →   9

● formic          acetic             propionic acid      →   15

25     Your answer is correct.

Let us now consider some *aromatic* carboxylic acids.

The simplest of these is benzoic acid. It is formed by the oxidation of benzaldehyde in the air, and therefore samples of benzaldehyde which are in contact with air invariably contain some benzoic acid.

**16**

Write the equation for the air oxidation of benzaldehyde to benzoic acid. For the sake of simplicity use [O] as the oxidant.     →   55

If you have difficulties, go instead to    63   .

26

$$H-\overset{\displaystyle Cl}{\underset{Cl}{\overset{|}{\underset{|}{C}}}}-C \overset{\displaystyle O}{\underset{OH}{\big/\big/}}$$     dichloroacetic acid

$$Cl-\overset{\displaystyle Cl}{\underset{Cl}{\overset{|}{\underset{|}{C}}}}-C \overset{\displaystyle O}{\underset{OH}{\big/\big/}}$$     trichloroacetic acid

Which is the strongest of the four carboxylic acids given below?

● Acetic acid              →   4

● Chloroacetic acid        →   17

● Dichloroacetic acid      →   43

● Trichloroacetic acid     →   34

**27**    No, NaOH is used.

Study the equation at  31  in detail.

**28**    A carbon atom with *four different groups* attached is asymmetric.

As an example, we will examine the three carbon atoms in lactic acid,

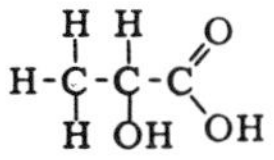

$$H-\overset{\overset{\displaystyle H}{|}}{\underset{\underset{\displaystyle H}{|}}{C}}-\overset{\overset{\displaystyle H}{|}}{\underset{\underset{\displaystyle OH}{|}}{C}}-C\overset{\nearrow O}{\underset{\searrow OH}{}}$$

The carbon atom to the left carries three H–atoms and the group

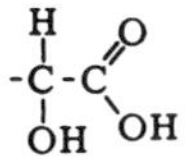

$$-\overset{\overset{\displaystyle H}{|}}{\underset{\underset{\displaystyle OH}{|}}{C}}-C\overset{\nearrow O}{\underset{\searrow OH}{}}$$

and is *not,* therefore, asymmetric.

The central carbon atom carries the following four groups,

     -H
     -CH₃
     -OH
     -COOH

i.e. four different ones; hence, it is *asymmetric.*

The carbon atom to the right *cannot* be asymmetric since it has only three groups attached:

     =O
     -OH  H
     CH₃-C-
         OH

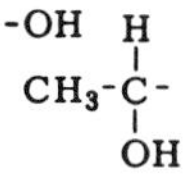

Repeat  44  .

**16**

|29|  $CH_3COOH + NaOH \rightarrow CH_3COONa + H_2O$

We now turn to the preparation of the fatty acids. The higher fatty acids can be isolated from vegetable or animal fats. The lower members of the series must, however, be prepared by chemical synthesis.

You already know a synthesis of carboxylic acids: the oxidation of a primary alcohol gives the corresponding aldehyde first, and then the acid.

Write the equation for the oxidation of ethanol to acetaldehyde and further to the carboxylic acid. The reaction can be carried out with e.g. $KMnO_4$, $HNO_3$, or $O_2$ as oxidants. For the sake of simplicity use an oxygen atom, [O], in your equation.

What is the reaction product called?

- Formic acid                    →    |3|

- Acetic acid                    →    |20|

- Propionic acid                 →    |12|

- I need the formula for acetaldehyde    →    |39|

**16**

|30|    Your answer is correct.

The systematic names of the carboxylic acids – methanoic acid, ethanoic acid, propanoic acid, etc., are not yet widely used.

We have already mentioned the occurrence of formic, acetic, and butyric acids. Where are they found?

a) Formic acid: ...................
b) Acetic acid: ...................
c) Butyric acid: ...................                    →    |38|

|31|    $NaOH + CO \rightarrow HCOONa$

Formic acid can be liberated from the salt by the addition of a strong acid, e.g. dilute $H_2SO_4$. Give the equation for this reaction!    →    |40|

---

**32**    No, CO is used.

See the equation at    31 .

---

**33**    Your answer is correct.

Now select the formulas corresponding to the following names:

maleic acid              chloroacetic acid

●  [maleic acid structure]        [chloroacetic acid structure: H-C-C-C with H, H, Cl, O, OH]        →    79

●  [maleic acid structure]        [structure: H-C-C with H, Cl, O, OH]        →    25

●  [structure with O, OH, C, C, O, OH]        [structure: H-C-C with H, Cl, O, OH]        →    13

---

**34**    Correct. – The more chlorine atoms in the neighborhood of the carboxyl function, the stronger the acid.

Consider now the *hydroxy*carboxylic acids, i.e. carboxylic acids carrying an additional hydroxyl group.

Lactic acid is an example. It is present in sour milk. Lactic acid is derived from propionic acid by adding an OH group at the central carbon atom.

What is the structure?    →    44

---

**35**        [structure: Cl-C-C with H, H, O, OH]

chloroacetic acid

A chlorine atom neighboring the carboxyl group increases the acidity. This effect is more pronounced when two or three chlorine atoms are present.

**16**

Give the structural formulas for
dichloroacetic acid and
trichloroacetic acid.    →    $\boxed{26}$

---

$\boxed{36}$    The first method of preparation of carboxylic acids, viz. oxidation of primary alcohols via the corresponding aldehydes, is generally applicable.

It is used on a large scale in industry, especially for the production of acetic acid from ethanol.

The second method, the oxidation of an aldehyde, is also generally applicable. This, too, is used in the chemical industry, e.g.:

$$\text{CH}_3\text{-C}\overset{\text{O}}{\underset{\text{H}}{\diagup}} + [\text{O}] \longrightarrow \text{CH}_3\text{-C}\overset{\text{O}}{\underset{\text{OH}}{\diagup}}$$

The third method, the reaction between CO and NaOH at elevated temperature and pressure, is a *unique* method for the synthesis of formic acid:

$$\text{NaOH} + \text{CO} \longrightarrow \text{HCOONa}$$

$$2\ \text{HCOONa} + \text{H}_2\text{SO}_4 \longrightarrow 2\ \text{HCOOH} + \text{Na}_2\text{SO}_4$$

**16**   The fourth method, fermentation, is a *unique* method for the production of acetic acid. Ethanol, itself formed by fermentation, is oxidized to acetic acid by microorganisms.

Return to   $\boxed{56}$   .

---

$\boxed{37}$    $\text{CH}_3\text{–COOH} \rightarrow \text{CH}_3\text{COO}^{\ominus} + \text{H}^{\oplus}$

A carboxylic acid reacts with a base to form a salt of the acid and water. Complete the equation for acetic acid:

$$\text{CH}_3\text{COOH} + \text{NaOH} \rightarrow \qquad\qquad\qquad \rightarrow \boxed{29}$$

---

$\boxed{38}$    a) Formic acid: ants and nettles.
b) Acetic acid: vinegar.
c) Butyric acid: rancid butter; perspiration.

Carboxylic acids of longer chain-length, particularly 16–18 carbon atoms, are constituents of fats. Hence, they are known as the *fatty acids*. Stearic acid, $\text{C}_{17}\text{H}_{35}\text{–COOH}$ is an example. (Note: stearic acid has a total of 18 carbon atoms!)

The fats will be dealt with in Program 23. At this point we can distinguish

a) the *lower* fatty acids – aliphatic carboxylic acids up to $C_4$. These are soluble in water;

b) the higher fatty acids – those with longer chains. The longer the chain the greater the insolubility in water.

Give an example of a *lower* and a *higher* fatty acid. Write their formulas and names.    →    46

---

**39**

$$CH_3-C\underset{H}{\overset{O}{\diagup}}\quad \text{acetaldehyde}$$

Return to    29    .

---

**40**    $2\ HCOONa + H_2SO_4 \rightarrow 2\ HCOOH + Na_2SO_4$

Which chemicals are used in the preparation of the sodium salt of formic acid?

● $NaOH + CO_2$       →    32

● $NaOH + CO$        →    48

● $Na_2CO_3 + CO$      →    27

**16**

---

**41**    Ethanol $\xrightarrow{\text{oxidation}}$ .................... $\xrightarrow{\text{oxidation}}$ acetic acid.

Write the name of the intermediate compound.    →    56

---

**42**    Your answer is incomplete. Maleic acid is *unsaturated* since it contains a C=C double bond.

$$\begin{array}{c} \text{H}_{\diagdown}\text{C}-\text{C}\overset{O}{\underset{\displaystyle \text{OH}}{\diagup}} \\ \quad \parallel \quad \diagdown \text{OH} \\ \text{H}^{\diagup}\text{C}-\text{C} \\ \qquad \underset{O}{\parallel} \end{array}$$

However, it is a *dicarboxylic* acid since it contains *two carboxyl groups*.

Thus, the compound is properly described as a ....................... ....................

Write the missing words.    →    50

| 43 |

No, repeat   | 35 |   carefully.

---

| 44 |

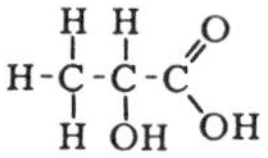

$$\underset{\overset{|}{\underset{H}{}}\;\overset{|}{\underset{OH}{}}}{H-\overset{H}{\underset{}{C}}-\overset{H}{\underset{}{C}}-C}\overset{\nearrow O}{\searrow_{OH}}$$   lactic acid

Lactic acid has an *asymmetric* carbon atom. What does that mean?

The central carbon atom in this compound is said to be *asymmetric* because it carries *four different* atoms or groups. These are:

   –CH$_3$
   –H
   –OH
   –COOH

Which of the following compounds contain asymmetric carbon atoms?

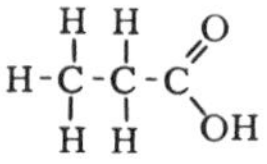

| (a) | (b) | (c) | (d) |
|-----|-----|-----|-----|
| propionic acid | chloroacetic acid | glycerol | 2–chloropropionic acid |

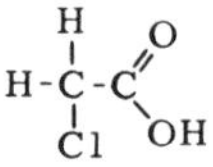

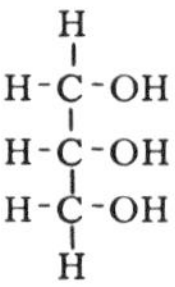

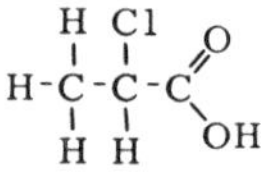

- (a), (c), and (d)    →   | 54 |

- (b) and (d)    →   | 71 |

- (d) only    →   | 62 |

- All four    →   | 18 |

- I don't know    →   | 28 |

---

| 45 |

Your answer is partially correct. However, one can also obtain acetic acid from wine. Continue at   | 48 |   .

16

---

**46**

Lower fatty acids:     formic acid,      HCOOH
     acetic acid,      $CH_3COOH$
     propionic acid,      $C_2H_5COOH$
     butyric acid,      $C_3H_7COOH$

Example of a higher fatty acid:
     stearic acid,      $C_{17}H_{35}COOH$

The carboxylic acids dissociate in aqueous solution in just the same manner as do inorganic acids: a proton and a carboxylate anion are formed, compare Program 6.

● Write the dissociation equation for
  acetic acid.            →    37

● I need further information.      →    22

---

**47**

fumaric acid

**16**

Fumaric acid is a *trans* compound.

Let us now consider the *halogenated* carboxylic acids.

Chloroacetic acid is formed by passing chlorine into glacial acetic acid (i.e. the undiluted acid). A hydrogen atom in the methyl group is replaced by chlorine.

Write the structure of chloroacetic acid.    →    35

---

**48**      Your answer is correct.

Dilute *acetic acid* is obtained from wine – hence the name *vinegar*. This is due to an enzyme-catalyzed air oxidation of the ethanol present in wine (acetic acid fermentation).

There is an intermediate oxidation product formed in this reaction. What is it? Check your answer at   56  , or, if you need help, go instead to   41   .

**49** Yes, maleic acid is a *cis* compound;
fumaric acid is a *trans* compound.

Give the name of this type of isomerism. → 57

---

**50** Maleic acid is an *unsaturated dicarboxylic acid*. Another unsaturated dicarboxylic acid with the same empirical formula also exists. Its name is *fumaric acid*.

Compare the two structural formulas:

maleic acid      fumaric acid

You may well be astonished that these two formulas describe two *different* acids.

We had learned that carbon-carbon bonds are flexible, that is, structural formulas which can be interconverted simply by stretching or rotation about C–C bonds are identical and do *not* represent isomeric molecules. For example, the two formulas

and

do not represent two isomers, but rather *one single compound,* 1,2-dichloroethane.

However, the above is only true for C–C *single* bonds. C=C double bonds, on the contrary, are rigid, and a rotation about the double bond requires high energy (it can take place on strong heating or in the presence of catalysts).

*Maleic acid* has a *cis* double bond and is said to be a *cis* compound because the two carboxyl groups are on the same side of the bond (Latin, *cis* = on this side).

*Fumaric acid* has a *trans* double bond and is said to be a *trans* compound because the two carboxyl groups are on different sides of the bond (Latin, *trans* = on the other side).

Now write the structural formulas for

cis-1,2-dichloroethene and
trans-1,2-dichloroethene    →    58

---

**51**      *Asymmetric carbon atom.*

Let us exercise some names again. Assign the proper names to the three acids below:

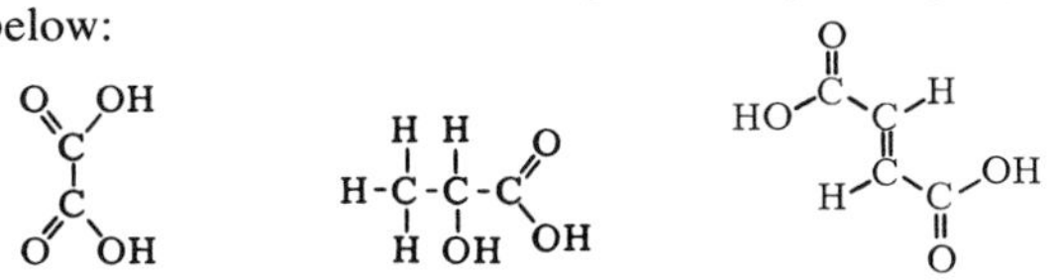

● oxalic          propionic          maleic acid     →      5

● oxalic          propionic          fumaric acid    →     14

● oxalic          lactic             fumaric acid    →     33

● butyric         lactic             fumaric acid    →     80

---

**52**

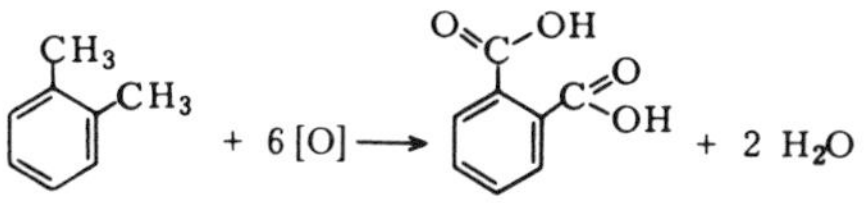

o−xylene                phthalic acid

Now give the structure of terephthalic acid.    →    59

---

**53**      You confused the three xylenes:

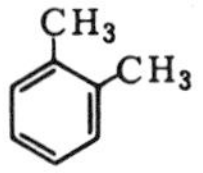

o−xylene 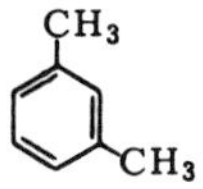               m−xylene 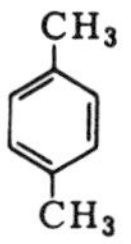               p−xylene

The oxidation of o-xylene gives an *ortho*-dicarboxylic acid.

The oxidation of p-xylene gives a *para*-dicarboxylic acid.

Go back to    73    .

**16**

**54**     No, neither propionic acid nor glycerol contain asymmetric carbon atoms.

The central carbon atom in propionic acid

$$H-\underset{\underset{H}{|}}{\overset{\overset{H}{|}}{C}}-\underset{\underset{H}{|}}{\overset{\overset{H}{|}}{C}}-C\overset{\nearrow O}{\underset{\searrow OH}{}}$$

does *not carry four different substituents,* but rather

     –CH₃
     –H
     –H
     –COOH

i.e. only three different ones.

Similarly, in glycerol,

$$\begin{array}{c} H \\ H-\overset{\overset{|}{}}{C}-OH \\ H-\overset{\overset{|}{}}{C}-OH \\ H-\overset{\overset{|}{}}{C}-OH \\ H \end{array}$$

the central carbon atom carries only three different substituents:

     –H
     –OH
     –CH₂OH
     –CH₂OH

Return to    44    .

---

**55**

benzaldehyde          benzoic acid

Aromatic carboxylic acids can also be obtained by oxidation of *methyl groups* attached to the aromatic ring.

Which methyl substituted aromatic hydrocarbon is required for this oxidative preparation of benzoic acid?

Give both the name and the formula.     →    64

| 56 | Acetaldehyde. |

We have described the following methods of preparation of carboxylic acids:

1. Oxidation of a primary alcohol (via the corresponding aldehyde).
2. Oxidation of an aldehyde.
3. Reaction between CO and NaOH at elevated temperature and pressure.
4. Oxidation of ethanol in wine by microorganisms.

Consider whether these four methods are generally applicable, or whether some of them are specific for the preparation of certain carboxylic acids. Then answer the question:

Which methods can be used for the preparation of acetic acid?

● Methods 1 and 2             →   45

● Methods 1, 2, and 4        →   78

● Methods 1 and 4             →   68

● Methods 3 and 4             →   21

● I need a brief summary of the
   four methods                   →   36

16

| 57 | *cis-trans isomerism.* |

The following mnemonic aid will help you remember that

> maleic acid is cis;
> fumaric acid is trans.

Write down the structural formula for fumaric acid.    →    47

| 58 | H-C-Cl<br>  ‖<br>H-C-Cl | Cl-C-H<br>    ‖<br>H-C-Cl |
|---|---|---|
| | *cis*−1,2−dichloroethene | *trans*−1,2−dichloroethene |

This type of isomerism is called *cis-trans isomerism.*

Is maleic acid a *cis* or a *trans* compound?

● *cis*                                    →   49

● *trans*                                  →   66

● I need further information       →   74

---

| 59 |

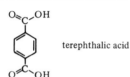

terephthalic acid

16

This terminates our description of individual carboxylic acids. The lower aliphatic acids are liquids (e.g. formic, acetic, propionic, and butyric acids); the higher ones are solids (e.g. stearic acid).

The aromatic carboxylic acids and the aliphatic dicarboxylic acids are colorless solids (e.g. oxalic acid, maleic and fumaric acids, benzoic, phthalic, and terephthalic acids).

Can you write the structures for all these compounds? If not, you should repeat this program!

You have learned that carboxylic acids form *salts* on reaction with strong bases. We will briefly describe the naming of these salts.

Give the equations for the reaction of the following acids with sodium hydroxide:

     a) formic acid
     b) acetic acid
     c) benzoic acid     →   70

| 60 |  *Oxalic acid*

It is necessary to learn the names and formulas for some further carboxylic acids. We limit the presentation to the most important ones, however.

| Previous programs described: | Analogous carboxylic acids are: |
| --- | --- |
| Unsaturated compounds | Unsaturated carboxyle acids (incorporating C=C bonds) |
| Halogen compounds | Halogenated carboxylic acids |
| Alcohols | Hydroxycarboxylic acids (having both OH and COOH groups). |

Let us now describe these three classes of carboxylic acids, starting with the *unsaturated* ones.

The following is an incomplete formula for an unsaturated dicarboxylic acid,

$$C-C\overset{\nearrow O}{\underset{\searrow O}{}}$$
$$\overset{\parallel}{C}-C\overset{\nearrow O}{\underset{\searrow O}{}}$$

in which the hydrogen atoms are missing. Complete the formula and check it at  | 67 | .

**16**

| 61 |     Yes, the oxidation of *o*-xylene gives an *ortho*-dicarboxylic acid, namely phthalic acid.
The oxidation of *p*-xylene gives a *para*-dicarboxylic acid, namely terephthalic acid.

Write out the equation for the oxidation of *o*-xylene.   →   | 52 |

| 62 |     Your answer is correct.

Compounds possessing asymmetric carbon atoms exist in two forms which are mirror images of each other. The relationship between them is the same as that between your left and right hands.

The two mirror image forms of lactic acid are shown schematically below. The asymmetric carbon atom is in the center, and the four different groups form the corners of the tetrahedron:

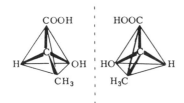

The two forms can be distinguished using *polarized light.* One form rotates the plane of polarized light to the right; the other to the left.

Thus, there are *two different* lactic acids:

        a) one which rotates the plane of polarized light to the *right,*
        b) one which rotates the plane of polarizid light to the *left.*

These two compounds are said to be *optical isomers* since they have the same empirical formula but influence polarized light in opposite ways.

Which condition must be met in order that a compound exists in two optically isomeric forms?

The compound must possess an ............... ...............

**16**    Write down the missing words.   →   51

---

63    Benzaldehyde is easily oxidized by air to benzoic acid:

$$\underset{\displaystyle\bigcirc}{\overset{\displaystyle O\diagdown_{C}\diagup^{H}}{}} \quad + [O] \longrightarrow$$

Complete the equation.   →   55

---

64    *Toluene,*   $\underset{\displaystyle\bigcirc}{CH_3}$

Give the equation for the oxidation of toluene to benzoic acid.   →   73

---

**65**

Decarboxylation.

What type of organic compound is formed on decarboxylation?  →   81

---

**66**

No, *maleic acid* is a *cis* compound.

Study   74   again.

---

**67**

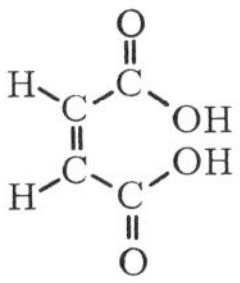

maleic acid

Did you have the correct formula?

What kind of acid is maleic acid?

- An unsaturated acid      →    42

- A dicarboxylic acid      →    76

- An unsaturated dicarboxylic acid      →    50

16

---

**68**

Your answer is partially correct. However, you forgot the most important acetic acid synthesis.
Go to   36   and study the first method in particular.

---

**69**

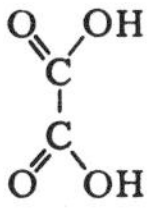

Give the name of this dicarboxylic acid.    →    60

| 70 | | | | |
|---|---|---|---|---|
| a) HCOOH | + NaOH | → | HCOONa | + $H_2O$ |
| b) $CH_3COOH$ | + NaOH | → | $CH_3COONa$ | + $H_2O$ |
| c) $C_6H_5COOH$ | + NaOH | → | $C_6H_5COONa$ | + $H_2O$ |

The sodium salt of formic acid is sodium *formate*.
The sodium salt of acetic acid is sodium *acetate*.
The sodium salt of benzoic acid is sodium *benzoate*.

Thus, the names of the carboxyl*ate* salts end with *ate* in the same way as e.g. sulf*ate*, or nitr*ate*.

Give the names and formulas for the following salts:

$C_6H_5COOK$, $(HCOO)_2Ca$, and $CH_3COOK$    →    77

---

71

No, chloroacetic acid does not have an asymmetric carbon atom!

The carbon atom to the left carries the following groups:

    –H
    –H
    –Cl
    –COOH,

that is, there are *not* four different groups. An asymmetric carbon atom is present *only* when it carries *four different* groups or atoms.

Return to    44 .

---

72    R–COOH → R–H + $CO_2$

In this reaction the carboxyl group is removed and replaced by a hydrogen atom.

What is this reaction called?    →    65

|73|

CH₃
[toluene structure] + 3 [O] ⟶ [benzoic acid structure] + H₂O

toluene          benzoic acid

The oxidation of the aliphatic hydrocarbons (alkanes) does not occur so readily. Furthermore, the oxidation of alkanes leads to mixtures of carboxylic acids which are difficult to separate.

Two important aromatic dicarboxylic acids are

phthalic acid                          terephthalic acid
(pronounced: "ftalic")                 (pronounced: "tereftalic")

[phthalic acid structure]              [terephthalic acid structure]

Both phthalic and terephthalic acid are used in the manufacture of synthetic polymers. The two compounds are obtained by air oxidation of the corresponding xylenes.

**16**

Which xylenes are used?

For the oxidation to phthalic acid:     For the oxidation to terephthalic acid:

● 　　*o*-xylene          *m*-xylene   →   |83|

● 　　*m*-xylene          *p*-xylene   →   |53|

● 　　*p*-xylene          *o*-xylene   →   |19|

● 　　*o*-xylene          *p*-xylene   →   |61|

|74|  Maleic acid is a *cis* compound. The two carboxyl groups are *on the same side* of the molecule.

[maleic acid structure]

Fumaric acid is the corresponding *trans* compound. Here, the two carboxyl groups are on *opposite sides* of the molecule:

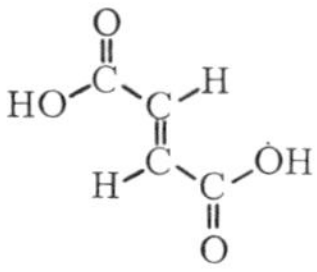

fumaric acid

What is the term used for this type of isomerism?

Write the name.   →   57

---

**75**   *Soaps are sodium or potassium salts of the higher fatty acids.*

Why do soaps not function well in hard water?

Check your written answer at   82   .

---

**76**   Your answer is incomplete. Naturally, maleic acid is a *dicarboxylic* acid,

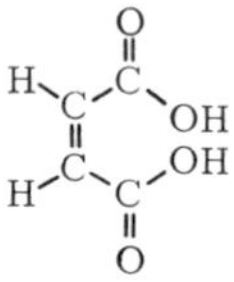

but is also contains a C=C double bond and is, therefore, *unsaturated*.

Thus, maleic acid is an .............. ...............

Write down the completed sentence.   →   50

---

**77**   $C_6H_5COOK$   potassium benzoate
$(HCOO)_2Ca$   calcium formate
$CH_3COOK$   potassium acetate

The salts of propionic acid are called *propionates*.
Those of stearic acid are *stearates*.

Write down and complete the following table:

| Name of acid | Name of salt |
| --- | --- |
| formic acid | .............. |
| acetic acid | .............. |
| .............. | propionate |
| .............. | stearate |
| benzoic acid | ..............  → 84 |

---

**78**     Your answer is correct.

Up until now we have limited the presentation to carboxylic acids carrying only *one* carboxyl group.

A molecule may, however, carry two, three, or more carboxyl groups.

Those possessing two carboxyl groups are called *dicarboxylic* acids. The simplest of these is *oxalic acid.* This compound is composed of two directly connected carboxyl groups.

Give the structural formula.    →    69

**16**

---

**79**     Your formula for maleic acid is correct, but the one for chloroacetic acid is wrong.

Repeat from    35    .

---

**80**     You gave the correct names for lactic acid and fumaric acid only.

Butyric acid is *not* a dicarboxylic acid. It has the formula $CH_3CH_2CH_2COOH$.

Return to    51    .

---

**81**     A hydrocarbon, R–H

This is the end of Program 16.

---

**82**     The answer is that insoluble calcium salts of the fatty acids are formed in hard (i.e. calcium carbonate rich) water.

Further reactions of the carboxylic acids will be treated in the next two programs. At this point we need consider the *decarboxylation* only.

The thermal *elimination of $CO_2$* from a carboxylic acid is called a decarboxylation.

Complete the decarboxylation equation

$$R-COOH \rightarrow R-H + \text{..............} \quad \rightarrow \quad \boxed{72}$$

---

| 83 |
|---|

You confused the three xylenes:

      $o-$xylene             $m-$xylene             $p-$xylene

The oxidation of *o*-xylene gives an *ortho*-dicarboxylic acid.

The oxidation of *p*-xylene gives a *para*-dicarboxylic acid.

Go back to   $\boxed{73}$   .

---

**16**

| 84 |
|---|

| Name of acid | Name of salt |
|---|---|
| formic acid | formate |
| acetic acid | acetate |
| propionic acid | propionate |
| stearic acid | stearate |
| benzoic acid | benzoate |

The sodium and potassium salts of the higher fatty acids are *soaps* (e.g. sodium and potassium stearates). They dissolve in water with the formation of foam.

In contrast, the *calcium* salts of the higher fatty acids are *insoluble* in water. Therefore, the use of soaps in hard water results in the precipitation of calcium salts of the fatty acids (hard water is rich in calcium carbonate) and the soaps lose their power as detergents.

What are soaps?

Describe the *chemical* nature of soaps (not their use as detergents) by completing the sentence:

Soaps are ..............   $\rightarrow$   $\boxed{75}$

# Program 17

## Carboxylic Acid Derivatives (I)

---

**1**     In Program 16 we described the carboxylic acids and some substituted derivatives. The latter were derived by altering the *hydrocarbon residues:*

- Replacement of H by a halogen atom gave the halogenated carboxylic acids (e.g. chloroacetic acid).
- Replacement of H by a hydroxyl group gave the hydroxycarboxylic acids (e.g. lactic acid).

We will now be concerned with derivatives formed by alteration of the functional group – *the carboxyl group.*

The most important are

     a) esters,
     b) acid chlorides,
     c) acid anhydrides,
     d) amides, and
     e) nitriles

Let us commence with the *esters.*

*Esters* are compounds of the type

$$R-C{\overset{O}{\underset{O-R'}{\diagdown}}}$$

in which the acid hydrogen atom has been replaced by a hydrocarbon residue, R'.

Pick out the ester among the following compounds:

●     $CH_3-CH_2-COONa$        →   $\boxed{20}$

●     $CH_3-\underset{OH}{CH}-C{\overset{O}{\underset{OH}{\diagdown}}}$     →   $\boxed{27}$

●     $CH_3-C{\overset{O}{\underset{O-C_2H_5}{\diagdown}}}$     →   $\boxed{13}$

●     $CH_3-\underset{Cl}{CH}-C{\overset{O}{\underset{OH}{\diagdown}}}$     →   $\boxed{8}$

---

**2** | Your answer to (a) is incorrect; the compound is *not* an acetic acid
ester.

Repeat　19 .

---

**3** | Formulate the expression for the law of mass action for the reaction

acid + alcohol ⇌ ester + water

(do not use chemical formulas but the general words, acid, alcohol, etc.).

$$\frac{[\text{ester}] \cdot \dotfill}{\dotfill \cdot \dotfill} =$$

→　10

---

**4** |

a) The ester　　$CH_3 - C\underset{O-C_2H_5}{\overset{O}{\diagdown}}$

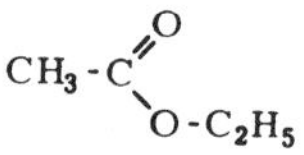

is a derivative of *acetic acid,*　$CH_3 - C\underset{OH}{\overset{O}{\diagdown}}$

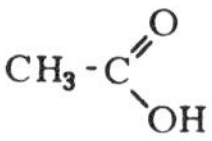

b) The hydrocarbon residue $-C_2H_5$ is the *ethyl* group.

The name of the ester is *acetic acid ethyl ester.*

Thus, the *names* of esters may be formed from

　1) the name of the parent acid plus
　2) the name of the alkyl group, and
　3) adding the word "ester".

Give the names of the following esters:

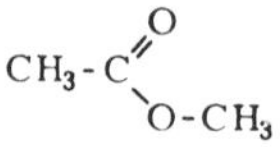

$CH_3 - C\underset{O-CH_3}{\overset{O}{\diagdown}}$　　　$H - C\underset{O-C_2H_5}{\overset{O}{\diagdown}}$

(a) 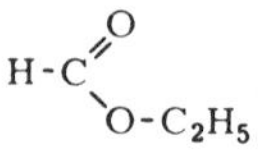　　　　　　　(b) 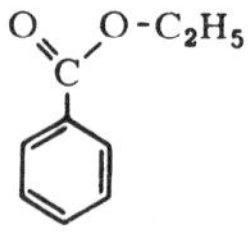　　　　　　　(c)　　　→　12

**5** Fats and aromas.

Esters of the lower acids and alcohols are liquids. The esters of longer chain-length alcohols are solids. All esters have little or no solubility in water. They are good solvents for fats, lacquers, and many other organic compounds.

Which of the following statements is true for acetic acid ethyl ester?

● It is solid and water insoluble          →     37

● It is liquid and water insoluble         →     15

● It is liquid and water soluble           →     47

---

**6** First write the equation for the hydrolysis of acetic acid ethyl ester by *water*.

Then replace water by sodium hydroxide.     →     43

---

**7** You are confusing acetyl chloride and benzoyl chloride. Repeat from 64 .

---

**8** The compound $CH_3-CH-C\overset{O}{\underset{OH}{}}$ is 2-chloropropionic acid.

$\underset{Cl}{}$

It is *not an ester* but a derivative of propionic acid in which a hydrogen atom *in the alkyl chain* has been replaced by chlorine.

Esters are compounds formed by derivatization of the *carboxyl group*.

Return to 1 .

---

**9** The following compound

$$CH_3-O \diagdown$$
$$\qquad SO_2$$
$$CH_3-O \diagup$$

is the dimethyl ester of sulfuric acid, or *dimethyl sulfate*.

Now write the structure of diethyl sulfate.     →     17

---

**17**

---

**10**  $$\frac{[\text{Ester}] \cdot [\text{Water}]}{[\text{Acid}] \cdot [\text{Alcohol}]} = K$$

The *equilibrium* concentrations in the esterification of one mole of acetic acid with one mole of ethanol are:

[Acid]  = 1/3 mol; [Alcohol]= 1/3 mol
[Ester]  = 2/3 mol; [H₂O[   = 2/3 mol

Inserting these into the law of mass action expression, we obtain:

$$\frac{\frac{2}{3} \cdot \frac{2}{3}}{\frac{1}{3} \cdot \frac{1}{3}} = K$$

Calculate $K$ and compare your result at [34] .

---

**11**    Your answer for (b) is wrong; there are *two* residues derived from propane:

1)  the propyl residue, $CH_3—CH_2—CH_2—$
2)  the isopropyl residue, $CH_3—\overset{|}{CH}—CH_3$

Therefore, both propyl and isopropyl *esters* exist.

You answer for (c) is also wrong; the compound is neither a phenyl ester nor a derivative of propionic acid.

Return to  [19] .

---

**12**

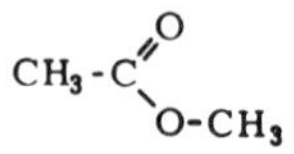

|  |  |  |
|---|---|---|
| acetic acid methyl ester | formic acid ethyl ester | benzoic acid ethyl ester |

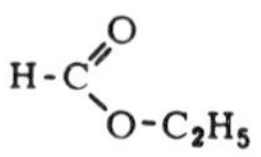

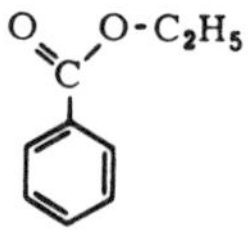

The names of esters can also be derived in the same manner as for salts:

formic acid ethyl ester is ethyl formate;
acetic acid ethyl ester is ethyl acetate;
benzoic acid ethyl ester is ethyl benzoate.

Remember that the salts of formic acid are called formates;
those of acetic acid are called acetates;
those of benzoic acid are called benzoates.

Complete the following list:

|  |  |  |
|---|---|---|
| acetic acid butyl ester | = | butyl acetate |
| formic acid methyl ester | = | .................... |
| .................... | = | propyl benzoate |
| .................... | = | ethyl acetate |
| benzoic acid methyl ester | = | .................... |
| .................... | = | phenyl acetate   →   $\boxed{21}$ |

---

$\boxed{13}$      Correct; the compound is an ester. Compare the structures of acids and esters:

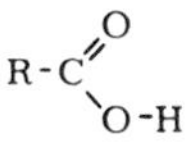

$$R-C\underset{O-H}{\overset{O}{\big<}} \qquad R-C\underset{O-R'}{\overset{O}{\big<}}$$

carboxylic acid 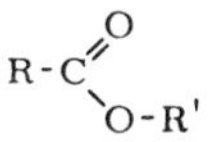        ester

$\boxed{17}$

An ester is a *derivative* of a carboxylic acid. It is formed by replacing the acidic hydrogen atom by a hydrocarbon residue, R'. R' will normally be an alkyl or aryl group.

Examine the formula

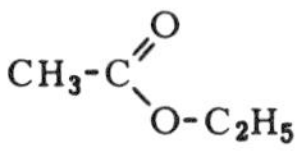

$$CH_3-C\underset{O-C_2H_5}{\overset{O}{\big<}}$$

and give

a) the name and structure of the acid from which this ester is derived;
b) the name and formula for the hydrocarbon residue which has replaced the
   carboxylic proton.   →   $\boxed{4}$

---

$\boxed{14}$      1) Removal of one or both of the reaction products from the equili-
             brium.
             2) Increasing the concentration of *one* of the starting materials.

This terminates our presentation of the synthesis of esters. We will now briefly describe their occurrence, properties, and applications.

Esters occur as *fats* in both animals and plants. *Fats* are the esters of the higher fatty acids (e.g. stearic acid) with *glycerol*. These will be described in more detail in Program 24.

Many natural *aromas* are due to esters of the lower carboxylic acids.

Thus, which natural products are esters?     →     | 5 |

---

| 15 |          Your answer is correct.

Esters are used as aromas, solvents, and softeners.

What are softeners?

*Softeners* are additives used to make hard polymers soft (e.g. polyvinyl chloride, PVC). This allows moulding, preparation of films, etc. Synthetic polymers will be described in Program 22.

Complete the sentence:

Hard polymers are softened by the addition of ...................     →     | 32 |

---

| 16 |          Your answer is wrong.

A catalyst can accelerate a chemical reaction, i.e. shorten the time needed to obtain *equilibrium*. Neither the amount nor the nature of the catalyst can change the *position* of the equilibrium.

Return to     | 35 |   .

---

| 17 |

$$C_2H_5-O \diagdown SO_2$$
$$C_2H_5-O \diagup$$

diethyl sulfate
(diethyl ester of sulfuric acid)

**Both dimethyl sulfate and diethyl sulfate are highly poisonous!**

The esters of nitric acid can be derived in a similar way. Construct the formula for ethyl nitrate.     →     | 25 |

**18**     Silver *formate cannot* give rise to an ester of *acetic* acid!

Complete the following:

the salts of formic acid are called ......................
the salts of acetic acid are called .........................
the salts of propionic acid are called ...................
the salts of benzoic acid are called ......................     →     26

---

**19**

benzoic acid methyl ester          acetic acid phenyl ester
or methyl benzoate                 or phenyl acetate

Assign the proper names to the following esters:

$$H-C\overset{O}{\underset{O-CH_2-CH_2-CH_3}{\diagdown}}$$

$$CH_3-C\overset{O}{\underset{O-CH\diagdown CH_3}{\diagup}}\overset{CH_3}{}$$

$$\overset{O}{\underset{C}{\diagdown}}\overset{O-C_2H_5}{\diagup}$$

(a)               (b)               (c)

**17**

● Propyl acetate      isopropyl acetate      ethyl benzoate      →     2

● Propyl formate      propyl acetate         phenyl propionate  →     11

● Propyl formate      isopropyl acetate      ethyl benzoate      →     28

● Propyl formate      propyl acetate         ethyl benzoate      →     36

---

**20**     $CH_3-CH_2-COONa$ is the sodium *salt* of propionic acid.

Go back to  1  and examine the general formula for an ester. Note that R'
designates an *organic* residue.

**21**   Compare:

|                          |   |                 |
|--------------------------|---|-----------------|
| formic acid methyl ester | = | methyl formate  |
| benzoic acid propyl ester| = | propyl benzoate |
| acetic acid ethyl ester  | = | ethyl acetate   |
| benzoic acid methyl ester| = | methyl benzoate |
| acetic acid phenyl ester | = | phenyl acetate  |

Give structures for the methyl ester of formic acid
and for ethyl acetate.  →   29

---

**22**   The concentration of the ester does *not* change once *equilibrium* has
been established (hence the name "equilibrium"). The number of molecules of
ester being *formed* (by esterification) equals the number *disappearing* by hydrolysis.

Suppose a certain esterification reaction is at equilibrium. Will the concentration
of the carboxylic acid change by continued heating of the mixture at reflux?

● Yes        →   39

● No         →   52

---

**17**

**23**   Your answer is correct.

If either the ester or the water formed (or both) are removed by distillation
during the reaction, the reverse process – hydrolysis – cannot take place. In this
case the equilibrium is continually shifted toward the ester, and the reaction
will go to completion.

The distillation of water or an ester from the reaction mixture is possible only if
the other reaction partners have *higher* boiling points. For example, the method
case the equilibrium is continuously shifted toward the ester, and the reaction
will go to completion.

|              |        |
|--------------|--------|
| ethanol      | 78 °C  |
| acetic acid  | 118 °C |
| water        | 100 °C |
| ethyl acetate| 77 °C  |

A second method of shifting the equilibrium toward the ester is to increase the
concentration of *one* of the *starting* materials.

For example, when one mole of acetic acid is treated with two moles of ethanol (i.e., an excess of the alcohol) the yield of ethyl acetate increases to 85%.

The increase in yield can be calculated with the aid of the law of mass action (details in Program 6) as described under $\boxed{3}$ .

What are the two methods used to shift the esterification equilibrium to the side of the ester?   $\rightarrow$   $\boxed{14}$

---

$\boxed{24}$   You forgot that two different residues are derived from propane:

    1)  the propyl residue, $CH_3—CH_2—CH_2—$
    2)  the isopropyl residue, $CH_3—\underset{|}{CH}—CH_3$

Therefore, there are also two kinds of esters:

    propyl and isopropyl esters.

Will a *propyl* or an *isopropyl* ester be formed in the reaction of 1-bromopropane with a silver carboxylate?

Return to   $\boxed{33}$  .

**17**

---

$\boxed{25}$   $CH_3–CH_2–O–NO_2$
    ethyl nitrate
    (ethyl ester of nitric acid)

We now turn to the *preparation* of esters. A general method is the treatment of a silver carboxylate (i.e., silver salt of a carboxylic acid) with an *alkyl halide*. The ester and silver halide are formed.

Complete the following equation and name all the compounds:

    $CH_3–COOAg + I–C_2H_5 \rightarrow$       +

(Note: $I–C_2H_5$ is the same as $C_2H_5–I$.)   $\rightarrow$   $\boxed{33}$

---

**26**      Compare:

The salts of formic acid are called *formates*;
The salts of acetic acid are called *acetates*;
The salts of propionic acid are called *propionates*;
The salts of benzoic acid are called *bezoates*.

Return to   33  .

---

**27**

The compound   $CH_3-\underset{\underset{\textstyle OH}{|}}{CH}-C\overset{\textstyle O}{\underset{\textstyle OH}{\diagup}}$   is *lactic acid*.

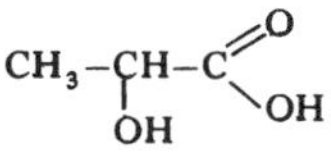

It is *not* an ester.

Lactic acid is derived from propionic acid by replacing an H-atom in the aliphatic chain by OH. Esters, on the contrary, are formed by derivatization of

the *carboxyl group*.

Return to   1  .

---

**28**      Your answer is correct.

Not only the carboxylic acids form esters. Esters of the inorganic acids are also known. For example, *diethyl sulfate* is an ester of *sulfuric acid*.

● Give the structural formula for
  diethyl sulfate.               →   17

● I need further information.      →   9

---

**29**

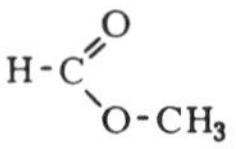

$H-C\overset{\textstyle O}{\underset{\textstyle O-CH_3}{\diagup}}$               $CH_3-C\overset{\textstyle O}{\underset{\textstyle O-C_2H_5}{\diagup}}$

formic acid methyl ester          acetic acid ethyl ester
or methyl formate                 or ethyl acetate

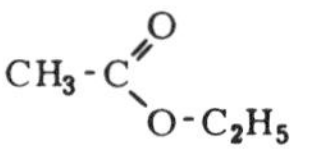

Compare these with your own formulas and correct any mistakes.

Now write the structures of methyl benzoate and acetic acid phenyl ester. Give *two* names for each compound.   →   19

| 30 |

The reaction started with **1** mol of acetic acid. When equilibrium was reached there was still 1/3 mol of acetic acid left.

How many moles of acetic acid had reacted?     →     | 38 |

---

| 31 |

Your answer is wrong.

The concentration of the ester *does not change* once equilibrium is attained (hence the name "equilibrium"). The number of molecules of ester being *formed* (by esterification) equals the number *disappearing* by hydrolysis.

Suppose a certain esterification reaction is at equilibrium. Will the concentration of the carboxylic acid change by continued heating of the mixture at reflux?

● Yes     →     | 39 |

● No     →     | 52 |

---

| 32 |

Hard polymers are softened by the addition of *softeners*.

The dioctyl ester of phthalic acid is a frequently used softener. Give the structural formula for this compound and check your answer at | 42 |    or, if you need help, go instead to | 73 |    .

**17**

---

| 33 |

$$CH_3\text{-}COOAg + I\text{-}C_2H_5 \longrightarrow CH_3\text{-}C\overset{O}{\underset{O\text{-}C_2H_5}{}} \quad + \quad AgI$$

silver acetate        ethyl iodide        ethyl acetate                silver iodide

The general equation is

$$R\text{-}COOAg + Hal\text{-}R' \longrightarrow R\text{-}C\overset{O}{\underset{O\text{-}R'}{}} + AgHal$$

where R and R' can be either different or identical.

For purely didactic purposes it is sometimes useful to "frame" the atoms or groups reacting with one another in a chemical reaction:

$$CH_3\text{-}COO^{\ominus} \boxed{Ag^{\oplus} + Hal\text{-}}R' \longrightarrow R\text{-}C\overset{O}{\underset{O\text{-}R'}{}} + AgHal$$

Which ester is formed by the treatment of 1-bromopropane with silver formate?

- Propyl acetate → ☐18☐

- Propyl formate → ☐41☐

- Isopropyl formate → ☐49☐

- I need further information → ☐24☐

---

☐34☐ The equilibrium constant, $K$, equals 4;

$$K = 4.$$

This value applies to the following equilibrium *only:*

acetic acid + ethanol $\rightleftarrows$ ethyl acetate + water

If other acids or alcohols are used, other values of $K$ will apply.

Knowing the value of $K$, we can determine the concentration of ester formed when *one* mole of acetic acid reacts with *two* moles of ethanol (i.e., 100% excess of ethanol).

If the concentration of ester equals x, then [water] = x also.

Hence, x mol of acid and x mol of alcohol have been used.

Acid:  starting amount:  1 mol
used:  x mol
present at equilibrium:  1–x mol

Alcohol:  starting amount:  2 mol
used:  x mol
present at equilibrium:  2–x mol

Hence:

$$\frac{[\text{ester}] \cdot [\text{water}]}{[\text{acid}] \cdot [\text{alcohol}]} = \frac{x \cdot x}{(1-x) \cdot (2-x)} = \frac{x^2}{2-3x + x^2} = 4$$

The solutions of this equation are

$$x_1 = 0.85$$
$$x_2 = 3.16$$

The second solution is chemically meaningless, since 3.16 mol of ester cannot be formed from one mole of acid. Therefore, the yield of ethyl acetate is 0.85 mol = 85%.

Go back to   23   .

---

**35**     Your answer is correct. 2/3 of a mole of acetic acid (and ethanol) is *consumed,* and 2/3 of a mole of ethyl acetate (and water) is *formed.*

$$CH_3COOH + C_2H_5OH \rightleftharpoons CH_3COOC_2H_5 + H_2O$$

At equilibrium:
          1/3             1/3           2/3                2/3 mol

Thus, the yield of the ester is only 2/3 of a mole = 66%. It would be desirable to displace the equilibrium toward the right, so that a yield higher than 66% is obtained.

One way of doing this is by preventing the reverse reaction, i.e. hydrolysis, from taking place.

How can this be done?

● By using a catalyst                                    →   16

● By removing one of the reaction products
   from the reaction mixture                            →   23

● By adding water                                       →   46

**17**

---

**36**     You forgot that two different residues can be derived from propane:

  1) the propyl residue, $CH_3$—$CH_2$—$CH_2$—
  2) the isopropyl residue, $CH_3$—$CH$—$CH_3$
                                          |
Therefore, two kinds of esters are also possible:

     propyl and isopropyl esters.

Return to   19   .

---

**37**      Acetic acid ethyl ester is formed from an acid and an alcohol which are both short chain molecules. Therefore, the ester is *liquid*.

Return to  [5] .

---

**38**      2/3 of a mole of acetic acid has reacted.

How much ester has now been formed?

Return to  [44] .

---

**39**      No; at equilibrium the amount of ester formed equals the amount disappearing by hydrolysis. Hence, the concentration of the acid does *not* change.

Repeat in detail from  [45] .

---

**40**

$$HCOOAg + CH_3\text{-}CH\text{-}CH_3 \longrightarrow H\text{-}C \overset{O}{\diagup} \underset{O\text{-}CH}{\diagdown} CH_3 + AgBr$$
$$\underset{Br}{|} \qquad\qquad \underset{CH_3}{\diagup}$$

isopropyl formate

Now give the equation for the reaction between silver formate and 1-bromo-propane. Then write the name of the ester formed.    →    [54]

---

**41**      Your answer is correct.

Now write the equation for the reaction between 1-bromopropane and silver formate (using structural formulas) and the name of the ester formed.    →    [54]

---

**42**

$$\text{C}\overset{O}{\overset{\|}{-}}\text{OC}_8\text{H}_{17}$$

dioctyl phthalate

$$\text{C}\underset{O}{\overset{\|}{-}}\text{OC}_8\text{H}_{17}$$

One of the most important reactions of esters is their cleavage back to the carboxylic acid and alcohol. As we already know, this is a hydrolysis reaction. This particular kind of hydrolysis is also referred to as a *saponification* (= soap formation, which is true for the alkaline hydrolysis of the higher fatty acid esters).

Ester + water $\rightleftharpoons$ acid + alcohol

The hydrolysis is normally carried out by boiling with water using a small amount of sulfuric acid as catalyst.

When one mole of ethyl acetate is boiled with one mole of water, only 1/3 of a mole of the ester is hydrolyzed. In other words, 1/3 of a mole of acetic acid and 1/3 of a mole of ethanol are formed. This is the situation at *equilibrium*.

Thus, the hydrolysis equilibrium is identical with the esterification equilibrium.

How can one achieve the complete hydrolysis of the ester?

1) Removal of .................... from the reaction mixture.
2) Increasing the concentration of ...................

Complete the two sentences.   →   $\boxed{50}$

---

$\boxed{43}$

$$CH_3 - C\!\!\overset{\displaystyle O}{\underset{\displaystyle OC_2H_5}{\diagdown}} \quad + \; NaOH \; \longrightarrow \; CH_3\text{-}COO^{\ominus}Na^{\oplus} \; + \; C_2H_5OH$$

ethyl acetate        sodium              sodium acetate        ethanol
                     hydroxide

The salt formation removes acetic acid from the equilibrium

$\boxed{17}$

ester + water $\rightleftharpoons$ acid + alcohol.

Therefore, the alkaline hydrolysis is *complete*.

This terminates our discussion of the esters. A second important group of carboxylic acid derivatives is the

   *acid chlorides.*

Acid chlorides are derived from the acids by replacing the hydroxyl group by Cl.

Give the structure of the acid chloride formed from acetic acid.   →   $\boxed{64}$

---

| 44 |

Concentrated sulfuric acid catalyzes the hydrolysis, i.e., it accelerates the attainment of equilibrium.

Let us now consider the esterification reaction *quantitatively*.

We start with a mixture of 60 g (1 mol) of acetic acid, 46 g (1 mol) of ethanol, and ca. 1 g concentrated $H_2SO_4$ as catalyst. The mixture is refluxed. Small samples (200 mg) are removed from time to time, and the concentration of acetic acid is determined by titration with an 0.1 N solution of sodium hydroxide. (Of course, the sulfuric acid is also titrated, but the amount is so small that it can be neglected. Furthermore, it is possible to calculate the amount of $H_2SO_4$ present in the 200 mg sample.)

The experiment will show that at the beginning of the reaction the concentration of acetic acid decreases rapidly. However, after prolonged boiling, no further change will take place. The reaction has reached equilibrium. At this point, 1/3 of a mole of acetic acid is present (33% of the original amount). The following graph gives the results of our titrations:

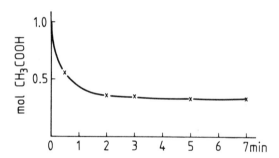

At the beginning of the reaction (0 min) there was one mole of acetic acid present. Already during the first minute the amount decreases drastically (to ca. 0.4 mol). Equilibrium is attained after ca. 5 min. After this time no concentration change is observed.

How much *ethyl acetate* is present at *equilibrium*?

● 1 mol          →     | 56 |

● 2/3 mol          →     | 35 |

● 1/3 mol          →     | 63 |

● I don't know          →     | 30 |

**45**     (a) Esterification; (b) hydrolysis

When a carboxylic acid is boiled with an alcohol (together with a catalytic amount of sulfuric acid), initially an *esterification* will take place. As the concentrations of ester and water increase, more and more of the ester will be hydrolyzed back to the acid and alcohol. After a certain time, the amount of ester *formed* will exactly equal the amount *disappearing* by hydrolysis.

The reaction is at *equilibrium*.

Having attained equilibrium, can we change the concentration of the ester by prolonged refluxing?

● Yes, the concentration increases          →     59

● No          →     52

● Yes, the concentration decreases          →     31

● I don't know          →     22

**46**     Addition of water has the opposite effect: the equilibrium

acid + alcohol ⇄ ester + water

is displaced toward the left; i.e. *hydrolysis* occurs.

The problem was, however, to find a method favoring the *esterification*.

Return to     35  .

**47**     No, esters are generally insoluble or only slightly soluble in water.

Return to     5  .

**48**
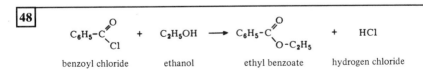

$$C_6H_5-C\underset{Cl}{\overset{O}{\diagup}} \quad + \quad C_2H_5OH \quad \longrightarrow \quad C_6H_5-C\underset{O-C_2H_5}{\overset{O}{\diagup}} \quad + \quad HCl$$

benzoyl chloride          ethanol          ethyl benzoate          hydrogen chloride

Thus, the reaction between an acid chloride and an alcohol gives an ..................     →     58

---

| **49** | You are wrong.

Isopropyl formate is formed from 2-bromopropane and silver formate.

Write out this equation using structural formulas. → 40

---

| **50** | 1) Removal of a *reaction product* (acetic acid or ethanol) from the reaction mixture.

   2) Increasing the concentration of *one of the starting materials*. The simplest is to add an excess of *water*.

A hydrolysis can easily be carried to completion by the addition of a base. In this case, the acid formed is removed from the equilibrium in the form of a salt.

Write the equation for the hydrolysis of ethyl acetate with sodium hydroxide. Give names of all the compounds and check your answer at 43 or, if you need help, go instead to 6 .

---

| **51** | No, you confused

   and

benzyl          phenyl

Be sure you understand the difference between these two residues. → 58

---

| **52** | Your answer is correct.

At equilibrium there is no concentration change. The catalyst (sulfuric acid or another strong acid) accelerates the *attainment* of equilibrium. It does not influence the *position* of the equilibrium.

Write your answer to the following question:

Why is a small amount of concentrated $H_2SO_4$ used in esterification reactions? → 44

---

**53**

Complete the equation:

$$CH_3-C \overset{O}{\underset{OH}{\big\langle}} \ + \ C_2H_5-OH \ \longrightarrow \ \dots\dots\dots \ + \ H_2O$$

acetic acid     . . . . . . . .     ethyl acetate

                or . . . . . . . .            →   **71**

---

**54**

$$HCOOAg \ + \ Br-CH_2-CH_2-CH_3 \ \longrightarrow \ H-C \overset{O}{\underset{O-CH_2-CH_2-CH_3}{\big\langle}} \ + \ AgBr$$

propyl formate
(formic acid propyl ester)

This reaction is of interest only on a laboratory scale due to the high cost of silver.

Other methods are used in *industry*. The most important is the esterification of a *carboxylic acid* with an *alcohol*.

(The term *esterification* implies the formation of an ester from an acid.)

Complete the general equation below, giving names of all the compounds:

$$R-C \overset{O}{\underset{OH}{\big\langle}} \ + \ R'-OH \ \longrightarrow \ \dots\dots\dots \ + \ \dots\dots$$

acid        alcohol        ester      . . . . . . . .

Check your answer at   **62**   or, if you need help, go instead to   **70** .

---

**55**

Your answer is correct.

Since esterification and hydrolysis are opposite and reversible processes, the equation should be written with a double arrow:

$$R-C \overset{O}{\underset{OH}{\big\langle}} \ + \ R'OH \ \rightleftharpoons \ R-C \overset{O}{\underset{OR'}{\big\langle}} \ + \ H_2O$$

Give the name of

     (a) the reaction proceeding toward the right: .....................
     (b) the reaction proceeding toward the left: .....................   →   **45**

| 56 |
|---|

　　One mole of ester cannot have formed since this would require the complete reaction between one mole of acid and one mole of the alcohol.

Although we did start with one mole each of the precursors, *the reaction does not proceed to completion.*

Hence, less than one mole of ester is formed.　　→　　44

| 57 |
|---|

*Acetic acid* and *hydrogen chloride* are formed:

$$CH_3-C\overset{O}{\diagup}_{\underline{Cl\ +\ H}-O-H} \longrightarrow CH_3-C\overset{O}{\diagup}_{OH} \quad + \quad HCl$$

　acetyl chloride　　water　　　　　acetic acid　　　hydrogen chloride

Many acid chlorides react with water already at room temperature (e.g. acetyl chloride). Others require heating (e.g. benzoyl chloride).

Alcohols and phenols also react with acid chlorides.

All these reactions can be expressed by the general equation:

$$R-C\overset{O}{\diagup}_{\underline{Cl\ +\ H}-O-R'} \longrightarrow R-C\overset{O}{\diagup}_{O-R'} \quad + \quad HCl$$

R' = H denotes the reaction with water.

R' = CH$_3$ denotes the reaction with methanol.

R' = ⬡ denotes the reaction with phenol.

Write the equation for the reaction between benzoyl chloride and ethanol, giving the names of all the compounds.　　→　　48

**17**

---

**58**      *Esters.*

Acid chlorides also form esters when they react with phenols. How would you prepare the phenyl ester of benzoic acid?

● From benzoyl chloride and benzyl alcohol    →    51

● From benzoyl chloride and phenol           →    67

● From benzyl chloride and phenol            →    77

● From benzyl chloride and benzyl alcohol    →    84

---

**59**      Your answer is wrong.

The concentration of the ester does *not* change once *equilibrium* has been established (hence the name "equilibrium"). The number of molecules of ester being *formed* (by esterification) equals the number *disappearing* by hydrolysis.

Suppose a certain esterification reaction is at equilibrium. Will the concentration of the carboxylic acid change by continued heating of the mixture at reflux?

● Yes      →    39

● No       →    52

**17**

---

**60**      (1) *Silver carboxylate + alkyl halide.*
           (2) *Carboxylic acid + alcohol.*
           (3) *Acid chloride + alcohol* (or phenol).

*Phosgene* is an industrially important acid chloride:

$$O=C\begin{matrix} \diagup Cl \\ \diagdown Cl \end{matrix}$$

From which acid is this chloride derived? You will see the answer when the two chlorine atoms are replaced by OH groups.    →    75

---

**61**      Your answer is wrong.

Study the equation at   69   once more. Give the name of the reaction proceeding from the *right* to the *left*.

---

**62**

$$R-C\overset{O}{\underset{OH}{\diagup}} \quad + \ R'-OH \longrightarrow R-C\overset{O}{\underset{O-R'}{\diagup}} \quad + \ H_2O$$

acid          alcohol          ester          water

Write the corresponding equation for the preparation of ethyl acetate. Give the names of all compounds involved. Check your result at [71] or, if you need help, go instead to [53] .

---

**63**    No; at the onset of the reaction there was *one* mole of acetic acid present. At equilibrium there was 1/3 of a mole still left.

How many moles of acetic acid have reacted?    →    [38]

---

**64**

$$CH_3-C\overset{O}{\underset{Cl}{\diagup}}$$

acetyl chloride

Acid chlorides are compounds of the general type:

$$R-C\overset{O}{\underset{Cl}{\diagup}}$$

The acid chloride formed from *acetic acid* is called ...................
The acid chloride formed from *benzoic acid* is called *benzoyl chloride.*

Write the structures and names of these two acid chlorides.    →    [72]

---

**65**    $CH_3-CH_2-CH_2-CH_2-OH$    +    $SOCl_2$    →
              1-butanol                              thionyl chloride

$CH_3-CH_2-CH_2-CH_2-Cl$    +    $SO_2$    +    HCl
       1-chlorobutane              sulfur dioxide     hydrogen chloride

This reaction presents a useful method for the preparation of alkyl chlorides from alcohols.

We now turn to the *chemical reactions* of *acid chlorides.* When they react with *water the carboxylic acid is regenerated.*

Write the equation for the reaction between acetyl chloride and water. What
are the products?

● Benzoic acid and hydrogen chloride          →          7

● Acetic acid and hydrogen chloride          →          57

● Acetic acid and chlorine          →          79

● I don't know          →          76

---

**66**

$$O = C \overset{Cl}{\underset{Cl}{\diagup}}$$

phosgene

**Phosgene is a highly poisonous gas. It is particularly dangerous because the
symptoms develop only hours after the exposure when treatment may be too
late!**

Like other acid chlorides, phosgene reacts with water and alcohols.

Give equations and names of the products for

(a) the reaction between phosgene and water;
(b) the reaction between phosgene and ethanol.          →          82

**17**

---

**67**          Your answer is correct.

Now give the equation for the synthesis of benzoic acid phenyl ester (including
the names of all reaction partners).          →          78

---

**68**          (1) *Silver acetate + methyl iodide* (or methyl bromide).
          (2) *Acetic acid + methanol.*
          (3) *Acetyl chloride + methanol.*

Describe the three preparative methods for esters. Fill in the *general names* of
the starting materials below:

          (1) ........................ + ........................
          (2) ........................ + ........................
          (3) ........................ + ........................          →          60

---

**69**

$$CH_3-C\underset{\displaystyle O-C_2H_5}{\overset{\displaystyle O}{<}} + H_2O \longrightarrow CH_3-C\underset{\displaystyle OH}{\overset{\displaystyle O}{<}} + C_2H_5-OH$$

ethyl acetate     water          acetic acid     ethanol

This cleavage of an ester with water is called a *hydrolysis.* (The alkaline hydrolysis of esters is also referred to as "saponification" = soap formation.)

Point out the reaction which is the *opposite* of ester hydrolysis:

● Esterification                    →     55

● Alcoholic fermentation          →     83

● Acetic fermentation              →     61

**70**     Let us formulate the equation as follows:

$$R-C\underset{\boxed{OH + H}-O-R'}{\overset{\displaystyle O}{<}} \longrightarrow \cdots\cdots\cdots + \cdots\cdots\cdots$$

acid          alcohol               ester          $\cdots\cdots\cdots$

Surely, you can now complete it.     →     62

**17**

**71**

$$CH_3-C\underset{\textbf{OH}}{\overset{\displaystyle O}{<}} + C_2H_5-\textbf{OH} \longrightarrow CH_3-C\underset{\displaystyle O-C_2H_5}{\overset{\displaystyle O}{<}} \quad + \textbf{ H}_2\textbf{O}$$

acetic acid     ethanol          ethyl acetate          water
                                (or acetic acid
                                ethyl ester)

The water is formed from the groups printed in boldface. Rewrite the above equation in such a way that you can "frame" the boldface groups in the starting materials.

● Check your result at                    80

● I need further information    →     85

**72**

CH₃-C(=O)Cl

O=C(Cl) (benzene ring)

acetyl chloride          benzoyl chloride

Note the difference between *benzoyl* and *benzyl!*

O=C(Cl) (benzene ring)          CH₂Cl (benzene ring)

benzoyl chloride          benzyl chloride

Acid chlorides are mostly formed by treatment of the corresponding carboxylic acid with *thionyl chloride,* $SOCl_2$. As an exercise complete the following equation:

$C_6H_5-C(=O)OH$  +  .........  ⟶  .........  +  ....  +  HCl

benzoic acid    thionyl chloride

(Give also the names of the reaction products.)   →   **81**

---

**73**

(benzene ring)C(=O)OH / C(=O)OH  +  $HO-C_8H_{17}$ / $HO-C_8H_{17}$  ⟶  ............  +  $H_2O$

phthalic acid        octanol              dioctyl phthalate

Complete the equation.   →   **42**

**17**

---

**74**

$CH_3-C(=O)OH$  +  $SOCl_2$  ⟶  $CH_3-C(=O)Cl$  +  $SO_2$  +  HCl

acetic acid    thionyl       acetyl       sulfur       hydrogen
               chloride      chloride     dioxide      chloride

Thionyl chloride effects the replacement of OH by Cl not only in *carboxylic acids,* but also in *alcohols.*

What, therefore, is the product of the reaction between 1-butanol and thionyl chloride? Give the names of all the compounds in your reaction equation.   →   **65**

| 75 |

*Phosgene* is the acid chloride of *carbonic acid*, $H_2CO_3$.

Write the structural formula for phosgene once more.   →   | 66 |

| 76 |

Complete the equation:

$$CH_3 - C \overset{O}{\underset{}{\diagup}} \boxed{Cl + H} \text{-O-H} \longrightarrow$$

Write the structures and the names of the products.   →   | 57 |

| 77 |

No, benzyl chloride,

$$CH_2\text{-}Cl$$

is *not* an acid chloride! Hence it does not form an ester with phenol.
Return to   | 57 |   .

| 78 |

| benzoyl chloride | phenol | phenyl benzoate | hydrogen chloride |

This reaction is carried out in the presence of a base. Therefore, free HCl is not actually formed.

You now know three methods of preparing esters:

(1)  silver carboxylate + alkyl halide
(2)  acid + alcohol
(3)  acid chloride + alcohol (or phenol).

From which starting materials can one prepare methyl acetate using these three methods?

(1):  ................... + ...................
(2):  ................... + ...................
(3):  ................... + ...................

Write the names of the materials needed and check your answers at   | 68 |   .

| 79 |

You made an error in your equation. Complete the following:

$$CH_3-C \overset{O}{\underset{\boxed{Cl \ + \ H}-O-H}{}} \longrightarrow$$

Give the structure and the name of the product.    →    | 57 |

---

| 80 |

$$CH_3-C \overset{O}{\underset{\boxed{OH \ + \ H}-O-C_2H_5}{}} \longrightarrow CH_3-C \overset{O}{\underset{O-C_2H_5}{}} + H_2O$$

The esterification is carried out by heating a mixture of

       46 g (1 mol) ethanol,
       60 g (1 mol) acetic acid, and
       1 g conc. $H_2SO_4$

together with some boiling chips in a 250 ml round-bottomed flask equipped with a reflux condenser. The mixture is refluxed for one hour.

(The concentrated sulfuric acid is a catalyst. Other strong acids can be used as well, e.g., HCl.)

A complete reaction will *not* take place under these conditions since the water formed reacts with the ester, regenerating the starting materials.

Give the equation for the reaction between ethyl
acetate and water.    →    | 69 |

**17**

---

| 81 |

     Compare:

$$C_6H_5-C \overset{O}{\underset{OH}{}} + SOCl_2 \longrightarrow C_6H_5-C \overset{O}{\underset{Cl}{}} + SO_2 + HCl$$

| benzoic acid | thionyl chloride | benzoyl chloride | sulfur dioxide | hydrogen chloride |

Apart from benzoyl chloride (a liquid), only *gases* are formed as by-products. This makes the present procedure very convenient since one only has to distill the acid chloride in order to free it from unreacted carboxylic acid and thionyl chloride.

The reaction is carried out by refluxing a mixture of 50 g of benzoic acid and 150 g of thionyl chloride. Due to vigorous evolution of $SO_2$ and HCl the reaction should be carried out in a hood. When no more gases are evolved, the excess thionyl chloride is distilled off, and then the benzoyl chloride (boiling point 194 °C) is distilled.

Give the equation for the preparation of acetyl chloride from acetic acid and thionyl chloride. Include the names of all reaction partners.    →    |74|

---

|82|

a)    $O=C\overset{Cl}{\underset{Cl}{}}$ + 2 $H_2O$ ⟶ $O=C\overset{OH}{\underset{OH}{}}$ + 2 HCl

                                    ↓

                          $CO_2$ + $H_2O$

The products are carbon dioxide, water, and hydrogen chloride.

b)    $O=C\overset{Cl}{\underset{Cl}{}}$ + $\begin{matrix}HO\text{-}C_2H_5\\HO\text{-}C_2H_5\end{matrix}$ ⟶ $O=C\overset{O\text{-}C_2H_5}{\underset{O\text{-}C_2H_5}{}}$ + 2 HCl

The products are diethyl carbonate and hydrogen chloride.

This is the end of Program 17.

---

**17**

|83|    Your answer is wrong.

Examine the equation at   |69|   and give the name of the process proceeding from the right *to the left.*

---

|84|    You made two errors:

(1) benzyl chloride is *not* an acid chloride; hence it gives no ester with phenol;

(2) you confused the two residues

     $CH_2-$          and         

     benzyl            phenyl

Repeat from   |57|   .

85

$$R-C \overset{O}{\underset{\boxed{OH} + \boxed{H}}{\Big\backslash}} -O-R' \longrightarrow R-C \overset{O}{\underset{O-R'}{\Big\backslash}} + H_2O$$

acid          alcohol                    ester          water

The reacting groups are "framed" in order to illustrate the origin of the water.

Return to    71   .

# Program 18

## Carboxylic Acid Derivatives (II)

---

**1**      Having considered esters and acid chlorides in the preceding program, we will now discuss

      acid anhydrides,
      amides, and
      nitriles.

The *anhydride* of a carboxylic acid is formed by the elimination of *one molecule of water* from *two molecules of the acid:*

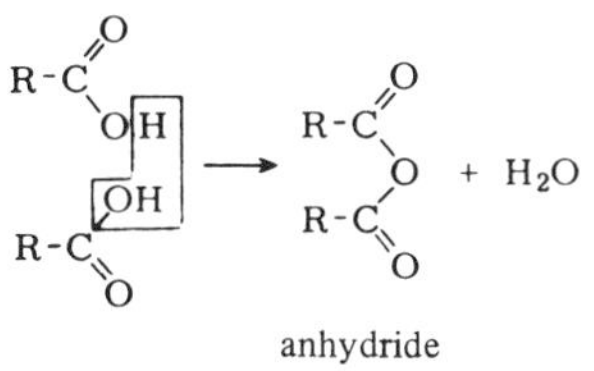

anhydride

Formulate with the aid of the above general equation the formation of acetic anhydride from two molecules of acetic acid.      →      $\boxed{8}$

---

**2**      Write the structural formulas for phthalic and terephthalic acids.

In phthalic acid the two carboxyl groups are *ortho* to each other.

In terephthalic acid they are *para*.

In order to form an anhydride, the two carboxyl groups must be near each other.

Now answer the question at    $\boxed{32}$    .

---

**3**     Since propionic acid is not a *dicarboxylic* acid, it does *not* form an *internal* anhydride. It can form an anhydride only by eliminating water from *two* acid molecules:

$$CH_3-CH_2-C\underset{OH}{\overset{O}{\lessgtr}} \qquad \qquad CH_3-CH_2-C\overset{O}{\lessgtr}$$
$$\longrightarrow \qquad \qquad O + H_2O$$
$$CH_3-CH_2-C\underset{O}{\overset{OH}{\lessgtr}} \qquad CH_3-CH_2-C\underset{O}{\lessgtr}$$

propionic acid          propionic anhydride

Repeat from   [25]   .

---

**4**     You only gave the correct name for formamide.

Return to   [33]   .

---

**5**

$$R\text{-}COO^{\ominus}NH_4^{\oplus} \xrightarrow{\text{heat}} R\text{-}C\underset{NH_2}{\overset{O}{\lessgtr}} + H_2O$$

ammonium acetate                    acetamide

Amides are often made from acid chlorides and ammonia. Try and formulate the preparation of *benzamide* from the corresponding acid chloride and ammonia.

Check your result at   [18]   , or, if you need assistance, go instead to   [28]   .

---

**6**     It is correct that $CH_3CN$ is acetonitrile, but HCN is not called "nitrile". A nitrile is a compound of the type

$$R–C≡N.$$

Repeat from   [30]   .

---

18

**7**    $P_2O_5$ is the anhydride of phosphoric acid, $H_3PO_4$.

We have already seen that anhydrides are dehydrating agents. Thus, *benzoic anhydride* was prepared from *benzoic acid* and *acetic anhydride*. The acetic anhydride reacts with the water liberated in the reaction, forming acetic acid.

Return to    52 .

---

**8**

acetic anhydride

As you will notice the derivation of an anhydride from a carboxylic acid is very simple – at least on paper. The names of the anhydrides are just as simple: they are formed by substituting the word "anhydride" for "acid".

Write the names for the anhydrides of

       propionic acid,
       benzoic acid, and
       phthalic acid.    →    16

**18**

---

**9**    Your answer is correct.

A second method for preparing nitriles consists of treating an alkyl halide with a cyanide. The halide atom is replaced by the cyanide group ($-C\equiv N$):

      R–Hal + NaCN ⟶ R–CN + NaHal

Using this method, propionitrile is formed by heating 1-iodopropane with sodium cyanide. Write the equation for this reaction, giving the names of all the compounds involved.    →    20

**10**    Terephthalic acid has the formula

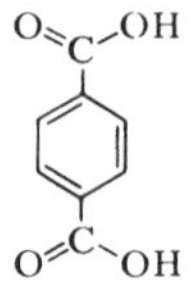

Does this compound form an internal anhydride?

- Yes              →   46

- No               →   23

- I don't know     →   2

---

**11**

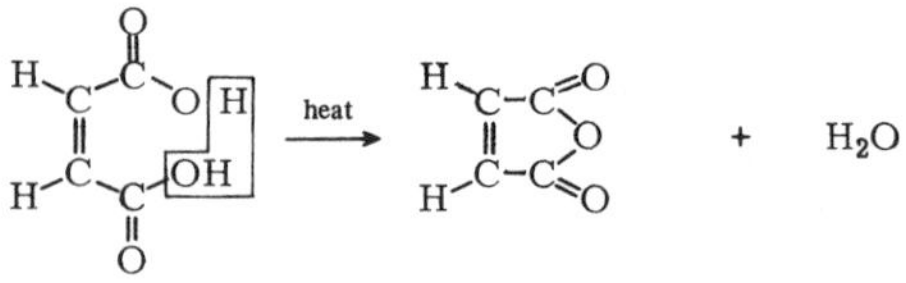

maleic acid          maleic anhydride   water

The water formed in this reaction must be removed by distillation as otherwise hydrolysis back to the acid occurs. The hydrolysis of anhydrides is a general reaction:

*Carboxylic acid anhydrides react with water to give carboxylic acids.*

Give the equation for the reaction between acetic anhydride and water.   →   19

---

**12**    You named all three amides wrongly. You must consider from which acids the amides are derived.

Return to   33  .

---

**13**    Compare:

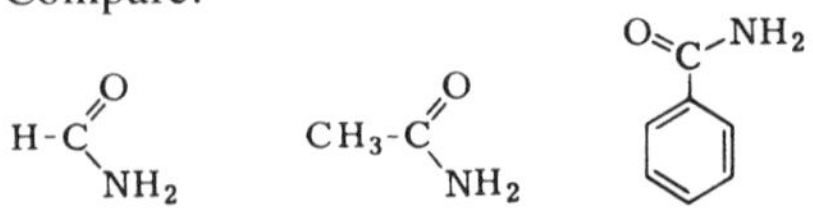

formamide          acetamide          benzamide

Most amides are colorless, crystalline compounds. *Formamide* is an exception: it is a liquid.

Amides can be obtained by heating the ammonium salts of the corresponding carboxylic acids:

$$R-COO^{\ominus}NH_4^{\oplus} \xrightarrow{\text{heat}} R-C\overset{\displaystyle O}{\underset{NH_2}{\diagdown}} + H_2O$$

This reaction, too, is reversible, and therefore the water formed must be removed, e.g. by distillation.

Which compound is formed by heating the ammonium salt of acetic acid?

Give the reaction equation and the names of the compounds involved.   →   5

---

**14**   Your answer is wrong. You probably confused maleic acid with fumaric acid.

Repeat from   25   .

---

**15**

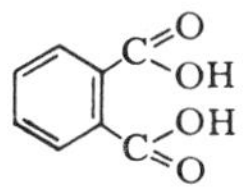

phthalic acid

Now give the structure and name of the anhydride formed from this acid.   →   32

18

---

**16**   *Propionic anhydride*
*benzoic anhydride*
*phthalic anhydride*

Write the structural formula for benzoic anhydride. It will be convenient to write two molecules of benzoic acid in such a way that you can "frame" the molecule of water to be eliminated.   →   25

17    Yes, both maleic and phthalic acids give the corresponding anhydrides on heating.

Complete the following equation writing the structures of the molecules:

maleic acid $\xrightarrow{\text{heat}}$ maleic anhydride + water    →    11

---

18

benzoyl          ammonia      benzamide      hydrogen
chloride                                     chloride

Note that the hydrogen chloride formed will react with surplus ammonia, giving ammonium chloride, $NH_4Cl$.

If you had the equation and all compound names correct, you are making good progress.

Let us now consider a third method for making amides:

*Amides are formed from carboxylic acid anhydrides by treatment with ammonia.*

Try and formulate the preparation of acetamide by this method. Give also the names of all the compounds involved. Check your answer at    26    or, if you need help, go instead to    34   .

**18**

---

19

acetic          acetic
anhydride       acid

Thus, in principle, acetic anhydride is capable of binding water. This property is taken advantage of in the *second method of preparing anhydrides,* namely, heating the acid together with acetic anhydride.

Which compounds are formed by heating a mixture of benzoic acid and acetic anhydride?    →    27

---

**20**

$$CH_3-CH_2-CH_2-I \;+\; NaCN \;\longrightarrow\; CH_3-CH_2-CH_2-CN \;+\; NaI$$

1-iodopropane     sodium       butyronitrile       sodium
               cyanide                              iodide

The carbon chain is *extended* by one carbon in this reaction.

The reaction also presents a useful way of obtaining *carboxylic acids* from alkyl halides: the nitriles can be hydrolyzed to the corresponding acids:

$$R-C{\equiv}N \;\xrightarrow[-NH_3]{+2\ H_2O}\; R-C{\overset{O}{\underset{OH}{\diagup}}}$$

Let us examine this reaction in more detail.

You will recall that nitriles can be made from amides. Give the general equation for the preparation of nitriles from amides.    →    |37|

---

**21**     *Potassium cyanide, KCN.*

**Hydrogen cyanide and its salts, the cyanides, are extremely poisonous.**

Assign the correct names to the following formulas:

       HCN                    $CH_3CN$

● cyanide                  acetonitrile          →    |56|

● nitrile                    acetonitrile          →    |6|

● nitrile of formic acid      nitrile of acetic acid     →    |54|

● hydrogen cyanide        acetonitrile          →    |9|

**18**

---

**22**     Your answer is correct.

Now write the names of the three amides.    →    |13|

---

**23**     Terephthalic acid,

.does *not* form an internal anhydride; the carboxyl groups are too far from each other.

In contrast, phthalic acid does form an internal anhydride:

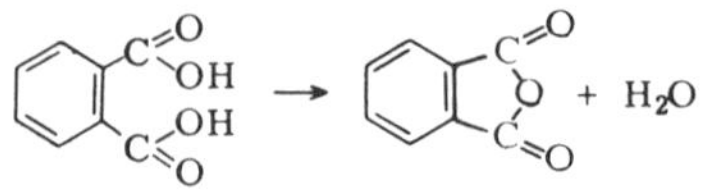

phthalic acid                    phthalic anhydride

Which of the following carboxylic acids forms an internal anhydride?

- Propionic acid        →     3
- Benzoic acid          →     40
- Fumaric acid          →     14
- Maleic acid           →     31

---

**24**     Your answer is only partially correct.

The two carboxyl groups in fumaric acid are *trans* to each other,

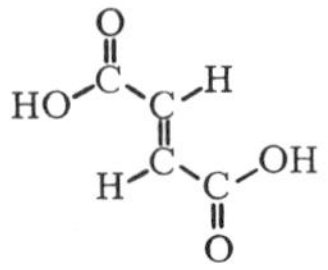

and therefore the distance between them is too large for the formation of an internal anhydride.

Go back to    31  .

---

**25**

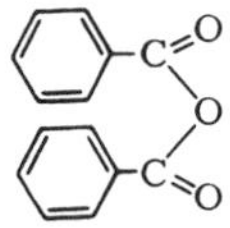

benzoic anhydride

The two carboxyl groups which react with each other to form an anhydride need not be in *different* molecules. Certain *dicarboxylic acids* can form *internal anhydrides*. Maleic acid is an example:

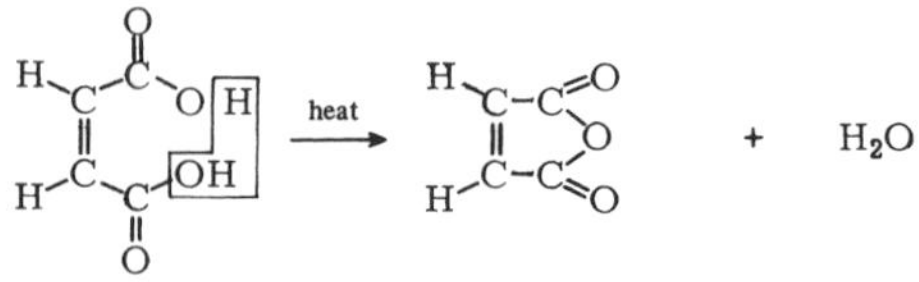

maleic acid              maleic anhydride

Phthalic acid also forms an internal anhydride.

Write the structure of *phthalic anhydride* and check your result at 32. (Phthalic acid is an *o*-dicarboxylic acid; if you have forgotten its formula, see 15.)

---

**26**

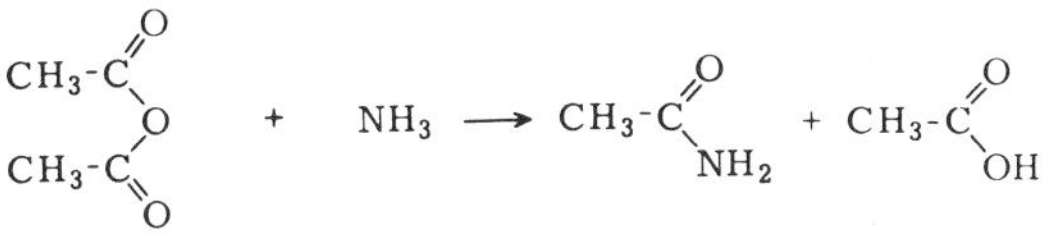


acetic anhydride  ammonia  acetamide  acetic acid

To recapitulate the three methods of preparing amides, complete the following:

(1) .................... $\xrightarrow{\text{heat}}$ amide + water

(2) .................... + .................... → amide + HCl

(3) carboxylic anhydride + .................... → amide + ..............

→ 36

---

**27** *Benzoic anhydride* and *acetic acid* are formed.



*Method 3:*

Acid anhydrides are formed by reacting an acid chloride with the salt of an acid:

acid chloride:

carboxylate:

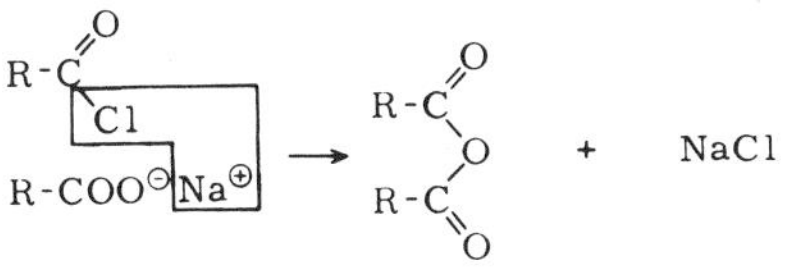


anhydride  sodium chloride

Which materials are needed for the preparation of acetic anhydride by this method? Write both the names and the structural formulas. → 35

---

**28**

$$R-C\overset{\displaystyle O}{\underset{\boxed{Cl \ + \ H}\overset{\displaystyle H}{\underset{\displaystyle H}{N}}}{\Big\|}} \longrightarrow R-C\overset{\displaystyle O}{\underset{\displaystyle NH_2}{\Big\|}} + HCl$$

Now write the analogous equation for the preparation of *benzamide*. Do not forget the names of the reaction partners.     →     18

---

**29**     No, $P_2O_5$ is not a catalyst.

$P_2O_5$ is the anhydride of phosphoric acid. Therefore, it will react with the water liberated in the reaction forming phosphoric acid. Since $P_2O_5$ is consumed in this process, it is not a catalyst:

   *catalysts* accelerate chemical reactions but are not consumed.

When acetamide is heated with $P_2O_5$, water is eliminated and phosphoric acid is formed. What is the other reaction product? Write the name of the compound.     →     38

---

**30**

$$3 \ CH_3-C\overset{\displaystyle O}{\underset{\displaystyle NH_2}{\Big\|}} \quad + \quad P_2O_5 \quad \longrightarrow \quad 3 \ CH_3-C\equiv N \quad + \quad 2 \ H_3PO_4$$

| acetamide | phosphorus pentoxide | acetonitrile | phosphoric acid |

The simplest nitrile, formonitrile, is derived from formic acid,

   $H-C\equiv N$

but is normally known as *hydrogen cyanide.*

Hydrogen cyanide is a weak acid. In other words, the dissociation equilibrium lies strongly to the left:

   $H-C\equiv N \leftrightharpoons H^{\oplus} + CN^{\ominus}$

The salts of this acid are the *cyanides,* e.g.

   NaCN, sodium cyanide.

Write the formula and the name of the potassium salt of hydrogen cyanide.
   →     21

| **31** | Your answer is correct. |

We shall now consider the preparation of *acid anhydrides*.

*Method 1:*

Certain dicarboxylic acids which can form internal anhydrides do so simply on heating. You already know two dicarboxylic acids that can form internal anhydrides. Which ones?

● Phthalic acid and fumaric acid      →      |24|

● Maleic acid and benzoic acid      →      |39|

● Maleic acid and phthalic acid      →      |17|

● Maleic acid and fumaric acid      →      |47|

---

| **32** |

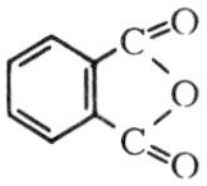

phthalic anhydride

An internal anhydride is formed *only* when the two carboxyl groups are *close enough* to react with each other.

Maleic acid is an example. Since this is a *cis*-dicarboxylic acid, the two carboxyl groups are on the same side of the molecule and can react with each other:

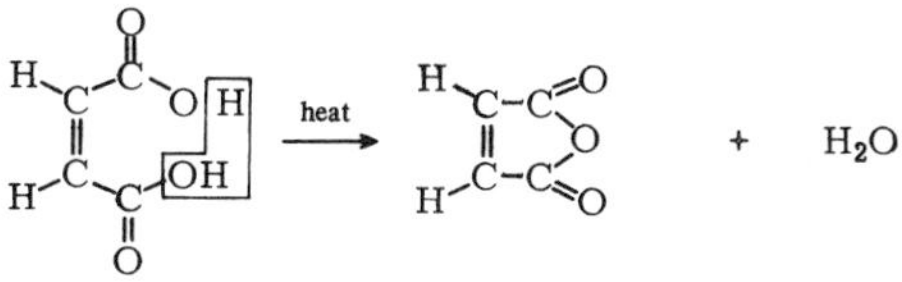

maleic acid          maleic anhydride

Fumaric acid is the corresponding *trans*-dicarboxylic acid. In this case the two carboxyl groups are on opposite sides of the molecule, and the distance between them is too great for an internal reaction to occur.

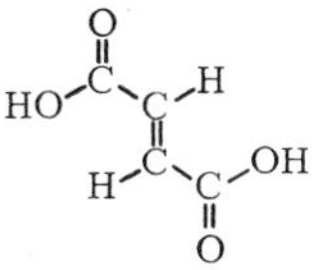

fumaric acid

**18**

Does terephthalic acid form an internal anhydride?

● Yes                                              →  46

● No                                               →  23

● I don't know                                     →  2

● I need the formula for terephthalic acid         →  10

---

33

$$H-C\overset{\displaystyle O}{\underset{\displaystyle NH_2}{<}} \qquad CH_3-C\overset{\displaystyle O}{\underset{\displaystyle NH_2}{<}} \qquad$$

O=C−NH₂ (c)

(a)                (b)                (c)

Now assign the proper names to the three compounds above.

|    (a)    |    (b)    |    (c)    |     |    |
|-----------|-----------|-----------|-----|----|
| ● Benzamide | formamide | acetamide | → | 12 |
| ● Formamide | benzamide | acetamide | → | 4  |
| ● Formamide | acetamide | benzamide | → | 22 |
| ● Acetamide | formamide | benzamide | → | 44 |

---

34

$$R-C\overset{O}{\underset{\underset{R-C\overset{O}{\underset{O}{}}}{O}}{<}} + NH_3 \longrightarrow R-C\overset{O}{\underset{NH_2}{<}} + R-C\overset{O}{\underset{OH}{<}}$$

Use this general equation to formulate the preparation of acetamide from the acid anhydride and ammonia. Give also the name of each compound.   →  26

**35**

Acetyl chloride:    $CH_3-C\overset{\displaystyle O}{\underset{\displaystyle Cl}{\big<}}$

$\longrightarrow$

sodium acetate:    $CH_3-COO^{\ominus}\ Na^{\oplus}$

Complete this scheme for the preparation of acetic anhydride.    $\rightarrow$    43

**36**

1) Ammonium salt of a carboxylic acid $\xrightarrow{\text{heat}}$ amide + water
2) Acid chloride + ammonia $\rightarrow$ amide + HCl
3) Acid anhydride + ammonia $\rightarrow$ amide + carboxylic acid

A special acid chloride described earlier is the *di*chloride of carbonic acid.

Give the name and formula for this compound.    $\rightarrow$    45

**37**

$R-C\overset{\displaystyle O}{\underset{\displaystyle NH_2}{\big<}} \xrightarrow{P_2O_5} R-C\equiv N + H_2O$

   amide          nitrile    water

Now write the general equation for the preparation of acid amides by heating the ammonium salt of the carboxylic acid.    $\rightarrow$    48

**18**

**38**    Yes, $P_2O_5$ is a powerful dehydrating agent, i.e. it binds water, forming phosphoric acid:

$$P_2O_5 + 3\ H_2O \rightarrow 2\ H_3PO_4$$

When *acetamide* is heated with $P_2O_5$, *acetonitrile* is formed.

Give the equation for this reaction and check it at    30    or, if you need help, go instead to    55    .

**39**    Your answer is only partially correct. Benzoic acid is not a dicarboxylic acid and does not, therefore, form an internal anhydride.

Return to    31    .

**40**   Benzoic acid does form an anhydride, but not an *internal* one. Internal anhydrides can be formed only from dicarboxylic acids.

Repeat from   25   .

---

**41**   $R-C\equiv N + 2\ H_2O \rightarrow R-COO^{\ominus}NH_4^{\oplus}$

nitrile     water     ammonium carboxylate

The hydrolysis of a nitrile is usually carried out with a base. In this case, $NH_3$ is evolved, and the alkali metal salt of the carboxylic acid is obtained after evaporation of the excess water:

$R-C\equiv N + H_2O + NaOH \rightarrow$

Complete this equation     $\rightarrow$   49

---

**42**

$$CH_3-C\!\!\begin{array}{c}{}^{\nearrow O}\\{}_{\searrow O}\end{array}\!\!\!\!\begin{array}{c}\\CH_3-C\!\!\begin{array}{c}\\{}_{\searrow O}\end{array}\end{array} + C_2H_5OH \longrightarrow CH_3-C\!\!\begin{array}{c}{}^{\nearrow O}\\{}_{\searrow O-C_2H_5}\end{array} + CH_3-C\!\!\begin{array}{c}{}^{\nearrow O}\\{}_{\searrow OH}\end{array}$$

acetic        ethanol        ethyl acetate        acetic acid
anhydride

**18**   As you will see, this reaction represents a fourth method of preparing esters.

Having discussed the acid anhydrides, we now turn to the *amides*.

An *amide* is derived from a carboxylic acid by replacing –OH by –NH₂. The general formula is

$$R-C\!\!\begin{array}{c}{}^{\nearrow O}\\{}_{\searrow NH_2}\end{array}$$

Write the formulas for the amides derived from formic, acetic, and benzoic acids.     $\rightarrow$   33

43

acetyl chloride: $CH_3-C\overset{O}{\underset{Cl}{\diagup}}$

sodium acetate: $CH_3-COO^{\ominus}Na^{\oplus}$

$\longrightarrow$ $\begin{matrix} CH_3-C\diagup^O \\ \diagdown O \\ CH_3-C\diagup \\ \diagdown O \end{matrix}$ + NaCl

acetic anhydride

We have now seen three methods of preparation of carboxylic acid anhydrides:

1) heating a dicarboxylic acid;

2) reacting a carboxylic acid with the anhydride of another acid;

3) treating an acid chloride with the salt of a carboxylic acid.

Which starting materials are required for the preparation of

phthalic anhydride by method 1,
benzoic anhydride by method 2,
acetic anhydride by method 3?

$\rightarrow$   51

44    You assigned the correct name to benzamide only. When you know the names of the acids, you can easily derive those of the amides.

Return to   33   .

**18**

45

$O=C\overset{Cl}{\underset{Cl}{\diagup}}$

phosgene

The reaction between this compound and ammonia gives *urea* – the diamide of carbonic acid:

$O=C\overset{Cl}{\underset{Cl}{\diagup}}$ + 2 $NH_3$ $\longrightarrow$ ..... + .....

phosgene    ammonia      urea      .....

Complete the equation.    $\rightarrow$   53

| 46 |

No, terephthalic acid does *not* form an internal anhydride.

The two carboxyl groups are *para* to each other and the distance between them is too large. You will easily see this by drawing the formula for terephthalic acid. Give the formulas for phthalic acid and phthalic anhydride also.    →    | 23 |

| 47 |

Your answer is only partially correct.

Fumaric acid is a *trans*-dicarboxylic acid,

$$HO-\overset{\overset{O}{\underset{\|}{}}}{C}-\underset{\underset{\|}{O}}{C}-H$$
$$H-\overset{\overset{}{}}{C}-\underset{\underset{\|}{O}}{C}-OH$$

and the distance between the two carboxyl groups is too long to permit the formation of an internal anhydride. Return to    | 31 |    .

| 48 |

$$R-COO^{\ominus}NH_4^{\oplus} \xrightarrow{\text{heat}} R-C\overset{O}{\underset{NH_2}{\diagdown}} + H_2O$$

ammonium salt                    amide        water

**18**

The last two equations discussed are *reversible*.

Write down the *reverse* equations, i.e.

1) nitrile → amide
2) amide → ammonium salt.    →    | 57 |

| 49 |

$$R-C\equiv N + H_2O + NaOH \longrightarrow R-COO^{\ominus}Na^{\oplus} + NH_3$$

nitrile                        sodium salt        ammonia
                                            (evaporates)

The free carboxylic acid is liberated from the sodium salt by treatment with a mineral acid:

$$RCOONa + HCl \longrightarrow R-C\overset{O}{\underset{OH}{\diagup}} + NaCl$$

Now complete the reaction scheme for the production of carboxylic acids *from alkyl halides:*

$$R\text{-Hal} \xrightarrow[-\text{NaHal}]{+\text{NaCN}} \quad \cdots \quad \xrightarrow{H_2O} \quad R\text{-C}\underset{NH_2}{\overset{O}{\diagup}} \xrightarrow[-\text{NH}_3]{+\text{NaOH}} RCOONa \xrightarrow{+\text{HCl}}$$

alkyl halide          $\cdots\cdots$          $\cdots\cdots\cdots$          sodium salt          $\cdots\cdots$

$\rightarrow$  $\boxed{58}$

---

$\boxed{50}$  *Ester* and *carboxylic acid.*

Now give the equation for the preparation of ethyl acetate from acetic anhydride.   $\rightarrow$  $\boxed{42}$

---

$\boxed{51}$

1) *Phthalic acid* $\xrightarrow{\text{heat}}$ phthalic anhydride.
2) *Benzoic acid + acetic anhydride* → benzoic anhydride.
3) *Acetyl chloride + sodium acetate* → acetic anhydride.

You already know one reaction of the anhydrides, namely the reaction with *water*. What product is formed?   $\rightarrow$  $\boxed{59}$

---

$\boxed{52}$

$$CH_3\text{-C}\underset{NH_2}{\overset{O}{\diagup}} \longrightarrow CH_3\text{-C}{\equiv}N \quad + \quad H_2O$$

acetamide          acetonitrile          water

The reaction is carried out by heating a mixture of acetamide and phosphorus pentoxide. The acetonitrile is distilled off as formed.

What is the function of phosphorus pentoxide ($P_2O_5$) in this reaction?
(Note: $P_2O_5$ is the anhydride of phosphoric acid.)

● $P_2O_5$ is a catalyst          $\rightarrow$  $\boxed{29}$

● $P_2O_5$ is a dehydrating agent          $\rightarrow$  $\boxed{38}$

● I don't know          $\rightarrow$  $\boxed{7}$

**18**

**53**

$$O=C\begin{smallmatrix}Cl\\Cl\end{smallmatrix} \ + \ 2\ NH_3 \longrightarrow O=C\begin{smallmatrix}NH_2\\NH_2\end{smallmatrix} \ + \ 2\ HCl$$

phosgene      ammonia         urea            hydrogen
                                              chloride

Urea is a product of metabolism present in urine. It serves to remove excess nitrogen from food (mainly proteins). Man excretes about 30 g of urea per day.

Urea is manufactured industrially from $CO_2$ and $NH_3$ at elevated pressure:

$$CO_2 \ + \ NH_3 \longrightarrow O=C\begin{smallmatrix}NH_2\\NH_2\end{smallmatrix} \ +$$

. . . . . . . .      . . . . . . . .      . . . . . . . . .      . . . . . . . . . .

Balance this equation and write the names of all the compounds involved.   →   |60|

---

**54**      It is true that HCN and $CH_3CN$ are the nitriles of formic and acetic acids, respectively, but you will find the proper names by studying   |30| attentively.

---

**55**      Note that **1** mol $P_2O_5$ reacts with **3** mol $H_2O$:

$$P_2O_5 + 3\ H_2O \rightarrow 2\ H_3PO_4.$$

Balance the equation for the preparation of acetonitrile starting from **3** mol acetamide and **1** mol $P_2O_5$.   →   |30|

---

**56**      $CH_3CN$ is acetonitrile, but the name "cyanide" implies a *salt* of HCN.

Repeat from   |30|   .

**18**

---

**57**

1)          $R-C{\equiv}N$ + $H_2O$ ⟶ $R-C\overset{O}{\underset{NH_2}{\diagup}}$

          nitrile                    amide

2)          $R-C\overset{O}{\underset{NH_2}{\diagup}}$ + $H_2O$ ⟶ $R-COO^{\ominus}NH_4^{\oplus}$

          amide                    ammonium carboxylate

These reactions illustrate the preparation of carboxylic acids from nitriles. The two reactions are usually carried out in a single procedure, that is, the amide is not isolated but directly hydrolyzed to the carboxylate.

Formulate the overall equation for the hydrolysis of a nitrile to an ammonium carboxylate   →   41

---

**58**

$R-Hal$ $\xrightarrow[-NaHal]{+NaCN}$ $R-C{\equiv}N$ $\xrightarrow{+H_2O}$ $R-C\overset{O}{\underset{NH_2}{\diagup}}$ $\xrightarrow[-NH_3]{+NaOH}$ $RCOONa$ $\xrightarrow{+HCl}$ $R-C\overset{O}{\underset{OH}{\diagup}}$

alkyl          nitrile          amide               sodium          carboxylic
halide                                              carboxylate      acid

These reactions allow the preparation of *carboxylic acids from alkyl halides.* Note that *an additional carbon atom is introduced into the carbon chain.*

This is the end of Program 18.

**18**

---

**59**     *Carboxylic acids.*

Anhydrides react with alcohols and phenols according to the equation

$\underset{R-C\diagdown_O}{\overset{R-C\diagup^O}{}}O$   +   $R'-OH$ ⟶ $R-C\overset{O}{\underset{O-R'}{\diagup}}$ + $R-C\overset{O}{\underset{OH}{\diagup}}$

anhydride          alcohol

Write the names of the products formed.   →   50

---

|60|

$$CO_2 \quad + \quad 2\,NH_3 \quad \longrightarrow \quad O=C{\Large\langle}^{NH_2}_{NH_2} \quad + \quad H_2O$$

carbon          ammonia          urea          water
dioxide

Ammonia is used in the manufacture of certain synthetic polymers.

Having presented the amides, we will now discuss the *nitriles.*

A *nitrile* is formed by the elimination of water from an *amide:*

$$R\text{-}C{\Large\langle}^{O}_{N\,H_2} \quad \longrightarrow \quad R\text{-}C{\equiv}N \; + \; H_2O$$

amide                    nitrile

In a similar manner write the equation for the preparation of *acetonitrile* from *acetamide*. Give the name of each compound involved.    →    |52|

# Program 19

## Sulfonic Acids

---

**1**   Sulfonic acids are carbon compounds containing the functional group

$-SO_3H$

in which sulfur is hexavalent and bonded to two oxygen atoms and an OH group. Develop the detailed structural formula for the $-SO_3H$ group and check your result at [10] .

---

**2**   Compare:

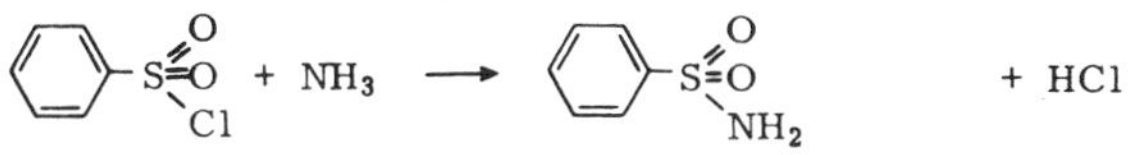

benzenesulfonyl
chloride

benzenesulfonamide

Sulfonamides often have bacteriostatic properties and some are used as drugs ("sulfa" drugs).

The names of sulfonyl chlorides and sulfonamides are easily derived from those of the sulfonic acids. Write the names of the following compounds:

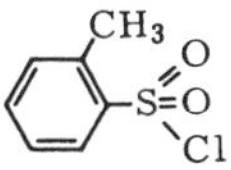

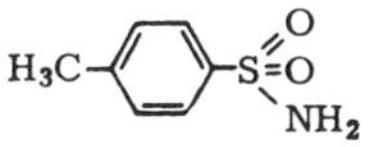

**19**

$\rightarrow$   [12]

---

---

3

CN
+ 2 H₂O  →  COOH + NH₃

benzonitrile                    benzoic acid

Thus, it is possible in two steps to *convert a sulfonic acid into a carboxylic acid:*

SO₃H                    CN                    COOH
—step 1→   —step 2→

How would you carry out the first step of this sequence?

● By heating the sulfonic acid with
  sodium hydroxide                    →  23

● By heating the sulfonic acid with
  sodium cyanide                    →  13

---

4

$$\left[ \begin{array}{c} \text{C}_6\text{H}_5-\text{S}^{=O}_{O} \end{array} \right]^{\ominus} \text{Na}^{\oplus} \qquad \left[ \begin{array}{c} \text{H}_3\text{C}-\text{C}_6\text{H}_4-\text{S}^{=O}_{O} \end{array} \right]^{\ominus} \text{K}^{\oplus}$$

The sulfonic acids and their derivatives may be designated as shown for sodium benzenesulfonate:

—SO₃Na

Derivatives of sulfonic acids are obtained in the same way as for carboxylic acids. Thus, sulfonic acid chlorides *(sulfonyl chlorides)* are obtained by replacing the OH group by Cl.

Write the structural formula for benzenesulfonyl chloride. Calculate its molecular mass (benzenesulfonic acid = 158; Cl = 35.5).

Which figure is correct?

● 158          →  21

● 176.5        →  14

● 193.5        →  24

● Neither      →  27

---

**5**

$$SO_3Na \quad + NaCN \longrightarrow \quad CN \quad + Na_2SO_3$$

This reaction is also carried out at elevated temperatures in the melt.

What is the name of the reaction product?

- Benzoic acid      →    ⟨19⟩

- Benzonitrile      →    ⟨15⟩

- Benzyl cyanide      →    ⟨25⟩

- I don't know      →    ⟨31⟩

---

**6**     The compound

$$H_3C-\!\!\langle\ \rangle\!\!-SO_3H$$

is a *sulfonic acid* derived from *toluene*. Hence its name is
*p*-toluene ....................    →    ⟨16⟩

---

**7**     Correct; a benzene ring substituent *directs the positioning* of another incoming substituent.

What is the effect of an *o, p*-director?

- It directs an incoming substituent to the
  *o*- and *p*-positions only          →    ⟨17⟩

- It directs an incoming substituent equally to
  all ring positions          →    ⟨33⟩

- I don't know          →    ⟨26⟩

**19**

---

**8**     Correct.

Both *o*- and *p*-toluenesulfonic acids are formed in the reaction of toluene with 1 mol $H_2SO_4$:

$$\overset{CH_3}{\langle\ \rangle}\!\!-SO_3H \quad \text{and} \quad H_3C-\!\!\langle\ \rangle\!\!-SO_3H$$

As further examples consider the reactions between

> (a) 1 mol $Cl_2$ and 1 mol benzenesulfonic acid;
> (b) 1 mol $Cl_2$ and 1 mol toluene.

Where do the chlorine atoms enter?

- In both cases in *ortho* and *para*      →   40

- In both cases in *meta*                →   43

- In *meta* for (a) and in *ortho, para* for (b)                →   35

- In *ortho, para* for (a) and in *meta* for (b)                →   49

---

**9**

and

naphthalene−1−sulfonic acid          naphthalene−2−sulfonic acid

Aromatic sulfonic acids are obtained by reacting the hydrocarbons with .................... or .................... . The reaction is called a ....................

Write the three missing words and check at   29   or, if you need help, go instead to   20   .

---

**19**

**10**

Sulfonic acids are compounds of the general formula

> $R{-}SO_3H$

in which an alkyl or aryl group is directly bonded to sulfur.

The *aromatic sulfonic acids* are particularly important and easy to make. Thus, benzenesulfonic acid is obtained by treating benzene with concentrated sulfuric acid at room temperature:

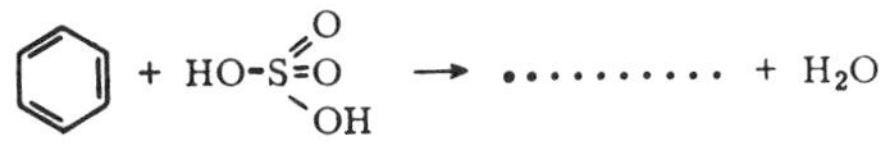

benzene     sulfuric acid

Complete this equation.    $\rightarrow$    20

---

**11**     Your answer is wrong. – The $CH_3$ group is *o*-, *p*-directing.

Return to    34   .

---

**12**

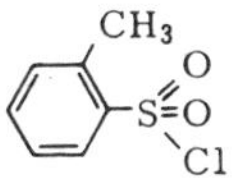

   *o*−toluenesulfonyl chloride

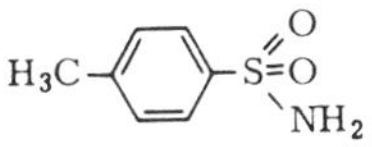

   *p*−toluenesulfonamide

Complete the following table:

| Name | Functional group (formula) | Properties; applications |
|---|---|---|
| Sulfonic acid | | acid/base properties? solubility in water? |
| Sulfonyl chloride | | used in the synthesis of............................? |
| Sulfonamide | | used as...................? |

                                          $\rightarrow$    22

---

**13**    Correct; the sulfonic acid group is replaced by CN on melting with sodium cyanide.

Let us now consider the *substitution rules* for aromatic compounds.

If toluene and benzenesulfonic acid are treated with chlorine, different substitution products are formed:

(a)    [structure: toluene] + $Cl_2$   →   [structure: o-chlorotoluene] + HCl

(b)    [structure: benzenesulfonic acid] + $Cl_2$   →   [structure: m-chlorobenzenesulfonic acid] + HCl

In reaction (a) the chlorine atom enters in the position *ortho* to the methyl group. In reaction (b) the chlorine atom enters in the position *meta* to the sulfonyl group.

These examples illustrate a general phenomenon: a substituent already on an aromatic ring determines the positioning of an incoming substituent. It *directs* the new substituent to either the *ortho* and *para* positions, or to the *meta* position.

> *ortho, para*-Directing substituents (*o-, p-directors*): alkyl, halogen, and OH groups.

> *meta*-Directing substituents (*m-directors*): sulfonic acid, carboxyl, and carbonyl groups.

Write down the names and formulas of the most important *o-, p-directors* and *m-directors*.    →    26

---

**14**     Your answer is correct; the molecular mass is 176.5.

Benzenesulfonyl chloride has the formula

[structure: benzenesulfonyl chloride]

Sulfonyl chlorides are used in the preparation of *sulfonamides*. The latter are analogous to the carboxylic acid amides.

Complete the equation:

[structure] + $NH_3$ →          + HCl

benzenesulfonyl             benzenesulfonamide
chloride                                         →   2

**19**

---

**15**        Your answer is correct.

The compound

CN
(benzene ring with CN)

is benzonitrile, and you have already learnt that this can be hydrolyzed to benzoic acid.

Complete the equation:

CN
(benzene ring with CN) + 2 H$_2$O  →        + NH$_3$

→  3

---

**16**

H$_3$C—(benzene ring)—SO$_3$H        *p*–toluenesulfonic acid

Sulfonic acids are very soluble in water.

Furthermore, *sulfonic acids are strong acids.*

Accordingly, salts of the sulfonic acids can be made. These are called *sulfonates.*

Write the structural formulas for sodium benzenesulfonate and potassium *p*-toluenesulfonate.   →   4

**19**

---

**17**        Your answer is correct.

What is the effect of a *m*-director?   →   34

---

**18**

SO$_3$Na                         OH
(benzene ring) + NaOH  →  (benzene ring) + Na$_2$SO$_3$

Phenol has been prepared industrially in this manner, but the process is now uneconomical. The most important modern procedure for the manufacture of phenol is the *cumene process,* which was discussed in Program 13. You will find a short repetition at 28 in the present program.

The sulfonic acid group is replaced by the CN group when a sulfonate is heated with sodium cyanide, NaCN, e.g.

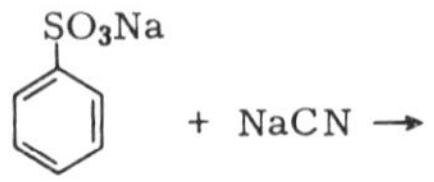

SO$_3$Na

+ NaCN →

Complete the equation.   →   $\boxed{5}$

---

$\boxed{19}$   No; benzoic acid is a carboxylic acid,

COOH

The carboxylic acids and their derivatives were discussed at length in the preceding program. Repeat $\boxed{5}$ .

---

$\boxed{20}$

benzene          sulfuric acid

$+ HO - S\overset{O}{\underset{OH}{}} \rightarrow \overset{O}{\underset{OH}{-S}} + H_2O$

This process is called *sulfonation*.

**19**

The reaction is normally accelerated by heating or by using oleum (a solution of SO$_3$ in concentrated sulfuric acid) in place of sulfuric acid itself.

The sulfonic acids are named simply by adding the term "sulfonic acid" to the name of the hydrocarbon.

What is the name of the sulfonic acid prepared from benzene according to the equation above?   →   $\boxed{30}$

---

$\boxed{21}$   Your answer is wrong. Please pay more attention to the question and the information given.

Return to $\boxed{4}$ .

| 22 | | |
|---|---|---|
| Name | Functional group | Properties; applications |
| Sulfonic acid | $-S\overset{O}{\underset{OH}{\overset{\parallel}{=}O}}$  or $-SO_3H$ | Strong acids; very soluble in water. |
| Sulfonyl chloride | $-S\overset{O}{\underset{Cl}{\overset{\parallel}{=}O}}$  or $-SO_2Cl$ | Used in the synthesis of sulfonamides. |
| Sulfonamide | $-S\overset{O}{\underset{NH_2}{\overset{\parallel}{=}O}}$  or $-SO_2NH_2$ | Used as drugs. |

The replacement of the sulfonic acid group by the hydroxyl group is an important reaction which occurs on heating with an excess of alkali, e.g. NaOH:

$$\underset{}{\overset{SO_3Na}{\bigcirc}} + NaOH \longrightarrow \underset{}{\overset{OH}{\bigcirc}} + Na_2SO_3$$

Write the names of the two products formed from the sodium salt of benzene-sulfonic acid.   →   32

---

**23**      You are mistaken. Sulfonic acids or their salts are converted into phenols by melting with sodium hydroxide:

$$\underset{}{\overset{SO_3Na}{\bigcirc}} + NaOH \longrightarrow \underset{}{\overset{OH}{\bigcirc}} + Na_2SO_3$$

**19**

In order to replace the $-SO_3H$ group by $-CN$, it is necessary to react with ...................

Complete the last sentence and then continue at   13  .

---

**24**      Your answer is wrong; you added the molecular mass of benzenesulfonic acid (158) and the atomic mass of chlorine (35.5).

Draw the structures of benzenesulfonic acid and benzenesulfonyl chloride, and you will see why your answer was wrong. Compute the correct molecular mass and then go to   14  .

**25**    Your answer is wrong; benzyl cyanide is

CH₂CN     CH₂–

;       is the *benzyl residue*

We have seen many other benzyl compounds previously, e.g.

CH₂Cl        CH₂OH

and

benzyl chloride     benzyl alcohol

Repeat  5  .

---

**26**

| *o-, p-Directors:* | alkyl, | –R |
| | halogen, | –Hal |
| | hydroxyl, | –OH |
| | | |
| *m-Directors:* | sulfonic acids, | –SO₃H |
| | carboxylic acids, | –COOH |
| | carbonyl groups, | ≻C=O |

Which substituent determines the outcome of a substitution reaction?

- The substituent already present on the aromatic ring    →    7

- The incoming substituent    →    36

- I don't know    →    13

---

**27**    Your answer is wrong. Draw the structure of benzenesulfonyl chloride beside the one of the sulfonic acid,

$$\text{(C}_6\text{H}_5)\text{–S}\overset{\text{O}}{\underset{\text{OH}}{=}}\text{O}$$

(158)

replacing OH by Cl. To obtain the molecular mass of the product, you must subtract the mass of OH (17) and add that of Cl (35.5). Write down the correct mass.    →    14

**28**     The *cumene process* starts from cumene (isopropyl benzene) which is oxidized to a peroxide. The latter is cleaved into phenol and acetone:

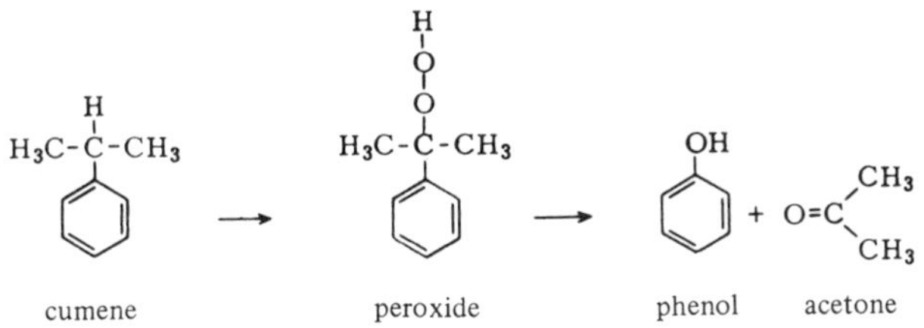

$H_3C$-$\overset{H}{\underset{|}{C}}$-$CH_3$     $\longrightarrow$     $H_3C$-$\overset{\overset{H}{|}\overset{O}{|}\overset{O}{|}}{C}$-$CH_3$     $\longrightarrow$     OH   +   $O=C\overset{CH_3}{\underset{CH_3}{}}$

cumene         peroxide      phenol    acetone

Return now to   |18| .

---

**29**     *Sulfuric acid* or *oleum.*
      *Sulfonation.*

Give the name of the compound:

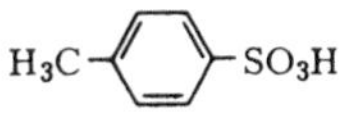

$H_3C$-⟨ ⟩-$SO_3H$

and check your answer at   |16| .

(If you need help, go instead to   |6| ).

---

**30**    ⟨ ⟩-$\overset{O}{\underset{OH}{S}}=O$   or   ⟨ ⟩-$SO_3H$

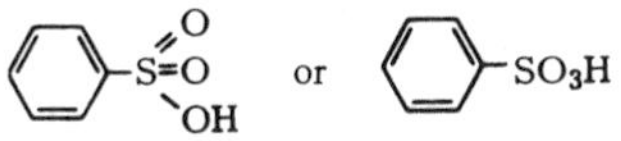

      benzenesulfonic acid

There are two possible naphthalenesulfonic acids, depending on the position of the $-SO_3H$ group as indicated below:

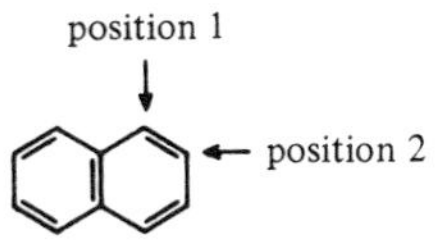

position 1
↓
⟨⟨ ⟩⟩ ← position 2

Draw the formulas for these two naphthalenesulfonic acids.    →   |9|

**19**

---

**31**     You should already know that the nitriles are named after the carboxylic acids, e.g.

CH$_3$CH$_2$COOH     CH$_3$CH$_2$–CN
propionic acid          propionitrile

By analogy give the name of the nitrile below:

benzoic acid          ⋯⋯⋯

→  15

---

**32**     *Phenol* and *sodium sulfite* are formed in the reaction between sodium benzenesulfonate and sodium hydroxide.

The reaction is carried out by heating a mixture of the reactants until it melts (ca. 300 °C). Thus, it is usual to talk of "reaction in the melt".

Write down the equation for the reaction between sodium benzenesulfonate and sodium hydroxide.   →   18

---

**33**     Your answer is wrong. It is important to understand the substitution rules. Therefore, repeat   13   carefully.

---

**19**

---

**34**     A *m-director* directs an incoming substituent into the *meta*-position only.

*m*-Directors are mostly unsaturated groups e.g.

-S=O         -C=O         C=O
 ‖  \          ‖  \          ‖
 O   OH        O   OH

sulfonic acid        carboxyl group        carbonyl groups

In order to predict the outcome of an aromatic substitution reaction, the following questions must be answered:

1) Which substituent is already present?
2) What kind of director is this?

*Example:*
1 mol toluene is treated with 1 mol conc. $H_2SO_4$. A sulfonic acid is formed. In which position does the sulfonic acid group enter?

- *ortho* → $\boxed{44}$

- *meta* → $\boxed{11}$

- *para* → $\boxed{51}$

- *ortho* and *para* → $\boxed{8}$

---

$\boxed{35}$ Your answer is correct; the $-SO_3H$ group directs the incoming substituent to the *m*-position; the methyl group is an *o*-, *p*-director. Thus, the reactions are

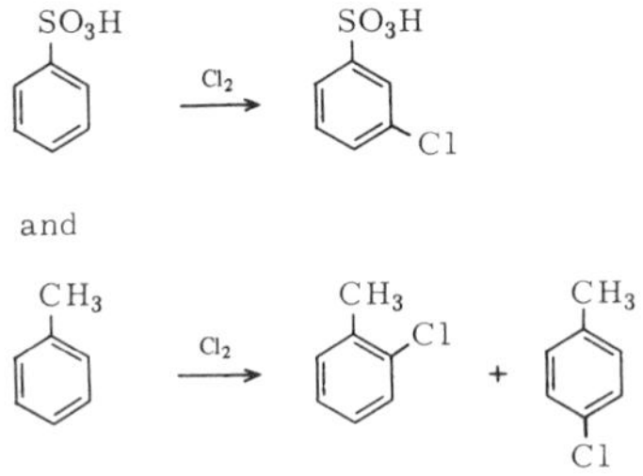

and

How many of the following substituents are *o*-, *p*-directing?

$-COOH, -Cl, -Br, -C_2H_5$.

**19**

- One → $\boxed{55}$

- Two → $\boxed{58}$

- Three → $\boxed{45}$

- All four → $\boxed{62}$

---

$\boxed{36}$ Your answer is wrong: the substituent already present on the aromatic ring directs the positioning of the incoming substituent.

Go back to $\boxed{13}$ .

---

| 37 | How can one prepare *m*-chlorobenzenesulfonic acid? |

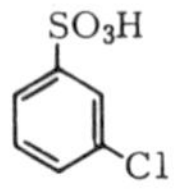

Possible routes:

    1) introduce the –SO₃H group first, then the Cl atom;
    2) introduce the Cl atom first, then the –SO₃H group.

The two reactions would proceed as follows:

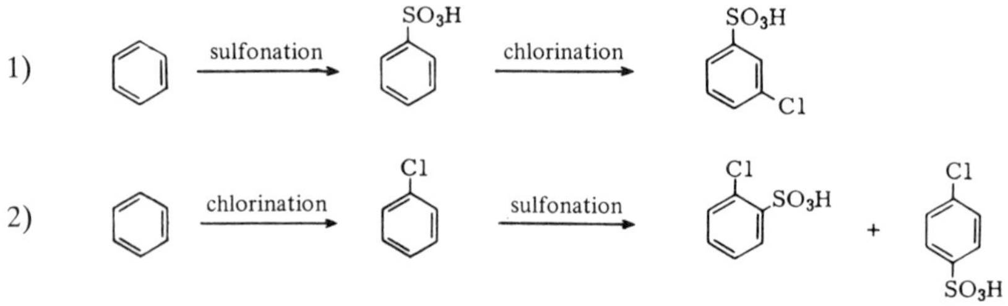

1)

2)

It is clear that only the first route gives the desired product.

Go now to ⌷48⌷ .

---

**19**

| 38 | Compare: |

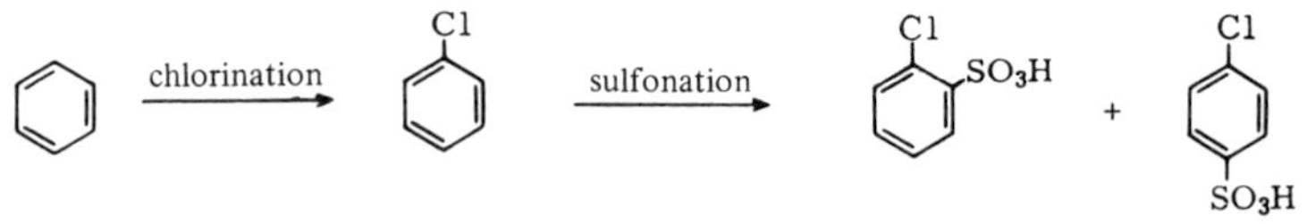

When *two* substituents are present on an aromatic ring, *both* exert a directing effect on the incoming substituent.

The following compound contains both a carboxyl and a hydroxyl group:

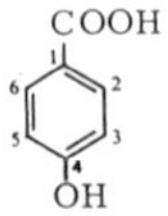

Where will the new substituent enter?

● Positions 2 and 3          →    53

● Positions 3 and 5          →    50

● Positions 5 and 6          →    57

● Positions 2 and 6          →    61

● I need more information     →    64

---

**39**    Compare:

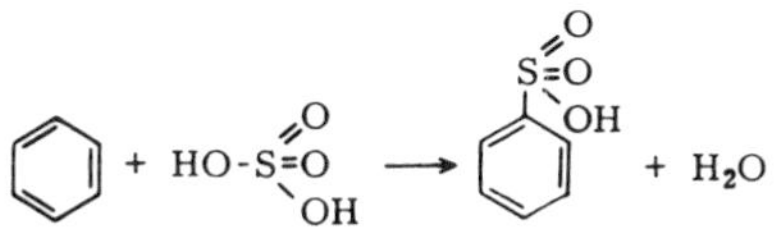

This reaction is reversible. Therefore, it is better written with a double arrow:

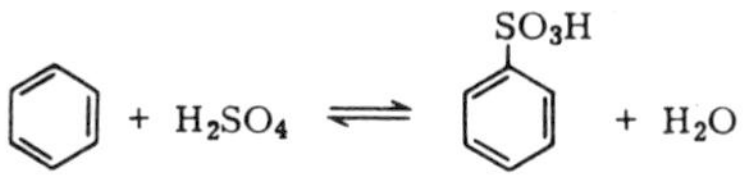

What are the products of the reaction between benzenesulfonic acid and water?

Check your answer at   52   .

**19**

---

**40**    Your answer is wrong. Read the explanation at   60   .

---

**41**    Your answer is wrong. Read the explanation at   37   .

---

**42**

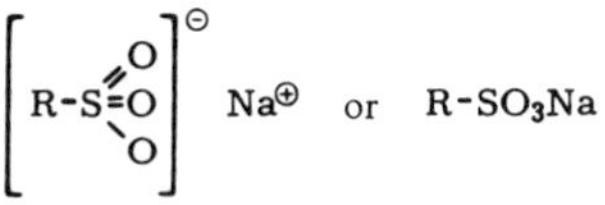

$$\left[ R - S \substack{\nearrow O \\ =O \\ \searrow O} \right]^{\ominus} Na^{\oplus} \quad \text{or} \quad R\text{-}SO_3Na$$

Alkanesulfonates with 12 to 18 carbon atoms are used as *detergents* and *emulsifiers*.

To recapitulate the preparation of these compounds, complete the following equation:

$$R-H \xrightarrow{\text{chlorosulfonation}} R- \hspace{2cm} \xrightarrow{\text{hydrolysis}} R-SO_3Na$$

alkane                  alkane-                sodium

                            sulfonyl             alkanesulfonate

                            chloride

→   56

---

**43**      Your answer is wrong. Read the explanation at   60 .

---

**44**      The methyl group is *ortho*- and *para*-directing. Hence, when toluene is sulfonated the $SO_3H$ group will enter in both the *ortho* and *para* positions.

Give the formulas and names of the two compounds formed.   →   8

---

**45**      Correct; only three of the four substituents mentioned are *o-, p-directors*. These are

       $-Cl, -Br, -C_2H_5$.

Now how many of the following substituents are *m*-directors?

       $-SO_3H$   $-CH_3$   $-OH$   $-C{\overset{\displaystyle O}{\underset{H}{\lessgtr}}}$   $-C{\overset{\displaystyle O}{\underset{R}{\lessgtr}}}$

● One         →   55

● Two         →   58

● Three      →   59

● Four        →   62

● All five     →   65

---

**46**      Chlorosulfonation is carried out with sulfur dioxide ($SO_2$) and chlorine.

This is the end of Program 19.

**47** Your answer is wrong. Read the explanation at 37 .

---

**48** Your answer is correct: benzene is first sulfonated, and then chlorinated.

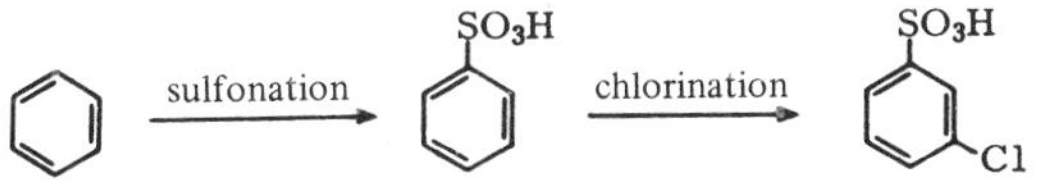

Which products are formed if the two reactions are carried out in the opposite sequence?

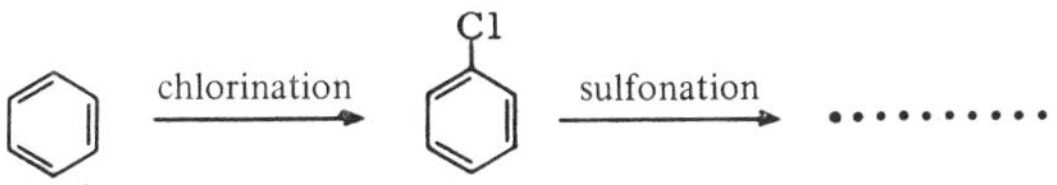

→ 38

---

**49** Your answer is wrong.

The sulfonic acid group is *m*-directing, so the reaction between benzenesulfonic acid and chlorine gives *m*-chlorobenzenesulfonic acid.

The methyl group is, however, *o*-, *p*-directing. Therefore, the reaction between toluene and chlorine gives *o*- and *p*-chlorotoluene.

Draw the structures of the three reaction products.  → 35

19

---

**50** Your answer is correct; a new substituent will enter in positions 3 and 5:

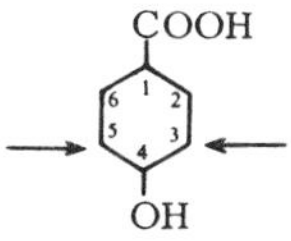

These positions are *ortho* with respect to the hydroxyl group and *meta* with respect to the carboxyl group.

Thus, in this example the directing effects of the two substituents reinforce each other.

This is not true for the following compound:

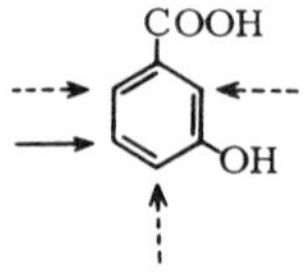

The introduction of a third substituent will lead to a mixture of products since the hydroxyl group directs toward the positions marked with dashed arrows whereas the carboxyl group directs toward that marked by a full arrow.

This terminates our discussion of the substitution rules.

Now complete the equation:

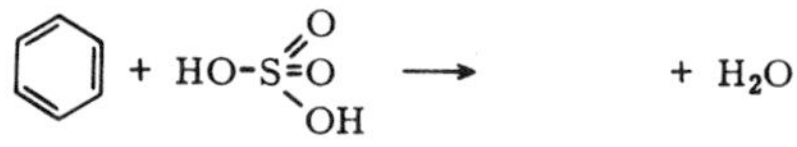

$$\bigcirc + HO-S\overset{O}{\underset{OH}{\overset{\|}{S}}}O \longrightarrow \qquad + H_2O$$

$$\rightarrow \boxed{39}$$

---

| 51 |    Your answer is wrong. The methyl group is an *o-*, *p*-director, So, which are the two products formed when toluene is sulfonated?

Write the formulas and the names.    →    $\boxed{8}$

---

| 52 |    The reaction between benzenesulfonic acid and water gives *benzene* and *sulfuric acid:*

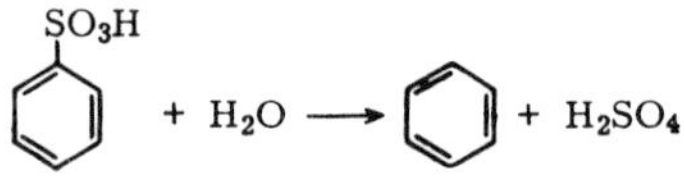

$$\underset{SO_3H}{\bigcirc} + H_2O \longrightarrow \bigcirc + H_2SO_4$$

This removal of the sulfonic acid group is carried out by prolonged boiling with water.

What are the products formed when *p*-toluenesulfonic acid is boiled with water?    →    $\boxed{63}$

---

| 53 |    Your answer is wrong. Read the explanation at    $\boxed{64}$ .

---

| **54** |

$$R-H + SO_2 + Cl_2 \longrightarrow R-S\overset{O}{\underset{Cl}{\lesseqgtr}}O + HCl$$

This reaction is termed a *chlorosulfonation.*

The sulfonyl chloride is hydrolyzed with sodium hydroxide to a salt of the sulfonic acid.

Write the formula for this salt. → |42|

| **55** |

Your answer is wrong. Repeat from |26| .

| **56** |

$$R-H \xrightarrow{\text{chlorosulfonation}} R-S\overset{O}{\underset{Cl}{\lesseqgtr}}O \xrightarrow{\text{hydrolysis}} \left[ R-S\overset{O}{\underset{O}{\lesseqgtr}}O \right]^{\ominus} Na^{\oplus}$$

alkane | alkanesulfonyl chloride | sodium alkanesulfonate

What are the reagents used in the chlorosulfonation?
Write the two names. → |46|

| **57** |

Your answer is wrong. Read the explanation at |64| .

| **58** |

Your answer is wrong.

Repeat from |26| .

**19**

| **59** |

Your answer is correct; only three of the five substituents mentioned are *m*-directors:

$$-SO_3H \quad -C\overset{O}{\underset{H}{\lesseqgtr}} \quad -C\overset{O}{\underset{R}{\lesseqgtr}}$$

A knowledge of the substitution rules is important in order to be able to prepare particular aromatic compounds. Consider the preparation of *m-chlorobenzenesulfonic acid* from *benzene* as an example.

In which order must the two substituents be introduced?

● First Cl, then SO₃H            →    41

● First SO₃H, then Cl            →    48

● The order is not important     →    47

● I don't know                   →    37

---

**60**    When

(a)  benzenesulfonic acid,  , and

(b)  toluene,

are chlorinated, different substitution patterns arise because in (a) the substituent already present (SO₃H) is *meta* directing, while in (b) the substituent (CH₃) is *o-, p*-directing. Now answer the question at ⃞8⃞ again.

---

**61**    Your answer is wrong. Read the explanation at ⃞64⃞ .

---

**62**    Your answer is wrong. Repeat from ⃞26⃞ .

---

**63**    *Toluene* and *sulfuric acid* are formed by boiling a mixture of *p*-toluenesulfonic acid and water.

Until now we have been concerned with *aromatic* sulfonic acids only, the reason being that these are more easily prepared then the *aliphatic* ones. The reaction between an alkane and sulfuric acid often gives a mixture of products which is difficult to separate.

However, the reaction between higher alkanes, SO₂, and Cl₂ leads to the introduction of the chlorosulfonyl group,

$$-S{\overset{\displaystyle \nearrow O}{\underset{\displaystyle \searrow Cl}{=}}}O$$

into the organic molecule. HCl is a by-product:

$$R\text{–}H + SO_2 + Cl_2 \rightarrow \quad \text{................} + HCl$$

alkane             alkanesulfonyl
chloride

Complete this equation    $\rightarrow$    $\boxed{54}$   .

---

$\boxed{64}$    There are two different substituents in the compound

The carboxyl group is *meta* directing, i.e., a new substituent will enter in either position 3 or 5.

The hydroxyl group is *ortho* and *para* directing, i.e., a new substituent can enter in positions 3, 5 or 1. However, since position 1 is already occupied by the carboxyl group, only positions 3 and 5 are available for the incoming substituent.

Thus, the directing effects of the two substituents reinforce each other.

Continue at    $\boxed{50}$   .

---

$\boxed{65}$    Your answer is wrong. Repeat from   $\boxed{26}$   .

# Program 20

## Organic Nitrogen Compounds (I)

**1** You already know some organic nitrogen compounds, e.g. the amides of carboxylic and sulfonic acids. We commence the present discussion with the

*nitro group,* $-N\overset{O}{\underset{O}{\diagup\!\!\!\diagdown}}$ or simpler $-NO_2$ .

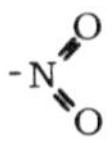

*Nitro compounds* are organic compounds containing this group. They are named by placing the prefix "nitro" in front of the name of the hydrocarbon. Thus, $CH_3-NO_2$ is *nitromethane.*

Name the following compounds:

$C_2H_5-NO_2$ and 

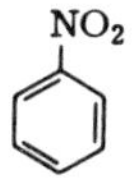

NO$_2$

 $\rightarrow$ 12

**2** The position of each functional group must be designated by a number.

For further information see 15 .

**3** Like toluene, phenol is very easily nitrated.

*Three* nitro groups are introduced into toluene on nitration. How many into phenol?

Return to 34 .

**20**

**4** You did not balance the equation correctly.

$$R-NO_2 + ..... H_2 \rightarrow R-NH_2 + 2 H_2O$$

Count the number of hydrogen atoms on the right and answer the question at 19 again.

**5** Your answer is wrong. Read the explanation at 15 .

---

**6**    You did not balance the equation correctly.

$$R-NO_2 + ..... H_2 \rightarrow R-NH_2 + 2 H_2O$$

Count the number of hydrogen atoms on the right and answer the question at ⌊19⌋ again.

---

**7**    Like toluene, phenol is very easily nitrated.

*Three* nitro groups are introduced into toluene on nitration. How many into phenol?
Return to ⌊34⌋ .

---

**8**    Your answer is wrong. Read the explanation at ⌊15⌋ .

---

**9**

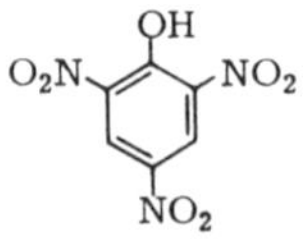

2,4,6-trinitrophenol

This compound is also known as *picric acid* (Greek, *pikros* = bitter) due to its taste and acidity. You will recall that phenols are normally *weak acids.* However, *the three nitro groups in picric acid strongly accentuate the acidity.*

This example shows that there can be a strong *interaction* between functional groups in a molecule. This should always be kept in mind when dealing with complicated compounds.

To recapitulate, which is the stronger acid, phenol or trinitrophenol?    →    ⌊20⌋

**20**

| 10 |

The mixture of *nitric acid* and *concentrated sulfuric acid* is known as "nitration acid" or "sulfuric-nitric acid".

Sulfuric acid need not be specified in nitration equations since it serves only to eliminate the water. Copy the following equation and write the name of each compound:

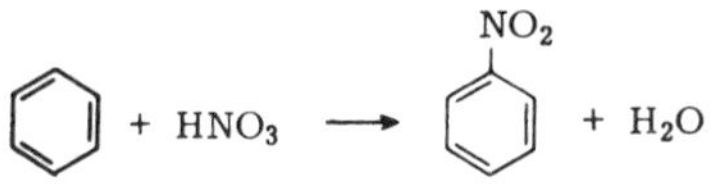

benzene

→   | 21 |

---

| 11 |

The reaction of 1 mol toluene with 1 mol nitric acid gives both *o*- and *p*-nitrotoluene: the methyl group is an *o*-, *p*-director.

The reaction of toluene with an *excess* of nitric acid gives trinitrotoluene.

Give the formula for the latter compound and check your result at | 25 | . If you have diffuculties, go instead to | 36 | .

---

| 12 |

$C_2H_5-NO_2$

nitroethane        nitrobenzene

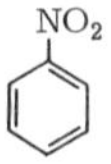

The *aromatic* nitro compounds are the most important; they are readily made by reacting the aromatic starting material with nitric acid. The process is called *nitration*.

**20**

*Complete the equation:*

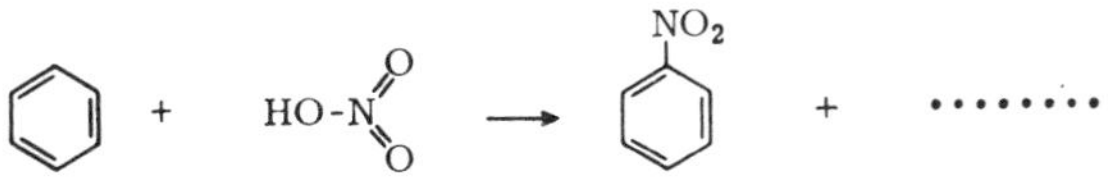

benzene        nitric acid        nitrobenzene

→   | 23 |

---

| 13 |

Your answer is wrong. Read the explanation at | 22 | .

| 14 | |
| --- | --- |
| Name | General formula |
| ammonia | $NH_3$ |
| primary amine | $R-NH_2$ |
| secondary amine | $R\!\!>\!NH$ |
| tertiary amine | $R\!\!>\!N-R$ |

In the above compounds the residues R may be either identical or different.

The names of the amines are formed from those of the hydrocarbon radicals by adding the suffix "amine". Thus,

$CH_3-NH_2$ is methylamine.

Is methylamine a

● primary,    →    28

● secondary, or    →    42

● tertiary amine?    →    46

---

15    The positions of substituents on the aromatic ring are indicated by numbers.

**20**

In trinitrotoluene the methyl group is given the number 1:

$$CH_3$$
$$O_2N \quad NO_2$$
$$NO_2$$

Now write the full name of this compound.    →    34

---

16    Your answer is wrong. Read the explanation at    22    .

**17**    Dimethylamine:

$$\underset{CH_3}{\overset{CH_3}{\diagdown}}NH \quad \left( or \quad \underset{H_3C}{\overset{H_3C}{\diagdown}}NH \right)$$

The abbreviated formula is $(CH_3)_2NH$ or $HN(CH_3)_2$.

*Trimethylamine* is derived from ammonia by replacing the three hydrogen atoms by methyl groups:

$$\underset{CH_3}{\overset{CH_3}{\diagdown}}CH_3-N \quad or \quad \underset{H_3C}{\overset{H_3C}{\diagdown}}H_3C-N \quad or \quad (CH_3)_3N \quad or \quad N(CH_3)_3.$$

Secondary and tertiary amines can carry different hydrocarbon residues. Examples are

    ethylmethylamine,
    ethyldimethylamine, and
    diethylmethylamine

Write the structural formulas for these three compounds.    →    29

---

**18**    Your answer is correct.

Until now we have dealt with *aliphatic* amines only. The simplest *aromatic* amine is

$NH_2$ (on benzene ring)    aniline.

Although this compound could be called phenylamine, it is generally known under the trivial name, *aniline*.

Write the formula for the compound formed by replacing one of the hydrogen atoms in the $NH_2$ group of aniline by methyl. The compound is called *methylaniline*.    →    47

20

| 19 |
|---|

*m*-Nitrobenzoic acid,

Nitro compounds are insoluble or only slightly soluble in water.

The most important *chemical reaction* of nitro compounds is their reduction to amines. The reduction can be carried out with various reducing agents, including hydrogen. The reduction with hydrogen is also known as hydrogenation.

The formation of an amine takes place according to the equation.

$$R-NO_2 + ..... H_2 \rightarrow R-NH_2 + 2 H_2O$$

How many moles of hydrogen are required per mole of nitro compound?

- 1            → ☐4
- 2            → ☐6
- 3            → ☐30
- 4            → ☐33
- I don't know  → ☐38

| 20 |
|---|

Trinitrophenol is a *stronger acid* than phenol itself: the three nitro groups augment the acidity of the OH group.

*Benzoic acid* is also nitrated by nitric acid. In which position of the aromatic ring will the nitro group enter?

- The *ortho* position           → ☐37
- The *meta* position            → ☐32
- The *para* position            → ☐40
- In both *ortho* and *para* position  → ☐43
- I don't know                   → ☐48

**20**

21

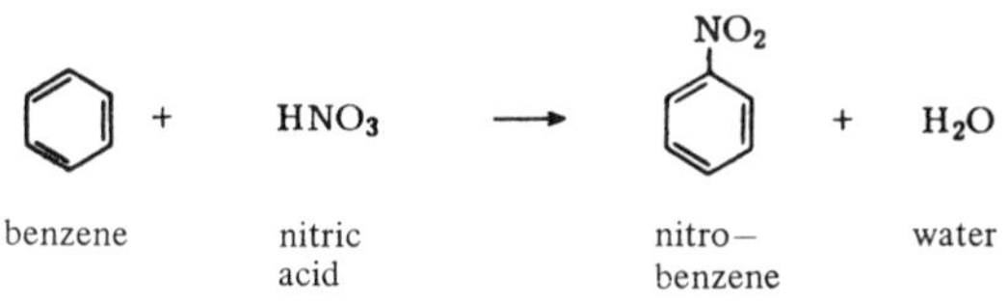

| benzene | nitric acid | nitro- benzene | water |

This reaction is carried out on an industrial scale.

Other aromatic compounds are nitrated in a similar manner. What is the product of the reaction of 1 mol toluene with 1 mol $HNO_3$?

- *o*-Nitrotoluene          →    31

- *m*-Nitrotoluene        →    35

- *p*-Nitrotoluene         →    39

- A mixture of *o*- and *p*-nitrotoluene    →    11

- I don't know             →    45

---

22    The reaction between phenol and an excess of nitric acid leads to the incorporation of *three* nitro groups in the position *ortho* and *para* to the OH group:

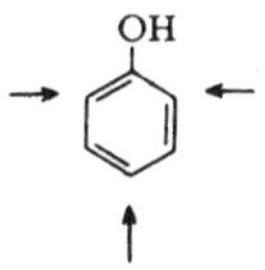

The product formed is a trinitrophenol – the precise name is 2,4,6-trinitrophenol.

Draw the formula for this compound.    →    9

**20**

---

23    *Water* is the by-product of nitration. The formation of water dilutes the nitric acid used, and the reaction would come to a standstill if concentrated sulfuric acid was not added in order to increase the acidity of the medium.

What, therefore, is the reaction medium used for nitrations?    →    10

| 24 |

Correct; the reaction between phenol and an excess of nitric acid gives trinitrophenol, more precisely called 2,4,6-trinitrophenol.

Draw the formula for this compound.    →    | 9 |

---

| 25 |

$CH_3$
$O_2N$        $NO_2$
$NO_2$

trinitrotoluene
(tri–nitro–toluene)

This name is not yet sufficiently precise. What is the correct designation?

● Trinitrotoluene              →    | 2 |

● 1,2,3-trinitrotoluene        →    | 5 |

● 2,4,6-trinitrotoluene        →    | 34 |

● 1,3,5-trinitrotoluene        →    | 8 |

● I need further information    →    | 15 |

---

| 26 |

Correct; there are four isomeric methylanilines. Three of them have the methyl group attached to the aromatic ring; the fourth is *N*-methylaniline.

Draw the four structures.    →    | 50 |

---

**20**

| 27 |

R
    N–R
R

tertiary amine

Complete the following table:

| Name | General formula |
| --- | --- |
| ammonia | $NH_3$ |
| primary amine | ................................................... |
| secondary amine | ................................................... |
| ............... amine | ................................................... |

Check your result at    | 14 |    or, if you need help, go instead to    | 41 |    .

| 28 | Correct; methylamine, $CH_3-NH_2$, is a *primary* amine. |

*Dimethylamine* is a *secondary* amine: two of the $NH_3$ hydrogens have been replaced by methyl groups.

Write the formula for dimethylamine.   →   | 17 |

---

| 29 |

Ethylmethylamine:
$$\begin{array}{c} H_3C \\ \phantom{H_3C} \diagdown \\ C_2H_5 \diagup \end{array} NH$$

ethyldimethylamine:
$$\begin{array}{c} H_3C \\ H_3C{-}N \\ C_2H_5 \end{array}$$

diethylmethylamine:
$$\begin{array}{c} H_3C \\ C_2H_5{-}N \\ C_2H_5 \end{array}$$

If you had any mistakes, repeat from | 41 |. If you had no mistakes, you can surely tell whether the following amines are primary, secondary, or tertiary:

| Ethylmethylamine | Ethyldimethylamine | Diethylmethylamine | | |
|---|---|---|---|---|
| ● primary | secondary | tertiary | → | 44 |
| ● secondary | secondary | secondary | → | 49 |
| ● secondary | tertiary | tertiary | → | 18 |
| ● tertiary | tertiary | tertiary | → | 53 |
| ● I don't know | | | → | 55 |

**20**

---

| 30 |   Correct. – The reduction of the $NO_2$ group to the $NH_2$ group requires *three* moles of hydrogen.

Give the general equation for the reduction of nitro compounds ($R-NO_2$) with hydrogen. Check your result at | 41 |   or, if you need help, go instead to | 19 | .

---

**31**        The methyl group is *o-, p*-directing. Hence, *two* compounds are formed in the nitration of toluene.

Which are the two compounds formed?    →    ⬚11⬚

---

**32**        Correct; the nitration of benzoic acid gives *m*-nitrobenzoic acid.

Draw the structure of this product:

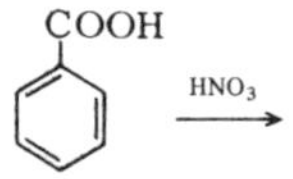

COOH
⬡   HNO₃
    ⟶

→    ⬚19⬚

---

**33**        Count the number of hydrogen atoms on the right-hand side of the equation:

$$R-NO_2 + \ldots H_2 \rightarrow R-NH_2 + 2\,H_2O.$$

How many moles are needed on the left in order to supply this number of hydrogen atoms? Return to ⬚19⬚ .

---

**34**        The three nitro groups are in positions, 2, 4, and 6, so the compound is 2,4,6-trinitrotoluene:

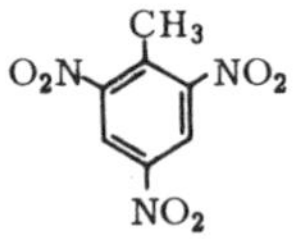

CH₃
O₂N ⬡ NO₂
    NO₂

This material, which is also known as TNT, is used as an explosive. The explosive power of atomic weapons is expressed in "tons TNT".

*Phenol,* too, is readily nitrated by nitric acid. How many nitro groups will enter the molecule when phenol is treated with an excess of nitric acid?

● One          →    ⬚3⬚

● Two          →    ⬚7⬚

● Three        →    ⬚24⬚

**20**

● Four                →    13

● Five                →    16

● I don't know        →    22

---

**35**    Your answer is wrong. Read the explanation at    45   .

---

**36**    The nitration of toluene with an excess of nitric acid leads to the incorporation of three nitro groups in the positions *ortho* and *para* to the methyl group:

Now write the structure of the trinitrotoluene formed    →    25

---

**37**    Your answer is wrong. Read the explanation at    48   .

---

**38**    Count the number of hydrogen atoms on the right-hand side of the equation:

$$R-NO_2 + \text{.....} \ H_2 \rightarrow R-NH_2 + 2 \ H_2O.$$

How many moles are needed on the left in order to supply this number of hydrogen atoms?

Return to    19   .

**20**

---

**39**    The methyl group is *o-*, *p*-directing. Hence, *two* compounds are formed in the nitration of toluene.

Which are the two compounds formed?    →    11

---

**40**    Your answer is wrong. Read the explanation at    48   .

---

**41**    $R-NO_2 + 3 H_2 \rightarrow R-NH_2 + 2 H_2O$

Compounds $R-NH_2$ possessing the *amino group* are called *amines*.

The amines are formally derived from ammonia by replacing one or more hydrogen atoms by hydrocarbon residues.

| $NH_3$ | $R-NH_2$ |
|--------|----------|
| ammonia | *primary* amine |

When only *one* hydrogen atom is replaced, the compound ($R-NH_2$) is called a *primary* amine.

When *two* hydrogen atoms are replaced, a *secondary* amine is obtained:

$$\begin{array}{c} R \\ {}^{\diagdown}NH \\ R^{\diagup} \end{array}$$

When all *three* hydrogen atoms are replaced, the result is a *tertiary* amine.

Give the general formula for a tertiary amine.   →   |27|

---

**42**    The general formulas for primary, secondary, and tertiary amines were given in tabular form at |14| . Compare these formulas with that of methylamine, and you will be able to answer the question.   →   |14| .

---

**43**    Your answer is wrong. Read the explanation at |48| .

**20**

**44**    Primary, secondary, and tertiary amines exist. The primary amines have the formula $R-NH_2$, i.e. the amine nitrogen is attached to only *one* hydrocarbon residue. This is not the case for the compounds mentioned at |29| . Repeat the problem at |29| .

---

**45**    The methyl group directs an incoming substituent to the *ortho* and *para* positions. In other words, *the positioning of an incoming substituent is determined by the one already present.*

Which two compounds, therefore, are formed in the nitration of toluene?   →   |11|

| 46 |
|---|

The general formulas for primary, secondary, and tertiary amines were given in tabular form at ⌐14⌐ . Compare these formulas with that of methylamine, and you will be able to answer the question.   →   ⌐14⌐

| 47 |
|---|

H⁀N⁀CH₃

NHCH₃

or

methylaniline

Since there are several different methylanilines, this compound is correctly called *N*-methylaniline. The prefix "*N*" indicates that the methyl group is attached to nitrogen.

Isomeric methylanilines are obtained by attaching the methyl group to a carbon atom in the ring. How many isomeric methylanilines can exist (including *N*-methylaniline)?

● Two          →   ⌐58⌐

● Three        →   ⌐63⌐

● Four         →   ⌐26⌐

● Five         →   ⌐67⌐

● I don't know   →   ⌐70⌐

| 48 |
|---|

The carboxyl group is a *m*-director. Hence the nitration of benzoic acid,

COOH

gives the *m*-nitro compound.

Write the formula for this compound.   →   ⌐19⌐

| 49 |
|---|

Your answer is wrong.

The three amines do not all carry the same number of alkyl residues.

Repeat   ⌐29⌐ .

**20**

---

| 50 |
The four isomeric methylanilines are:

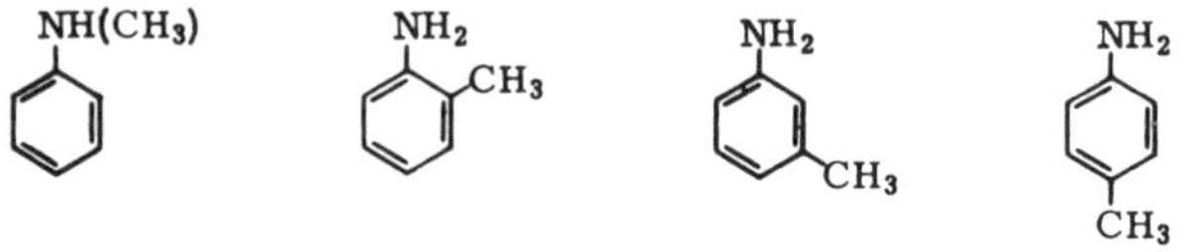

The three compounds methylated in the ring have particular trivial names: they are called *o-*, *m-*, and *p*-toluidine.

You have learned that the positions of substituents may also be expressed by *numbers*, rather than the symbols *o-*, *m-*, and *p-*.

Write the appropriate names of the three toluidines and check your result at [88] . If you have difficulties, see [72] .

---

| 51 |

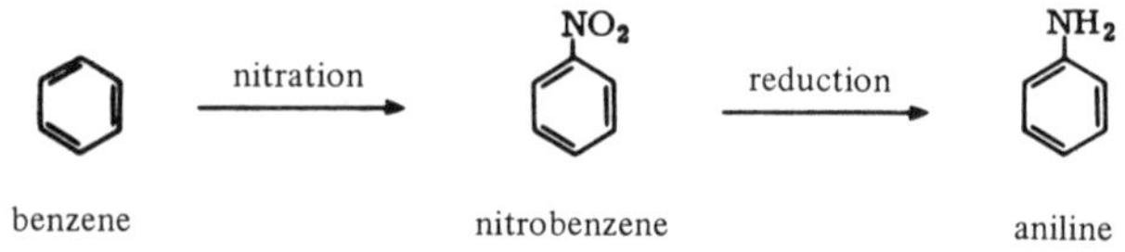

benzene          nitrobenzene          aniline

Two methods are used for the *reduction of nitrobenzene to aniline* in the chemical industry:

   1) Reduction with *hydrogen* gas and a *catalyst*.
   2) *Reduction with nascent hydrogen,* generated from iron and hydrochloric acid.

*Oxides of iron* are formed as by-products in the second reaction. These may be yellow, brown, red, or black, depending on the reaction conditions, and are used as *pigments*. Pigments are insoluble compounds used as colorants in paints, synthetics, concrete, etc.

How is the nascent hydrogen generated?

From ................... and ...................

And what is the by-product?   →   [64]

---

| 52 |

$CH_3I + NH_3 \rightarrow CH_3-NH_2 + HI$

iodide methyl     methylamine

The reaction is not as simple as formulated here. Some of the alkylamine will react with the alkyl halide, giving *secondary* and *tertiary* amines as by-products.

What are the possible by-products in the reaction between methyl iodide and ammonia? Write their formulas and check your answer at $\boxed{71}$ or, if you need help, go instead to $\boxed{81}$ .

---

$\boxed{53}$ Your answer is wrong.

The three amines do not all carry the same number of alkyl residues.

Repeat $\boxed{29}$ .

---

$\boxed{54}$

$$NH_2 + 2\ CH_3Br \longrightarrow N(CH_3)_2 + 2\ HBr$$

This is a *methylation* reaction; aniline has been *methylated*. It is a special case of an *alkylation* reaction. Using various alkyl halides and amines, similar *N*-alkylations can be carried out.

It is possible to prepare primary amines in another way so that no secondary or tertiary amines are formed as by-products. This reaction is called *Hofmann's degradation*. It is a somewhat complicated reaction whereby an acid amide is converted to a primary amine:

$$R-C{\overset{O}{\underset{NH_2}{}}} \longrightarrow R-NH_2$$

The reaction is called a *degradation* because the *product has one carbon atom less than the starting material*.

**20**

Without going into mechanistic details, let us consider two examples of this reaction.

Which amines are formed by Hofmann's degradation of (a) acetamide, (b) propionamide? $\rightarrow$ $\boxed{74}$

---

$\boxed{55}$ Primary, secondary, and tertiary amines exist.

You will find the general formulas at $\boxed{14}$ .

---

| **56** |
|---|

Correct; the reduction (hydrogenation) of nitro compounds gives *primary* amines:

$$R-NO_2 + 3 H_2 \rightarrow R-NH_2 + 2 H_2O$$

What kind of amines are generally prepared in this way?

● Aliphatic amines        →        80

● Aromatic amines        →        68

| **57** |
|---|

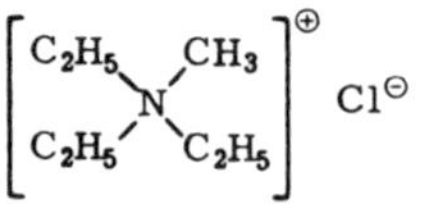

$$\left[ \begin{array}{c} C_2H_5 \quad CH_3 \\ N \\ C_2H_5 \quad C_2H_5 \end{array} \right]^{\oplus} Cl^{\ominus}$$

methyltriethylammonium chloride

Give the name of the following compound:

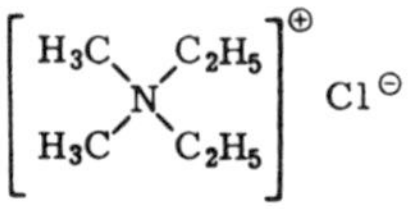

$$\left[ \begin{array}{c} H_3C \quad C_2H_5 \\ N \\ H_3C \quad C_2H_5 \end{array} \right]^{\oplus} Cl^{\ominus}$$

→        69

| **58** |
|---|

Your answer is wrong.

Read the explanation at        70        .

**20**

| **59** |
|---|

$$H_3C \diagdown_N \diagup CH_3$$

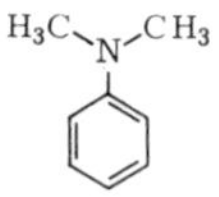

N, N–dimethylaniline

Not only trisubstituted but also tetrasubstituted amines are known. The latter are the *quaternary ammonium salts.*

You already know an inorganic ammonium compound, namely ammonium chloride (see Volume I):

$$NH_3 + HCl \rightarrow NH_4Cl$$

We can write the structure of NH$_4$Cl as follows:

$$\left[\begin{array}{c} H \quad H \\ N \\ H \quad H \end{array}\right]^{\oplus} Cl^{\ominus}$$

Quaternary ammonium salts are tetraalkyl (or aryl) derivatives of the ammonium ion:

$$\left[\begin{array}{c} R \quad R \\ N \\ R \quad R \end{array}\right]^{\oplus} Cl^{\ominus}$$

Write down the corresponding formula for R = CH$_3$.    →    |73|

---

|60|    True, an organic nitro compound does not react with dilute aqueous acid, nor does the acid make it water soluble.

Supposing our unknown compound is an amine, what will happen on treatment with acid?

● The compound dissolves     →    |77|

● Nothing happens     →    |85|

---

|61|    The question was about the name of the product, not the starting material.

Repeat    |84|  .

**20**

---

|62|    Correct; the methyl group is an *o*-, *p*-director, so no *m*-nitrotoluene is formed in the nitration of toluene. Accordingly, *m*-toluidine cannot be obtained by nitrating toluene followed by reduction.

Now consider the preparation of aliphatic amines. These are obtained by alkylation of ammonia with alkyl halides, e.g.:

$$C_2H_5Br + NH_3 \rightarrow C_2H_5-NH_2 + HBr$$

Write the corresponding equation for the reaction of ammonia with methyl iodide.    →    |52|

---

---

**63**   Your answer is wrong.

There are three different *ring*-substituted methylanilines (in these the methyl group is bonded to carbon). There is a fourth compound where the methyl group is at nitrogen.

All in all there are *four* methylanilines.

Draw the four formulas.   →   50

---

**64**   Nascent hydrogen is generated from *iron* and *hydrochloric acid. Iron oxides* are obtained as by-products. The latter may be yellow, brown, red, or black and are used for coloring paints, synthetics, or concrete.

What are such coloring materials called?   →   76

---

**65**   Correct; the compound is

$$\left[\begin{array}{c} H_3C \diagdown \diagup H \\ N \\ H_3C \diagup \diagdown H \end{array}\right]^{\oplus} Cl^{\ominus}$$

dimethylammonium chloride

Write the formula for tetramethylammonium bromide also.   →   89

---

**66**   The reaction of ammonium salts with bases is known from inorganic chemistry. An example is:

$$\left[\begin{array}{c} H \diagdown \diagup H \\ N \\ H \diagup \diagdown H \end{array}\right]^{\oplus} Cl^{\ominus} + NaOH \longrightarrow NH_3 + NaCl + H_2O$$

Trimethylammonium chloride reacts analogously with sodium hydroxide forming trimethylamine:

$$\left[\begin{array}{c} H_3C \diagdown \diagup CH_3 \\ N \\ H_3C \diagup \diagdown H \end{array}\right]^{\oplus} Cl^{\ominus} + NaOH \longrightarrow$$

Complete the equation.   →   92

**20**

---

**67** Your answer is wrong.

Read the explanation at 70 .

---

**68** Correct.

The reduction of nitro compounds is of importance in the preparation of *aromatic* amines.

Aniline is manufactured in this manner, starting from benzene which is first nitrated and then reduced.

benzene         nitrobenzene         aniline

Complete this equation.    → 51

---

**69** Diethyldimethylammonium chloride.

In the examples given above the counterion was always chloride. Similarly, quaternary ammonium bromides, iodides, and sulfates are known. They are named in the same way as described for the chlorides.

We now turn to the *preparation of amines:*

You already know the *reduction (hydrogenation) of nitro compounds,* $R–NO_2$.

**20**

What kind of amines can be prepared in this way?

● Primary, secondary, and tertiary     → 79

● Primary only     → 56

● Secondary only     → 83

● Tertiary only     → 86

● I don't know     → 91

| 70 |
Whenever there are *two* substituents on an aromatic ring, there will be *three isomeric compounds.*

You already know the three xylenes from Program 9:

*o*—xylene          *m*—xylene          *p*—xylene

Similarly, there are three isomeric benzenes carrying a methyl and an amino group. However, as we already know, there is a fourth isomeric compound, in which the benzene ring carries a methyl-substituted amino group:

Thus, in all there are four isomeric methylanilines.

Draw the four formulas.     →     | 50 |

---

| 71 |

Dimethylamine and trimethylamine

**20**

are the by-products in the reaction between ammonia and methyl iodide,

$$CH_3\text{--}I + NH_3 \rightarrow CH_3\text{--}NH_2 + HI.$$

As it is difficult to separate the mixtures of primary, secondary, and tertiary amines formed, the reaction between ammonia and alkyl halides is of little importance in the preparation of *primary* alkyl amines.

When the reaction with ammonia is carried to completion, however, *tertiary* amines are obtained in good yields. As an example, the reaction between 1 mol $NH_3$ and 3 mol methyl bromide gives trimethylamine. Similarly, aniline reacts with two moles of methyl bromide, giving *N,N*–dimethylaniline.

Write the equation for the latter reaction.     →     | 54 |

| 72 |    In systematic nomenclature the carbon atoms in a benzene ring are numbered 1 to 6:

and the positions of substituents are also designated by numbers, e.g.

NH₂

CH₃

1,2-toluidine or
*o*-toluidine

Give the corresponding names for *m*- and *p*-toluidine and draw the formulas for these compounds.   →   |88|

---

| 73 |

$$\left[\begin{array}{c} H_3C \diagdown \quad \diagup CH_3 \\ N \\ H_3C \diagup \quad \diagdown CH_3 \end{array}\right]^{\oplus} \quad Cl^{\ominus}$$

The compound is tetramethylammonium chloride.

Give the formula for methyltriethylammonium chloride also.   →   |57|

---

| 74 |

$$CH_3-C \diagup^{O}_{\diagdown NH_2} \longrightarrow CH_3-NH_2$$

acetamide          methylamine

$$C_2H_5-C \diagup^{O}_{\diagdown NH_2} \longrightarrow C_2H_5-NH_2$$

propionamide       ethylamine

What is this reaction called?   →   |84|

If you forget the name, go back to   |54| .

**20**

75

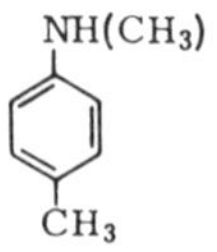

NH(CH₃)

CH₃

*N*—methyl—*p*—toluidine

This compound is a *secondary* amine.

The compound

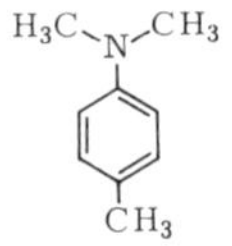

H₃C—N—CH₃

CH₃

is a *tertiary* amine. It is a dimethyl-*p*-toluidine, or more precisely *N,N*–dimethyl-*p*-toluidine (*N,N* because both methyl groups are at N).

What is the formula for *N,N*-dimethylaniline?     →     59

---

76      *Pigments.*

The toluidines are aromatic amines. Some of them are prepared from toluene by nitration followed by reduction. However, *one* of the toluidines is not obtainable in this manner. Which one?

- *o*-Toluidine      →     87

- *m*-Toluidine      →     62

- *p*-Toluidine      →     90

- I don't know      →     95

Note: To answer the question consider the compounds formed on nitration of toluene.

---

77      True; amines dissolve in the form of salts in dilute aqueous acid.

We have now seen how amines can be distinguished from nitro compounds. Further chemical properties of the amines will be discussed in the following program.

This is the end of Program 20.

---

**78**

$$\left[ \begin{array}{c} H_3C \quad\! CH_3 \\ N \\ H_3C \quad\! CH_3 \end{array} \right]^{\oplus}_{2} \quad SO_4^{2\ominus}$$

tetramethylammonium sulfate

You now know that amines form salts. Consequently, some of the equations previously given for the alkylation of amines were incomplete: a free amine, for example, cannot exist in the presence of HCl, as written in the equation:

$$\begin{array}{c} H_3C \\ \quad NH \; + \; CH_3Cl \; \longrightarrow \\ H_3C \end{array} \quad\quad \begin{array}{c} H_3C \\ \quad N\text{-}CH_3 \; + \; HCl \\ H_3C \end{array}$$

instead a trimethylammonium *salt* is formed:

$$\begin{array}{c} H_3C \\ \quad NH \; + \; CH_3Cl \; \longrightarrow \\ H_3C \end{array} \quad \left[ \begin{array}{c} H_3C \quad\! CH_3 \\ N \\ H_3C \quad\! H \end{array} \right]^{\oplus} Cl^{\ominus}$$

The free amines can be liberated from the salts by basification, e.g. addition of NaOH. Trimethylammonium chloride will give trimethylamine and sodium chloride.

Write the equation for this neutralization reaction, and check your result at 92 . If you need help, go instead to 66 .

---

**79**    Your answer is wrong.

Read the explanation at 91 .

**20**

---

**80**    No, *aliphatic* nitro compounds are *difficult* to obtain. Accordingly, the reduction of nitro compounds is of little use in the preparation of aliphatic amines.

This reduction is important in the preparation of ................... amines.

Complete the sentence.    →  68

---

**81**    The initial reaction between ammonia and methyl iodide can be written as follows:

$$CH_3\text{-}I \; + \; NH_3 \rightarrow CH_3\text{-}NH_2 \; + \; HI.$$

Dimethylamine is formed as a by-product because some of the methylamine reacts further with methyl iodide:

$$CH_3\text{-}NH_2 \;+\; CH_3I \;\longrightarrow\; \begin{matrix} H_3C \\ \\ H_3C \end{matrix}\!\!\!\diagdown\!\!\!\diagup\, NH \;+\; HI$$

Dimethylamine can react further to form trimethylamine.

Thus, two by-products, dimethylamine and trimethylamine are formed.

Write the formulas for the two by-products.   →   $\boxed{71}$

---

$\boxed{82}$

aniline

Aniline is a liquid boiling at ca. 200 °C. It is slightly soluble in water.

*Aromatic amines* are important intermediates in the chemical industry. They are used, for example, in the preparation of dyes.

Let us now consider a method which permits a distinction between nitro compounds and amines to be made.

Suppose you have an unknown liquid organic compound. You know that it is pure, contains nitrogen, and is insoluble in water.

Treat this compound with dilute aqueous hydrochloric acid. What would happen if it is a nitro compound?

● It would dissolve       →   $\boxed{93}$

● Nothing would happen   →   $\boxed{60}$

---

$\boxed{83}$    Your answer is wrong.

Read the explanation at · $\boxed{91}$ .

| 84 |    *Hofmann's degradation:*

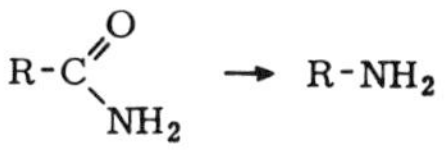

$$R-C\overset{\displaystyle O}{\underset{\displaystyle NH_2}{\diagup}} \quad \longrightarrow \quad R-NH_2$$

The reaction is called a "degradation" because the product has one carbon atom less than the starting material.

Let us now consider further *chemical properties* of the amines.

*Amines form salts with acids.* Like ammonia, amines are *bases*. The following reactions are typical:

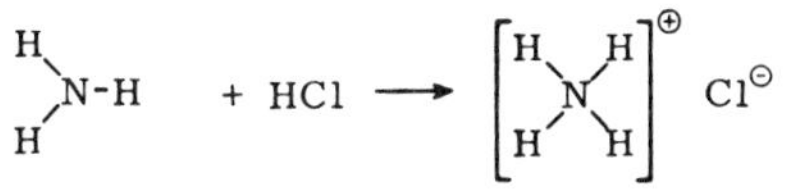

$$\overset{\displaystyle H}{\underset{\displaystyle H}{\diagdown}}N\text{-}H \quad + \text{ HCl} \quad \longrightarrow \quad \left[\overset{\displaystyle H \quad H}{\underset{\displaystyle H \quad H}{N}}\right]^{\oplus} Cl^{\ominus}$$

ammonia                              ammonium chloride

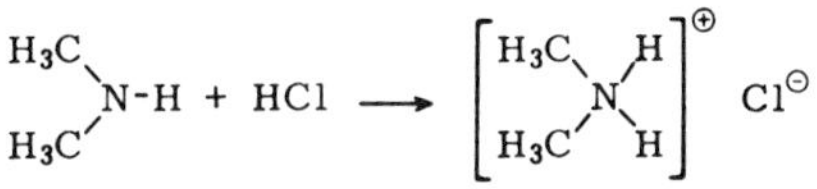

$$\overset{\displaystyle H_3C}{\underset{\displaystyle H_3C}{\diagdown}}N\text{-}H \text{ + HCl} \quad \longrightarrow \quad \left[\overset{\displaystyle H_3C \quad H}{\underset{\displaystyle H_3C \quad H}{N}}\right]^{\oplus} Cl^{\ominus}$$

dimethylamine

Give the name of the product of the last reaction.

● Dimethylamine                    → [61]

● Dimethylammonium chloride        → [65]

● Methylammonium chloride          → [94]

---

| 85 |    No, amines are bases and form salts with acids. Accordingly, an amine which is itself insoluble in water will dissolve in the form of a salt upon treatment with aqueous acid.    → [77]

**20**

---

| 86 |    Your answer is wrong.

Read the explanation at [91] .

---

| 87 |    Your answer is wrong.

Read the explanation at [95] .

---

88

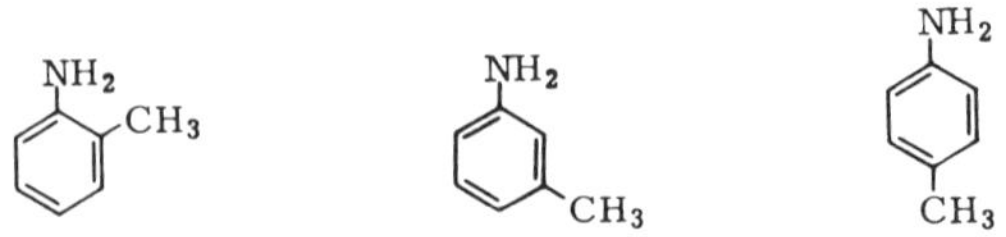

NH₂

.CH₃

o–toluidine
or 1,2–toluidine

NH₂

CH₃

m–toluidine
1,3–toluidine

NH₂

CH₃

p–toluidine
1,4–toluidine

As with aniline, the amine hydrogens can be replaced by methyl groups in the above compounds.

Give the formula for *N*-methyl-*p*-toluidine.     →     75     .

89

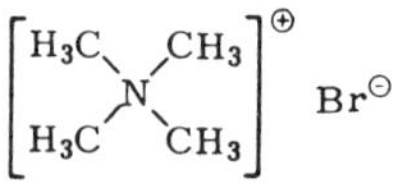

$$\left[ \begin{array}{c} H_3C \quad CH_3 \\ N \\ H_3C \quad CH_3 \end{array} \right]^{\oplus} \quad Br^{\ominus}$$

tetramethylammonium bromide

Amine salts of other acids are also known, e.g. the following salt of a dibasic acid:

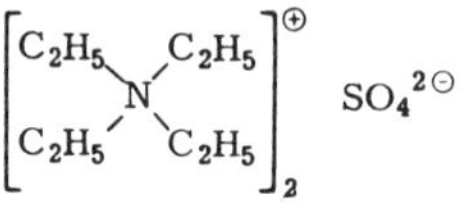

$$\left[ \begin{array}{c} C_2H_5 \quad C_2H_5 \\ N \\ C_2H_5 \quad C_2H_5 \end{array} \right]_2^{\oplus} \quad SO_4^{2\ominus}$$

tetraethylammonium sulfate

Give the name of the compound

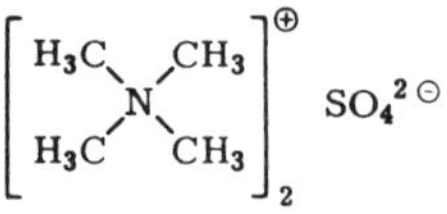

$$\left[ \begin{array}{c} H_3C \quad CH_3 \\ N \\ H_3C \quad CH_3 \end{array} \right]_2^{\oplus} \quad SO_4^{2\ominus}$$

→     78

90     Your answer is wrong.

Read the explanation at     95     .

**20**

---

**91**     The reduction of nitro compounds to amines takes place according to
the general equation

$$R–NO_2 + 3 H_2 \rightarrow R–NH_2 + 2 H_2O$$

Note that *only primary amines* are obtainable.

What is the main importance of this reaction?

● The preparation of aliphatic amines     →     80

● The preparation of aromatic amines     →     68

---

**92**     Compare:

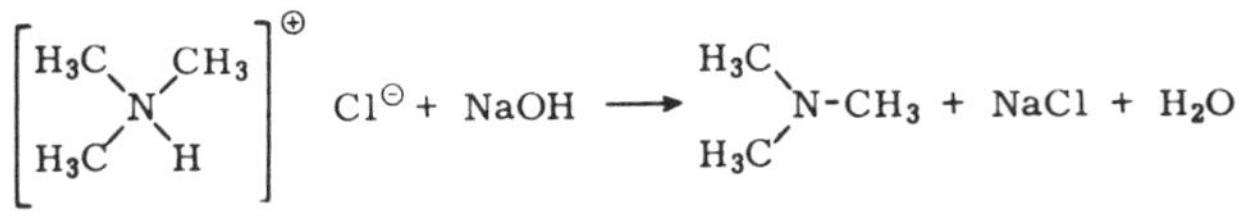

$$\begin{bmatrix} H_3C & CH_3 \\ & N \\ H_3C & H \end{bmatrix}^{\oplus} Cl^{\ominus} + NaOH \longrightarrow \begin{matrix} H_3C \\ \\ H_3C \end{matrix} N\text{-}CH_3 + NaCl + H_2O$$

trimethylammonium          trimethylamine
chloride

The most significant *physical properties* of amines are:

the lower aliphatic amines (e.g. methylamine, dimethylamine, trimethylamine,
and ethylamine) are gases at room temperature. They are soluble in water.

The higher aliphatic amines are liquid or solid and insoluble in water.

The simplest aromatic amine is a liquid boiling at ca. 200 °C. It is slightly soluble
in water.

**20**

What is the name of this aromatic amine?

Write the formula and the name.     →     82

---

**93**     No, a water insoluble nitro compound will also be insoluble in aqueous
acid. The nitro compounds are not basic and do not, therefore, form salts with
acids. Hence, nothing will happen when the nitro compound is treated with
dilute hydrochloric acid.     →     60

---

---

**94** Study **84** more carefully. The product in question has, as you will notice, *two* methyl groups.

---

**95** Since a methyl group is an *o-, p-*director the nitration of toluene gives a mixture of *ortho-* and *para-*nitrotoluene,

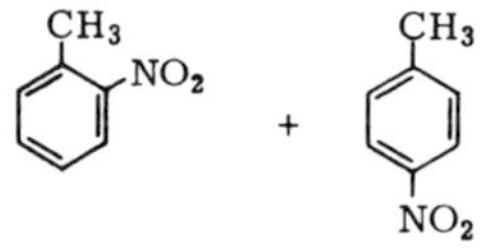

Since *m*-nitrotoluene is *not* formed in this reaction, you will see that one of the toluidines is inaccessible in this way. Modify your answer to the question at **76** .

---

**20**

# Program 21

## Organic Nitrogen Compounds (II)

---

**1**  We will now describe some further reactions of amines.

You will recall that acid chlorides react with ammonia giving *amides:*

$$R-C\overset{\displaystyle O}{\underset{Cl}{\diagup}} + NH_3 \longrightarrow R-C\overset{\displaystyle O}{\underset{NH_2}{\diagup}} + HCl$$

Primary amines react in the same way as ammonia:

$$R-C\overset{\displaystyle O}{\underset{Cl}{\diagup}} + H_2N-R \longrightarrow \cdots\cdots\cdots + HCl$$

Complete the equation.   →  13

---

**2**

aniline     phosgene     phenyl isocyanate

Isocyanates are very reactive compounds, used for example in the manufacture of certain synthetic polymers. This will be described in more detail in Program 22.

Another important reaction takes place when an amine is treated with *nitrous acid,* $HNO_2$. The nitrogen atom in nitrous acid is trivalent. Write the structural formula for this compound.   →  18

**21**

---

**3**  Compare:

$$R-NH_2 + \overset{Cl}{\underset{Cl}{\diagdown}}C=O \longrightarrow R-NH-\underset{\underset{Cl}{|}}{C}=O \longrightarrow \cdots\cdots\cdots$$

I

Compound I is not stable; it eliminates a molecule of HCl.

What is the end product?   →  14

---

| 4 |

$$NaNO_2 + HCl \rightarrow HNO_2 + NaCl$$

sodium nitrite          nitrous acid

The nitrous acid reacts with the amine giving a diazonium salt. This reaction should be carried out at a temperature below +5 °C since the diazonium salt can decompose. The solution is, therefore, cooled by the addition of ice.

Write the overall equation for this reaction:

$$R–NH_2 + HNO_2 + HCl \rightarrow \ldots\ldots\ldots\ldots + 2\ H_2O$$

$$\rightarrow \quad \boxed{16}$$

| 5 |

*Methanol* and *ethanol* are formed in the reaction of nitrous acid with methylamine and ethylamine, respectively.

Primary *aromatic* amines also react with nitrous acid:

$$R–NH_2 + HO–N=O \xrightarrow{-H_2O} R–NH–N=O \rightarrow R–N=N–OH$$

$$\text{I} \qquad\qquad\qquad \text{II}$$

In this case the product II does *not* decompose; in the presence of acids (e.g. hydrochloric acid) it forms a stable salt:

$$R–N=N–OH + HCl \rightarrow [R–N\equiv N]^{\oplus}\ Cl^{\ominus} + \ldots\ldots\ldots\ldots$$

What is the by-product?   $\rightarrow$   $\boxed{21}$

**21**

| 6 |

Correct; an amide is formed.

The reaction between ethylamine and acetyl chloride is a *acylation*.

Let us summarize:

| Reaction | Possible Reagents | Reaction Products |
|---|---|---|
| alkylation | alkyl chlorides | sec. or tert. amine |
| acylation | acid chlorides | acid amide |

Alkyl chlorides are *alkylating agents*. Acid chlorides are *acylating agents*.

Other alkyl halides can be used as well, e.g. the bromides or iodides. Acylations can also be carried out with acid anhydrides in place of the acid chlorides:

$$CH_3-NH_2 \quad + \quad \begin{matrix} CH_3-C{\overset{O}{\diagup}} \\ CH_3-C{\diagdown_O} \end{matrix}O \longrightarrow \quad + \quad \begin{matrix} CH_3-C{\overset{O}{\diagup}}_{NH-CH_3} \quad \text{amide} \\ CH_3-C{\overset{OH}{\diagup}}_{O} \quad \text{acetic acid} \end{matrix}$$

methylamine          acetic anhydride

Write the corresponding equation for the reaction between ethylamine and acetic anhydride.   →   |17|

---

| 7 |

$$\begin{matrix} R \\ \diagdown \\ NH \\ \diagup \\ R \end{matrix} + HO\text{-}N{=}O \longrightarrow \begin{matrix} R \\ \diagdown \\ N\text{-}N{=}O \\ \diagup \\ R \end{matrix} + H_2O$$

The important fact to remember is that *secondary amines react with nitrous acid,* water being eliminated.

Do tertiary amines react in the same way?

● Yes                              →   |25|

● No                               →   |20|

● I don't understand the question     →   |32|

---

| 8 |

$$CH_3-C{\overset{O}{\diagup}}_{Cl} + CH_3\text{-}NH_2 \longrightarrow CH_3-C{\overset{O}{\diagup}}_{NHCH_3} + HCl$$

$$CH_3-C{\overset{O}{\diagup}}_{Cl} + \begin{matrix} CH_3 \\ \diagdown \\ NH \\ \diagup \\ CH_3 \end{matrix} \longrightarrow CH_3-C{\overset{O}{\diagup}}{\underset{\diagdown_{CH_3}}{N}}_{CH_3} + HCl$$

**21**

Do tertiary amines form acid amides in the same manner?

● Yes                              →   |29|

● No                               →   |19|

● I need further information          →   |40|

---

**9**　　The initial reaction between a primary amine and nitrous acid leads to the elimination of water:

$$R–NH_2 + HO–N=O \rightarrow R–NH–N=O + H_2O$$

I

The product I (a nitrosoamine) is not stable but rearranges to II through migration of a hydrogen atom from N to O concomitant with a shift of the double bond:

$$R–NH–N=O \rightarrow R–N=N–OH$$

I　　　·　　II

The compound II is also unstable; it decomposes to an alcohol (R–OH) and
................　　→　　22

---

**10**　　Correct; the temperature of a diazotization reaction is kept at or below +5 °C since the diazonium salts decompose at higher temperatures.

The instability of diazonium salts leads to many important reactions. Some examples follow.

When an aqueous solution of a benzenediazonium salt is heated, nitrogen is evolved and the corresponding phenol is formed, e.g.:

This reaction is analogous to one already seen for a particular class of aliphatic amines with nitrous acid. Which class of amines?

- Primary　　　　　　　　　　→　　41

- Secondary　　　　　　　　　→　　47

- Tertiary　　　　　　　　　　→　　50

- I need further information　　→　　53

---

| 11 |
|---|

Correct; the N-atom number 1 is tetravalent as in an ammonium ion. Therefore, it carries the positive charge.

The name "diazonium salt" comes from "two" (di..) and "nitrogen" (azo...), i.e. they are compounds where two nitrogen atoms are attached to an organic moiety.

From this information derive the general formula for diazonium salts.   →

| 23 |
|---|

| 12 |
|---|

$$C_6H_5\text{-}NH_2 \quad + \quad HNO_2 \quad + \quad HCl \quad \longrightarrow \quad [C_6H_5\text{-}N{\equiv}N]^{\oplus} \; Cl^{\ominus} \quad + \quad 2\,H_2O$$

The diazonium salt formed is called *benzenediazonium chloride*. It is not necessary to remember the names of individual diazonium salts, however. For simplicity, we may speak of the diazonium salt derived from a particular amine.

The diazonium salt derived from *o*-toluidine has the formula

Write the structure of the diazonium salt derived from *p*-toluidine.   →   | 34 |

| 13 |
|---|

Compare:

$$R\text{-}C\!\!\begin{array}{c}{\scriptstyle O}\\{}\end{array}\!\!\diagdown_{Cl} + NH_2\text{-}R \longrightarrow R\text{-}C\!\!\begin{array}{c}{\scriptstyle O}\\{}\end{array}\!\!\diagdown_{NHR} + HCl$$

**21**

*Substituted amides* (i.e. amides carrying a substituent at nitrogen) are formed in the reaction between primary amines and acid chlorides.

The reaction with secondary amines

$$HN\!\!\diagup^{R}_{\diagdown R}$$

is analogous. Formulate the general equation.   →   | 24 |

14      $R-N=C=O$

Compounds of this type are called *isocyanates*. The isocyanato group has the formula

$-N=C=O$

The reaction between aniline and phosgene is an example of the formation of isocyanates.

The product is called *phenyl isocyanate.*

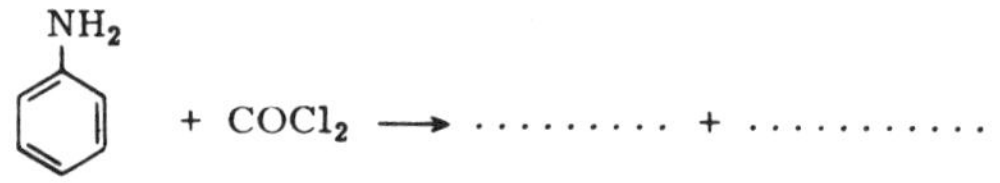

$+ COCl_2 \longrightarrow \cdots\cdots\cdots + \cdots\cdots\cdots$

Complete the equation.    $\rightarrow$    2

---

15      Alkylation and acylation of amines:

| Reaction | Reagents | Reaction Product |
|---|---|---|
| alkylation | alkyl halides | sec. and tert. amine |
| acylation | acid chlorides or anhydrides | acid amide |

Primary amines undergo an interesting reaction with phosgene, $COCl_2$ (the dichloride of carbonic acid).

The first step of the reaction is an acylation of the amine with liberation of one molecule of HCl.

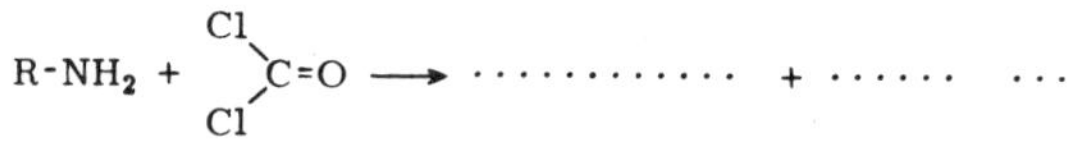

$$R\text{-}NH_2 + \underset{Cl}{\overset{Cl}{\diagdown}}C=O \longrightarrow \cdots\cdots\cdots + \cdots\cdots\cdots$$

Complete the equation.    $\rightarrow$    3

---

16      $R-NH_2 + HNO_2 + HCl \rightarrow [R-N\equiv N]^{\oplus}Cl^{\ominus} + 2\ H_2O$

This type of reaction is called a *diazotization*. The amine is *diazotized,* and a *diazonium* salt is formed.

Do you remember the reaction conditions? Complete the following:

**21**

1) The amine is dissolved in an excess of ...................

2) The nitrous acid is generated by the dropwise addition of an aqueous solution of ...................

3) The reaction temperature should be maintained at ca. ....................°C.

→   27

---

**17**   Compare:

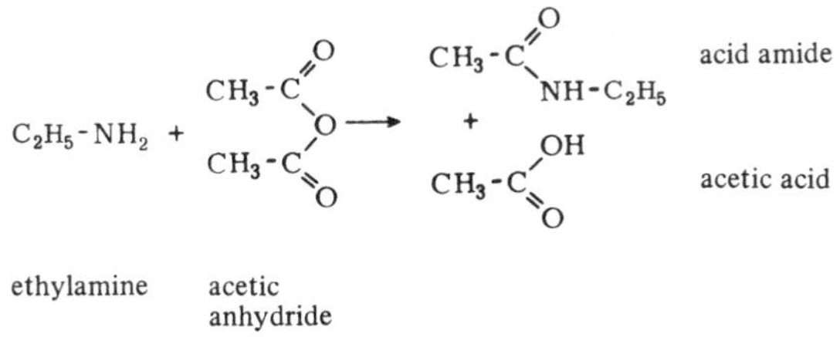

$C_2H_5-NH_2$ + acetic anhydride → acid amide + acetic acid

ethylamine    acetic anhydride

Which of the following compounds are *alkylating agents?*

a) Methyl chloride
b) Ethyl chloride
c) Ethyl bromide
d) Ethyl iodide
e) Acetyl chloride
f) Acetic anhydride

● a) and b)      →   37

● a), b), c), and d)      →   28

● e) and f)      →   31

● All six      →   35

**21**

---

**18**   $H-O-N=O$

nitrous acid

Primary, secondary, and tertiary amines can be distinguished with the aid of their differing reactions with nitrous acid.

Consider first the secondary amines:

$$\begin{array}{l} R \\ \phantom{R}\diagdown \\ \phantom{RR}NH \ + \ HO\text{-}N\text{=}O \ \longrightarrow \\ \phantom{R}\diagup \\ R \end{array}$$

nitrous acid

The reaction takes place with elimination of water. Knowing this, derive the structure of the product and write the complete equation.  →  $\boxed{7}$

---

$\boxed{19}$     Correct; since tertiary amines lack an N–H function, they do not react with acid chlorides.

The reaction between an amine and an acid chloride is called an *acylation*. The amine is *acylated* in such a reaction. The related *alkylation* reactions were described in the previous program. Compare these two types of reaction:

alkylation: $R\text{-}NH_2 \ + \ R\text{-}Cl \ \longrightarrow \ R\text{-}NH\text{-}R \quad +HCl$

acylation: $R\text{-}NH_2 \ + \ R\text{-}C\overset{\displaystyle O}{\underset{\displaystyle Cl}{\big<}} \ \longrightarrow \ R\text{-}NH\text{-}\overset{\displaystyle O}{\underset{}{C}}\text{-}R \ + \ HCl$

In an alkylation reaction an *alkyl* group is attached to the amine nitrogen, in an acylation reaction the *acyl* group is attached.

What is the outcome of the acylation reaction between ethylamine and acetyl chloride?

- A reaction does not take place     →  $\boxed{30}$

- A secondary amine is formed     →  $\boxed{33}$

- An amide is formed     →  $\boxed{6}$

- I need more information     →  $\boxed{36}$

---

$\boxed{20}$     Correct; since tertiary amines lack an N–H function, they do not react with nitrous acid.

Primary amines do, however, possess an N–H group, and the initial reaction with nitrous acid takes place just as for secondary amines:

$$R\text{–}NH_2 + HO\text{–}N\text{=}O \rightarrow \ldots\ldots\ldots\ldots + H_2O$$

Complete the equation   →  $\boxed{9}$

---

| 21 | *Water* is the by-product: |

$$R-N=N-OH + HCl \rightarrow [R-N\equiv N]^{\oplus} Cl^{\ominus} + H_2O$$

Compounds of the type

$$[R-N\equiv N]^{\oplus} X^{\ominus}$$

are called *diazonium salts*. This name is derived from the French word for nitrogen (azot); the prefix "di" denotes two nitrogen atoms. The suffix "onium" signifies that the molecule is positively charged like an amm*onium* ion.

$$[R-N\equiv N]^{\oplus} X^{\ominus} \qquad \begin{bmatrix} H & H \\ & N \\ H & H \end{bmatrix}^{\oplus} X^{\ominus}$$

diazonium salt                    ammonium salt

Consequently, these compounds exist only as salts, where $X^{\ominus}$ is the counterion.

Let us number the two nitrogen atoms in the diazonium ion as follows:

$$\overset{1 \quad 2}{[R-N\equiv N]^{\oplus}}$$

Which of the two nitrogen atoms carrries the positive charge?

● Number 1          →          11

● Number 2          →          38

● I don't know      →          44

**21**

| 22 | *Nitrogen.* |

The equation is repeated below:

$$R-NH_2 + HO-N=O \rightarrow R-NH-N=O \rightarrow$$
$$R-N=N-OH \rightarrow R-OH + N_2$$

Which alcohols will be formed in the reactions between methylamine and ethyl-amine with nitrous acid?          →          5

| 23 | $[R-N\equiv N]^{\oplus} X^{\ominus}$ |
|---|---|

If your formula was incorrect, go back to  [5] .

The *chloride* ion is the most common counterion in diazonium salts, i.e., $X^{\ominus} = Cl^{\ominus}$.

To carry out the reaction, the amine is first dissolved in an excess of hydrochloric acid. Since the free nitrous acid is unstable, it is generated by the slow addition of a solution of sodium nitrite in water. The sodium nitrite reacts with the hydrochloric acid present in excess:

$$NaNO_2 + HCl \rightarrow \text{................} + \text{................}$$

| sodium | nitrous |
|---|---|
| nitrite | acid |

Complete the equation.    →    [4]

---

| 24 | |
|---|---|

$$R-C\overset{O}{\underset{Cl}{<}} + HN\overset{R}{\underset{R}{<}} \longrightarrow R-C\overset{O}{\underset{\underset{R}{N}}{<}}R + HCl$$

Now give the equation for the reactions between acetyl chloride,

$$CH_3-C\overset{O}{\underset{Cl}{<}}$$

and (a) methylamine, (b) dimethylamine.    →    [8]

---

| 25 | Your answer is wrong. |
|---|---|

Read the explanation at  [32] .

---

| 26 | *Ethers* are the products of alcohol alkylation. |
|---|---|

*Esters* are the products of alcohol acylation.

Now complete the following table for the alkylation and acylation of *amines:*

| Reaction | Possible Reagent | Reaction Product |
|---|---|---|
| Alkylation | ................................ | ................................ |
| Acylation | ................................ | ................................ |

21

Check your result at ☐15 or, if you need help, go instead to ☐6 .

---

**27** Diazotization reactions (i.e. preparation of diazonium salt solutions) are carried out as follows:

1) The amine is dissolved in an excess of *hydrochloric acid.*

2) The nitrous acid is generated by the dropwise addition of a solution of *sodium nitrite.*

3) The reaction temperature should not be higher than + *5 °C.*

Give as an example the equation for the diazotization of aniline:

$$C_6H_5-NH_2 + \ldots\ldots\ldots\ldots + HCl \rightarrow \ldots\ldots\ldots\ldots + \ldots\ldots\ldots\ldots$$

                    nitrous  acid                              →   ☐12

---

**28** Correct; methyl chloride, ethyl chloride, ethyl bromide, and ethyl iodide are *alkylating* agents.

Which acid derivatives can be used as *acylating* agents? Mention two classes of compounds and check your answer at ☐39 or, if you need help, repeat from ☐6 .

---

**29** As tertiary amines,

$$\begin{array}{c} R \\ \diagdown \\ N\text{-}R \\ \diagup \\ R \end{array}$$

*do not possess an amine hydrogen atom,* they cannot react with acid chlorides.   →   ☐40

**21**

---

**30** A reaction does in fact occur between acetyl chloride and ethylamine. The general equation for such reactions is given at ☐1 .

---

**31** Carboxylic acid derivatives such as the chlorides and the anhydrides are not alkylating but *acylating* agents.

Examine the question posed at ☐17 in more detail.

---

**32**          Secondary amines react with nitrous acid with elimination of water:

$$\begin{array}{c} R \\ \diagdown \\ \diagup \quad NH \\ R \end{array} + HO-N=O \longrightarrow \begin{array}{c} R \\ \diagdown \\ \diagup \quad N-N=O \\ R \end{array} + H_2O$$

Since tertiary amines do not possess an amine hydrogen atom,

$$\begin{array}{c} R \\ \diagdown \\ \diagup \quad N-R \\ R \end{array}$$

they cannot react in the same manner with nitrous acid. They do not, in fact, react at all.    →    [20]

---

**33**          Carboxylic acid *amides* are the products of the reactions between acid chlorides and amines. Examples are given at [8] .

---

**34**

$$N\equiv N] \overset{\oplus}{} Cl^{\ominus}$$

[benzene ring with CH₃ substituent]

Some diazonium salts can be isolated as solids. Normally, this is not done, however, because most diazonium salts are *highly explosive* and shock sensitive. The harmless *aqueous solutions* are used in chemical reactions. Since diazonium salts are also unstable in solution, low temperatures are required.

**21**     Three experimental conditions necessary for the preparation of diazonium salts are given below. Which of the three conditions is due to the instability of the diazonium salts?

● The amine is dissolved in an excess of
   hydrochloric acid.                    →    [45]

● The nitrous acid is generated by slow
   addition of a sodium nitrite solution.    →    [48]

● The reaction temperature is ca. + 5 °C.    →    [10]

---

---

**35**     Not all of the compounds given are alkylating agents. Some of them are *acylating* agents!

Repeat   [17]   .

---

**36**     By now you should be able to answer this question. Examples of the reaction between amines and acid chlorides are given at   [8]   .

---

**37**     Not only alkyl *chlorides* are alkylating agents. The same is true of other alkyl halides, e.g. the bromides and iodides.

Examine the question at   [17]   once more.

---

**38**     Your answer is wrong; admittedly, the question was a difficult one. You will find the explanation at   [44]   .

---

**39**     Acid *chlorides* and *anhydrides* are acylating agents.

Examples are acetyl chloride and acetic anhydride.

Not only amines, but also other compounds possessing reactive hydrogen atoms are susceptible to alkylation and acylation, e.g. alcohols:

Alkylation: $R-OH + R-Cl \longrightarrow R-O-R + HCl$

Acylation: $R-OH + R-C \overset{O}{\underset{Cl}{\big<}} \longrightarrow R-C \overset{O}{\underset{O-R}{\big<}} + HCl$

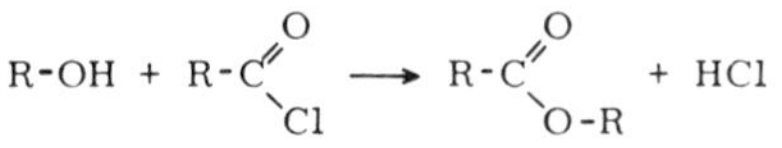

You already know these two reactions and can easily, therefore, complete the following statements:

**21**

a) The alkylation of alcohols gives ....................
b) The acylation of alcohols gives ....................   →   [26]

---

**40**     From the formulas for primary, secondary, and tertiary amines,

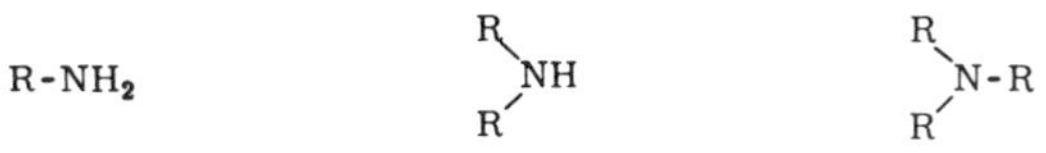

$R-NH_2$     $\overset{R}{\underset{R}{\big>}}NH$     $\overset{R}{\underset{R}{\big>}}N-R$

it is evident that only the primary and secondary ones possess amine hydrogen atoms (N–H groups). Therefore, only the latter react with acid chlorides to give amides.

Do tertiary amines form amides with acid chlorides?

● Yes     →     $\boxed{29}$

● No     →     $\boxed{19}$

---

$\boxed{41}$     Your answer is correct.

The reaction between primary aliphatic amines and nitrous acid results in nitrogen elimination and the formation of alcohols. The related transformation of aromatic amines into phenols requires higher temperatures.

The reaction is carried out by diazotization in dilute sulfuric acid and subsequent boiling of the diazonium salt solution.

This reaction makes it possible to prepare phenols starting from aromatic nitro compounds according to the scheme:

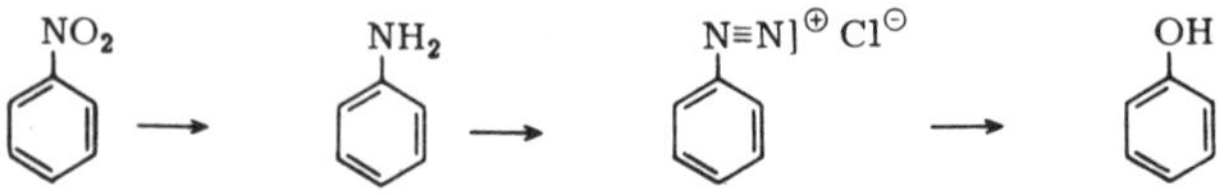

If the diazonium salt is heated with an aqueous solution of potassium iodide, KI, rather than just water, an aryl iodide is formed in place of the phenol:

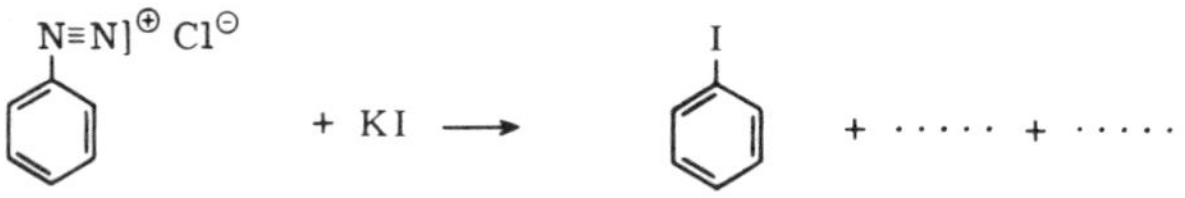

Write the complete equation.     →     $\boxed{54}$

**21**

---

$\boxed{42}$     Correct; the coupling reaction of phenols takes place in *alkaline* solution.

Usually, the coupling between phenols or anilines with diazonium salts takes place in the position *para* to the OH or $NR_2$ group.

Substituted phenols and anilines also couple with diazonium salts.

As an example consider the reaction between salicylic acid and benzenediazonium chloride:

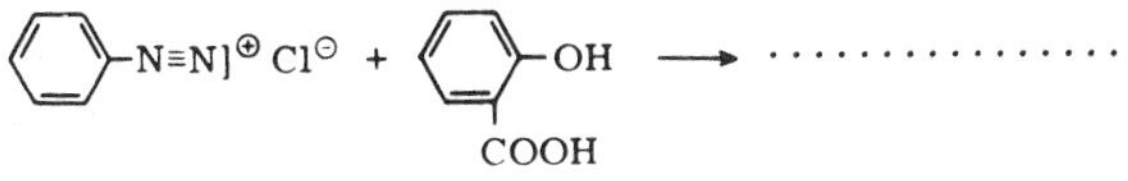

salicylic acid

Complete the equation.   →   52

---

**43**

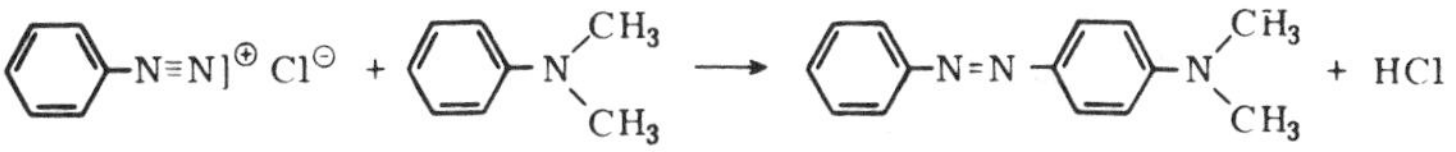

Under which reaction conditions is this reaction carried out?

● Alkaline solution     →   59

● Acid solution     →   55

● Neutral solution     →   65

● I don't know     →   56

---

**44**     In the formula for a diazonium salt,

$$\overset{1\quad 2}{[R\text{–}N\equiv N]^{\oplus}}\ X^{\ominus}$$

the number 1 nitrogen atom is tetravalent, number 2 trivalent. The same is true for ammonium salts and amines, respectively:

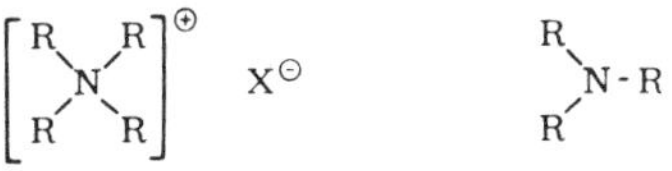

quaternary           tertiary amine
ammonium salt

**21**

Tetravalent nitrogen compounds are, of necessity, positively charged.
Accordingly, *ammonium* compounds exist only as *salts*. Amines are neutral compounds.

The situation is quite similar to that in a diazonium salt. Which of the two nitrogen atoms carries the positive charge?

Write down the answer.   →   11

**45**      The reaction conditions selected by you have no bearing on the instability of diazonium salts.

Due to the thermal instability of diazonium salts, even in aqueous solution, *the reaction temperature should be no higher than + 5 °C.*   →   $\boxed{10}$

---

**46**      $-OH$ and $-I$.

The *coupling reaction* is one of the most important reactions for diazonium salts. It takes place with phenols and aromatic amines as illustrated for phenol in the following example:

Complete the analogous reaction with an amine:

→   $\boxed{58}$

---

**47**      Water and not nitrogen is eliminated in the reaction between *secondary* aliphatic amines and nitrous acid.

You will find a summary of the reactions of different amines with $HNO_2$ at   $\boxed{53}$   .

---

**48**      The reaction conditions selected by you have no bearing on the instability of diazonium salts.

Due to the thermal instability of diazonium salts, even in aqueous solution, *the reaction temperature should be no higher than + 5 °C.*   →   $\boxed{10}$

---

**49**      $[C_6H_5-N\equiv N]^\oplus \; CN^\ominus \xrightarrow{\text{Cu(I)CN}} C_6H_5-CN + N_2$

This reaction is also a *Sandmeyer reaction.*

The Sandmeyer reaction is synthetically important. It allows the introduction of several different groups into an aromatic ring starting from the readily accessible nitro compounds (obtained by nitration). The nitro compounds are reduced to amines, and these are diazotized before subjection to the Sandmeyer reaction:

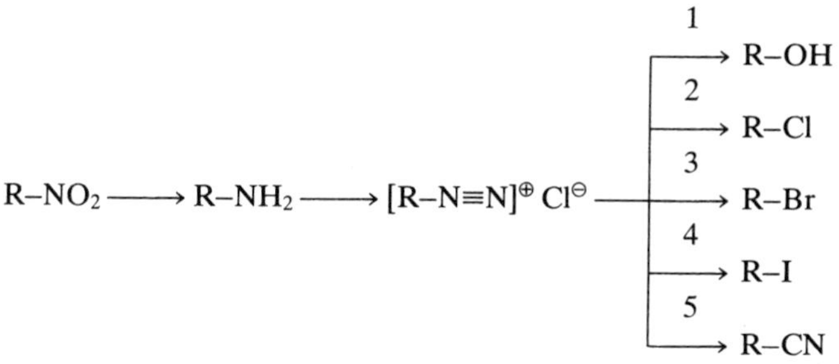

$$R-NO_2 \longrightarrow R-NH_2 \longrightarrow [R-N{\equiv}N]^{\oplus}\, Cl^{\ominus}
\begin{array}{l}
\xrightarrow{\;1\;} R-OH \\
\xrightarrow{\;2\;} R-Cl \\
\xrightarrow{\;3\;} R-Br \\
\xrightarrow{\;4\;} R-I \\
\xrightarrow{\;5\;} R-CN
\end{array}$$

Which of the five Sandmeyer reactions above require the use of Cu(I) as a catalyst?

- All five                    →   63
- Numbers 1 and 4            →   66
- Numbers 2, 3, and 5        →   60
- Numbers 2, 3, and 4        →   61
- I don't know              →   71

---

**50**    No; – tertiary amines do not react with $HNO_2$.

See the summary at   53 .

---

**51**    The bis-azo compound formed from diazotized benzidine and salicylic acid is:

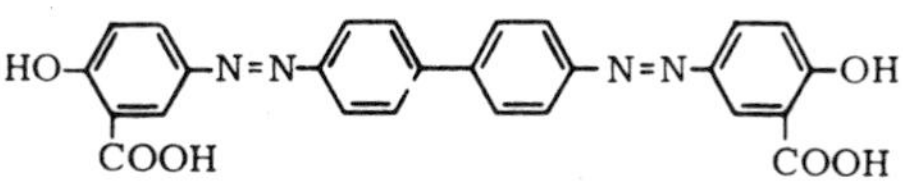

HO–⟨ ⟩–N=N–⟨ ⟩–⟨ ⟩–N=N–⟨ ⟩–OH
　　COOH　　　　　　　　　　　COOH

Let us now consider the quantitative determination of nitrogen in organic compounds. The compound to be analyzed is mixed with an oxidizing agent and is completely combusted in a special apparatus.

Under these conditions, what are the products formed from the carbon and hydrogen atoms in the organic compound?   →   70

**21**

52

$$\text{Ph}-N\equiv N]^{\oplus} Cl^{\ominus} + \text{Ph}-OH \longrightarrow \text{Ph}-N=N-\text{Ph}-OH + HCl$$

COOH

COOH

salicylic acid

The *azo compounds* formed in coupling reactions have important applications as *dyes*.

Consider how one can prepare the following azo dye:

$$\text{CH}_3-\text{Ph}-N=N-\text{Ph}-NH_2$$

- From diazotized aniline by coupling
  with *o*-toluidine                    →     67

- From diazotized *o*-toluidine by coupling
  with aniline                          →     62

- I cannot answer the question          →     69

53     The reactions between amines and nitrous acid are summarized below.

*Primary* aliphatic amines give the corresponding alcohols:

$$R\text{-}NH_2 + HO\text{-}N=O \longrightarrow R\text{-}OH + N_2 + H_2O$$

*Secondary* aliphatic amines react with elimination of water:

$$\begin{array}{c} R \\ \phantom{x} \end{array} NH + HO\text{-}N=O \longrightarrow \begin{array}{c} R \\ \phantom{x} \end{array} N\text{-}N=O + H_2O$$

*Tertiary* aliphatic amines do not react with $HNO_2$.     →     41

**21**

54

$$N\equiv N]^{\oplus} Cl^{\ominus}$$

$$\text{Ph} + KI \longrightarrow \text{Ph}-I + N_2 + KCl$$

Attempts to prepare bromo- and chlorobenzene using analogous procedures give very bad yields. The reactions are, however, catalyzed by Cu(I) salts, and good yields of bromo- and chlorobenzene are obtained in the presence of Cu(I)Br and Cu(I)Cl, respectively. The catalytic action of copper(I) salts was discovered by *Sandmeyer* who gave his name to the reaction.

Write down and complete the equations for the following two Sandmeyer reactions:

$$C_6H_5-N\equiv N^{\oplus} \ Cl^{\ominus} \xrightarrow{Cu(I)Cl} C_6H_5Cl + N_2$$

$$C_6H_5-N\equiv N^{\oplus} \ Br^{\ominus} \xrightarrow{\hspace{1cm}} \dots\dots\dots + N_2 \ \rightarrow \ \boxed{64}$$

---

**55**      Correct; the diazonium coupling of *amines* is carried out in acid solution.

Other compounds couple in *basic* solution. Which class of compounds?    →    $\boxed{42}$

---

**56**      See the general rule for coupling reactions of phenols and amines at   $\boxed{68}$ .

---

**57**

| Organic compound (e.g. aniline) | $\xrightarrow[\text{(with oxidant)}]{\text{combustion in a}\atop \text{CO}_2\text{ stream}}$ | combustion gases $(CO_2, H_2O, N_2)$ |

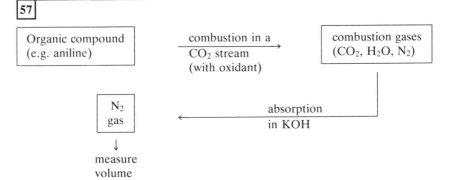

$N_2$ gas $\quad\xleftarrow[\text{in KOH}]{\text{absorption}}$

↓
measure volume

The *qualitative* detection of nitrogen in organic compounds is carried out by destructive heating with a piece of sodium. Nitrogen, carbon, and sodium combine to form sodium cyanide, NaCN, which can be identified by a color reaction.

**21**

This is the end of Program 21.

---

**58**

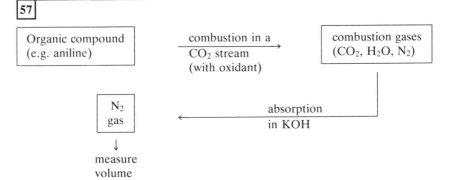

Without looking back at   $\boxed{46}$ , give the equation for the coupling of phenol with benzenediazonium chloride. Check your result at   $\boxed{68}$ .

---

59  You are mistaken. See the general rule for coupling reactions of phenols and amines with diazonium salts at  68  .

60  Correct; a Cu(I) catalyst is needed in the following *Sandmeyer reactions:*

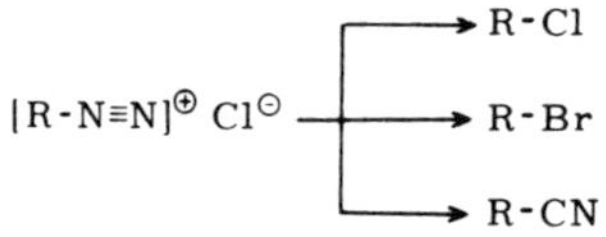

$$[R\text{-}N{\equiv}N]^{\oplus} Cl^{\ominus} \begin{cases} \rightarrow R\text{-}Cl \\ \rightarrow R\text{-}Br \\ \rightarrow R\text{-}CN \end{cases}$$

There are two other substituents which can be introduced in similar reactions *without* the need of a Cu(I) catalyst. Which substituents?  →  46

61  Your answer is wrong.

Read  71  in detail.

62  Your answer is correct:

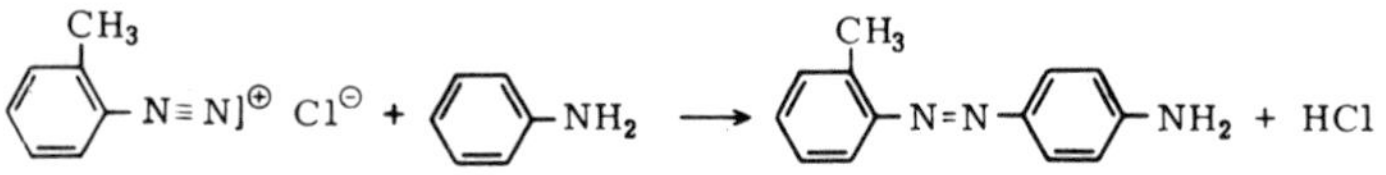

diazotized
o–toluidine

aniline

More complicated azo dyes containing two azo groups can be prepared from benzidine,

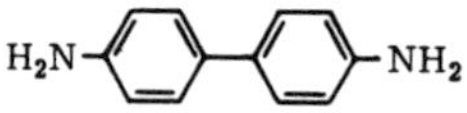

**21**

Diazotation gives a bis-diazonium salt,

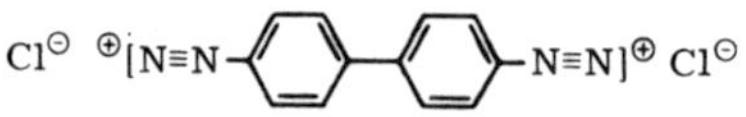

which couples with two molecules of salicylic acid,

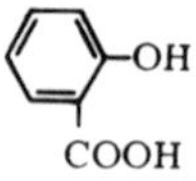

giving a *bis-azo compound* ("bis" for "two").

Draw the formula for the compound formed.  →  51

---

**63**     Your answer is wrong.

Repeat from   |41|  .

---

**64**

$$C_6H_5-N\equiv N^{\oplus} \; Cl^{\ominus} \xrightarrow{Cu(I)Cl} C_6H_5Cl + N_2$$

$$C_6H_5-N\equiv N^{\oplus} \; Br^{\ominus} \xrightarrow{Cu(I)Br} C_6H_5Br + N_2$$

Aromatic nitriles can also be prepared in this manner using copper(I) cyanide as a catalyst.

Complete the equation

$$C_6H_5-N\equiv N^{\oplus} \; CN^{\ominus} \xrightarrow{Cu(I)CN} \ldots\ldots\ldots\ldots\ldots + N_2$$

and give the general name of the reaction.    →   |49|

---

**65**     Your answer is wrong; the coupling reaction does not occur in neutral solution.

Continue at   |46|  .

---

**66**     On the contrary, *no catalyst* is needed for these two reactions!

Read   |71|   carefully.

**21**

---

**67**     The two components would react as follows:

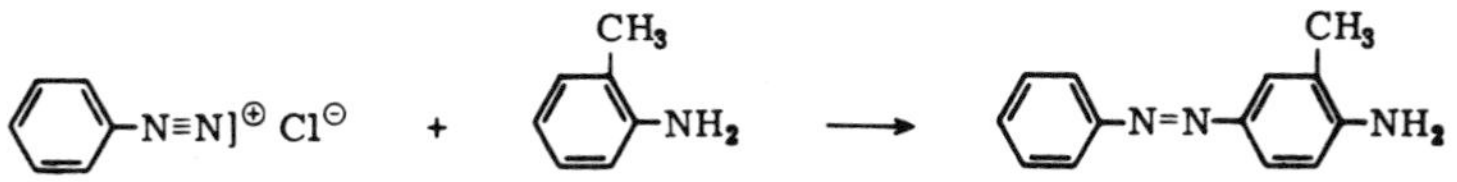

Does this lead to the desired compound?    →   |52|

---

**68**

$$\text{⟨⟩-N≡N]}^{⊕} \text{Cl}^{⊖} + \text{⟨⟩-OH} \longrightarrow \text{⟨⟩-N=N-⟨⟩-OH} + \text{HCl}$$

The following general rules apply to diazonium coupling reactions:

- *phenols* couple in *alkaline* solution;
- *amines* couple in *acidic* solution.

It is easy to remember these rules since:

- phenols *dissolve* (and couple) in alkaline solution, whereas
- amines *dissolve* (and couple) in acidic solution.

Now write the equation for the coupling reaction between benzenediazonium chloride and *N,N*-dimethylaniline:

$$\text{⟨⟩-N}\begin{smallmatrix} \text{CH}_3 \\ \text{CH}_3 \end{smallmatrix}$$          →          43

---

**69**          The azo compound,

$$\overset{\text{CH}_3}{\text{⟨⟩}}\text{-N=N-⟨⟩-NH}_2$$

has to be prepared from a diazonium salt and an aromatic amine (aniline).

$$[\text{R-N≡N}]^{⊕} \text{ Cl}^{⊖} + \text{⟨⟩-NH}_2$$

**21**

Which diazonium salt would you use?     →     52

---

**70**          *Carbon dioxide* and *water* are formed.

The nitrogen present in the organic compound is liberated as nitrogen gas in this reaction.

The combustion is carried out in a stream of $CO_2$ in the presence of an oxidant. The gases formed are passed through a solution of potassium hydroxide in which $CO_2$ and $H_2O$ are absorbed (i.e. both the $CO_2$ formed by combustion and the carrier gas). $N_2$ remains as a permanent gas whose volume permits one to calculate the amount of nitrogen originally present in the organic molecule.

The following scheme summarizes the analytical procedure. Copy it down and give the formulas for the gases formed in the combustion reaction.

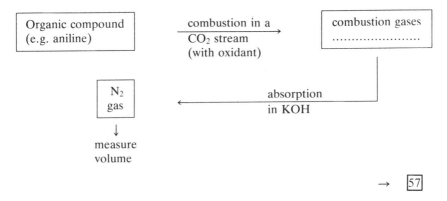

→   57

---

71    The following examplary diazonium salt reactions were presented above:

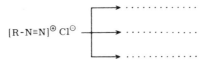

The formation of phenols and iodides (reactions 1 and 4) take place at elevated temperature *without* the use of a catalyst. The three other reactions require the corresponding Cu(I) compounds as catalysts. Complete the following scheme for the *catalyzed* reactions only:

$[R-N\equiv N]^{\oplus} Cl^{\ominus}$ ┤
  → . . . . . . . . . . . .
  → . . . . . . . . . . . .
  → . . . . . . . . . . . .

**21**

→   60

# Program 22

## Synthetic Polymers

---

**1**     The time is past when synthetic polymers were used as substitutes for other, "natural", materials. Today, synthetics are commonplace as construction materials, fibers, films, paints, etc.

A detailed discussion of this interesting field would take a separate volume. The present program aims at presenting the most important basic principles.

The phenomenon of *polymerization* has been described already. What is the product of a polymerization called?

Write the name.    →    13

---

**2**    Correct; the simplest unsaturated compound is ethene or *ethylene*,

$$\begin{array}{c} H \quad H \\ | \quad\;\; | \\ C = C \\ | \quad\;\; | \\ H \quad H \end{array}$$

The first preparation of polyethylene (1935) required very high pressures and temperatures. It took 20 years to discover that much lower pressures and temperatures will suffice in the presence of certain catalysts. The polymers prepared in these two ways have different properties.

Draw a fragment of a polyethylene molecule consisting of just three units of ethylene. Compare your formula with that at   16   or, if you need help, go instead to   27 .

---

**3**    1,6-Hexamethylenediamine has the formula:

$$H_2N-CH_2-CH_2-CH_2-CH_2-CH_2-CH_2-NH_2$$

**22**

The polycondensation is carried out by heating a mixture of this diamine with adipic acid at temperatures above 200 °C. The macromolecule nylon is formed by elimination of water:

Draw the formula for the reaction product which is a fragment of the macromolecule. Compare at $\boxed{15}$ .

---

$\boxed{4}$    Nylon-6,6 is prepared from adipic acid,

$$HOOC–CH_2–CH_2–CH_2–CH_2–COOH$$

and 1,6-hexamethylenediamine,

$$H_2N–CH_2–CH_2–CH_2–CH_2–CH_2–CH_2–NH_2.$$

Nylon-6,6 and nylon-6 are used as synthetic materials, particularly in the form of fibers. Perlon® is a trade-mark for nylon-6.

In contrast to polyacrylonitrile, polyamides are spun from the melt rather than a solution. The molten polyamide emanating from fine pinholes solidifies to fibers on cooling.

An important *polyester,* also used as a synthetic fiber, is prepared by polycondensing methyl terephthalate and ethylene glycol. Try and construct the formulas for the starting materials and check your result at $\boxed{22}$ .

If you need help, go instead to $\boxed{34}$ .

---

**22**

$\boxed{5}$    Vinyl chloride has the formula

   or    $CH_2{=}CHCl$

*Polyvinyl chloride* (PVC) is a very important material used, for example, in floor coatings, tubing, containers, bottles, sheets and films.

Styrene (vinylbenzene) also forms an important polymer. If you have forgotten the formula for styrene but know that of vinyl chloride, then you can surely construct the formula for vinylbenzene. Check your formula at $\boxed{20}$ .

---

**6**

$$\begin{array}{c} \\ \bigcirc \!\!\!\!-\!\!C_{\phantom{O}}^{\phantom{O}O}\!\!-\!OR \\ \phantom{\bigcirc}-\!C\!\!-\!OR \\ \phantom{\bigcirc OR}O \end{array}$$

Phthalic esters are used as *plasticizers,* i.e. compounds that soften polymers.

Many polymers are used in the manufacture of *synthetic fibers.* Polyacrylonitrile is an important example. It is made from acrylonitrile (vinyl cyanide).

Try and construct the formula for acrylonitrile.

Check your answer at [19] , or, if you need help, go instead to [31] .

---

**7**      Homopolymer of B:

B–B–B–B–B–B–B–B....

1:1 copolymer of A and B:

(I)      A–B–A–B–A–B–A–B–A–B....

(II)      A–A–B–A–B–B–B–A–B–A....

The regular copolymer (I) is only one possibility. The irregular composition (II) is more frequently obtained.

The polymers considered so far were all *linear.* The polymeric material is composed of strands of molecules which are partly tangled, partly parallel to each other:

**22**

The fine structure of polymeric materials has been intensely studied in recent years. This research is of importance not only from a theoretical point of view, but also for the use and processing of the materials.

Write down the names of the four polymers considered so far. Check your answer at [17] or go to [24] for a recapitulation.

| 8 |
|---|

*Propylene* (= propene) is the starting material for the preparation of *polypropylene.*

Propylene is a three-carbon compound containing a single and a double bond.

Now give the formula for propylene and check your result at ⟨18⟩ .

| 9 |
|---|

Correct; amides are formed from carboxylic acids and *primary* or *secondary* amines.

In order to obtain a *poly*amide, a *di*carboxylic acid is reacted with a *di*amine:

Elimination of water leads to the formation of a macromolecule.

Nylon was the first polyamide to be manufactured on a large scale. "Nylon" was originally a trade mark but is now used as the generic name of all polyamides.

The original nylon is made from *adipic acid* and *hexamethylenediamine.*

Adipic acid is a straight-chain aliphatic dicarboxylic acid with a total of six carbon atoms.

Give the formula for this compound.    →    ⟨21⟩

| 10 |
|---|

The methylene group has the formula $-CH_2-$. There are *six* methylene groups in 1,6-hexamethylenediamine (hexa = six):

$$-CH_2-CH_2-CH_2-CH_2-CH_2-CH_2-$$

An amino group is attached to each end of the chain, i.e. in positions 1 and 6.

Complete the formula for hexamethylenediamine.    →    ⟨3⟩

**22**

**11**   You do not understand the subject properly!

A tertiary amine,

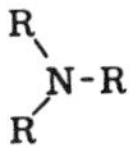

R
 \
  N-R
 /
R

has no hydrogen atoms attached to nitrogen and cannot, therefore, react with a carboxylic acid.

Amides are formed only from *primary* and *secondary* amines.   → ⑨

---

**12**

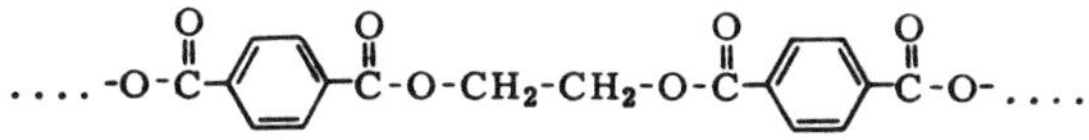

.... -O-C(=O)—⬡—C(=O)-O-CH₂-CH₂-O-C(=O)—⬡—C(=O)-O- ....

Continuation of the reaction results in the formation of a macromolecule.

The copolymer formed from dimethyl terephthalate and ethylene glycol is used exclusively in the manufacture of fibers. Diolene®, Terylene®, and Trevira® are trade marks.

A polycarbonate with important applications has the following formula:

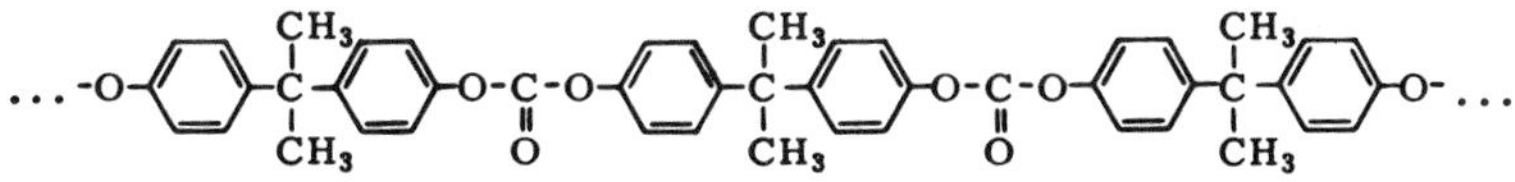

... -O—⬡—C(CH₃)(CH₃)—⬡—O-C(=O)-O—⬡—C(CH₃)(CH₃)—⬡—O-C(=O)-O—⬡—C(CH₃)(CH₃)—⬡—O- ...

It is marketed in Germany under the name of Makrolon®, and in the USA as Lexan®.

What kind of polymer is this?

- A polyamide      →   42
- A polyester       →   37
- A polyether      →   46
- I don't know     →   51

**22**

13  *A macromolecule.*

The molecular mass of a macromolecule can vary between ten thousand and a hundred thousand.

The three chemical reactions used in the preparation of polymeric materials are:

(a) Polymerization
(b) Polycondensation
(c) Polyaddition.

The first of these, the *polymerization* reaction, takes place using unsaturated starting materials. Which is the simplest unsaturated compound?

● Ethane          →    23

● Ethylene (ethene)  →    2

● I don't know      →    26

14  Correct; caprolactam is a seven-membered ring:

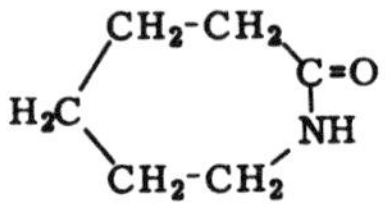

Polymerization to a polyamide can occur when the caprolactam ring opens:

caprolactam                    polyamide

This polyamide is known as nylon-6, and is formed from a single six-carbon precursor.

Do you recall the starting materials used in the preparation of nylon-6,6?

Write down their names and check your answer at   4  .

If you are uncertain, repeat as from   3  .

**15**    Nylon has the formula:

$$\ldots -NH-\overset{O}{\underset{}{\overset{\|}{C}}}(CH_2)_4\overset{O}{\underset{}{\overset{\|}{C}}}-NH(CH_2)_6NH-\overset{O}{\underset{}{\overset{\|}{C}}}(CH_2)_4\overset{O}{\underset{}{\overset{\|}{C}}}-NH(CH_2)_6NH-\overset{O}{\underset{}{\overset{\|}{C}}}-\ldots$$

This polyamide has a molecular mass of 25000–50000. Since both starting materials (adipic acid and hexamethylenediamine) have six carbon atoms, the polymer is called nylon-6,6.

Polyamides of this type are obtained by reacting two precursors having *two* amino groups and *two* carboxyl groups, respectively.

Other starting materials may be used, however. For example, a compound having both an amino and a carboxyl group may also form a polyamide. 6-Aminohexanoic acid (aminocaproic acid) is one example.

Which two end groups are needed to complete the formula for aminocaproic acid?

$$-CH_2-CH_2-CH_2-CH_2-CH_2-$$

● Two $NH_2$-groups            →    [32]

● Two COOH-groups          →    [38]

● One $NH_2$-group and one COOH-group    →    [25]

---

**16**    The following fragment of polyethylene is derived from three molecules of ethylene:

$$\ldots..CH_2-CH_2-CH_2-CH_2-CH_2-CH_2\ldots..$$

Polyethylene has a molecular weight of about 100000. It is one of the most widely used plastics and a great variety of modified products is produced in the world. All kinds of containers, bottles, sheets, tubes, and "polyethene" wrapping material are but a few examples of the uses for polyethylene.

**22**

There are two ways of preparing polyethylene. Depending on the method in question one speaks of

                .................... polyethylene and
                .................... polyethylene.

The first production of polyethylene required very high pressures and temperatures. Low pressure polymerization was achieved by use of special ............ → [29]

$\boxed{17}$
Polyethylene
polypropylene
polyvinyl chloride
polystyrene

These polymers are so-called *thermoplastics,* i.e. they soften on heating and can therefore be moulded.

In the fabrication of plastic materials the powdered or granular polymer is softened by heating and then forced, under pressure, into the mould where it is allowed to cool. When the material has hardened the mould is opened and the desired article is retrieved. Beakers, dishes, switches, combs, radio cabinets, etc., are fabricated in this manner.

Thermoplastics can also be drawn, in the heat, to films or tubes.

In order to soften hard polymers for better processing, so-called plasticizers (softeners) are added.

The esters of phthalic acid with higher alcohols are widely used as plasticizers.

Give the general formula for these esters, derived from an alcohol R–OH. Check your result at $\boxed{6}$ or, if you need help, go instead to $\boxed{30}$ .

---

$\boxed{18}$
Propylene:

$$CH_3-CH=CH_2.$$

Polypropylene has the formula

$$\ldots-\underset{\underset{CH_3}{|}}{CH}-CH_2-\underset{\underset{CH_3}{|}}{CH}-CH_2-\underset{\underset{CH_3}{|}}{CH}-CH_2-\cdots$$

The properties of polypropylene are similar to those of polyethylene, but the softening point is higher.

*Polyvinyl chloride* is another important plastic obtained by the polymerization of an olefin, namely vinyl chloride.

● Give the formula for vinyl chloride        →   $\boxed{5}$

● I need further information        →   $\boxed{28}$

---

**19**     *Acrylonitrile* (vinyl cyanide): $CH_2=CH-CN$.

The fabrication of fibers of *polyacrylonitrile* is achieved by dissolving the polymer in a suitable solvent; jets of the solution are allowed to enter a heated chamber where the solvent evaporates, leaving the fiber. (Simple though this may sound, the development of synthetic fibers took years of research as an entirely new technology was required.)

*Dralon*® and *Orlon*® are two of the trade names of polyacrylonitrile fibers frequently found on textiles. (The ® signifies a registered trade mark.)

Let us now turn to the second chemical reaction leading to macromolecules: the *polycondensation*.

In a polycondensation reaction, a molecule of water (or another small molecule) is eliminated in each condensation step. For example, *polyamides* are obtained in such reactions. Polyamides are polymeric *carboxylic acid amides*. How are such compounds formed?

● From a carboxylic acid and an alcohol                    →    36

● From a carboxylic acid and a primary or secondary amine  →    9

● From a carboxylic acid and a tertiary amine              →    33

● I don't know                                            →    40

---

**20**     Styrene (vinylbenzene)

The polymerization of this compound gives *polystyrene,* a water clear, rather brittle material.

**22**

Polymers prepared from one single type of monomer units are called *homopolymers.* One can also obtain polymers by reacting two or more *different* unsaturated starting materials. Such polymers are called copolymers, and the process is a copolymerization.

The properties and potential applications of synthetic macromolecules can be varied in a controlled manner depending on the composition of these copolymers. Synthetic materials can be "tailor-made" for particular applications. Styrene is often one of the components of copolymers.

Denoting two monomer units A and B, we can formulate a homopolymer as follows:

$$A–A–A–A–A–A–.....................$$

Formulate the analogous homopolymer formed from B alone and a copolymer formed from a 1:1 mixture of A and B.    →   $\boxed{7}$

---

$\boxed{21}$     $HOOC–CH_2–CH_2–CH_2–CH_2–COOH$

adipic acid

This *dicarboxylic acid* is one of the components used in the preparation of nylon.

The other component is a *diamine,* 1,6-hexamethylenediamine. This is a linear aliphatic diamine possessing six carbon atoms.

Give the formula for 1,6-hexamethylenediamine and check your result at $\boxed{3}$ , or, if you need help, go instead to $\boxed{10}$ .

---

$\boxed{22}$

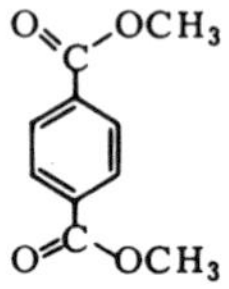

dimethyl terephthalate      ethylene glycol

*Methanol is eliminated* in the polycondensation reaction between these two compounds:

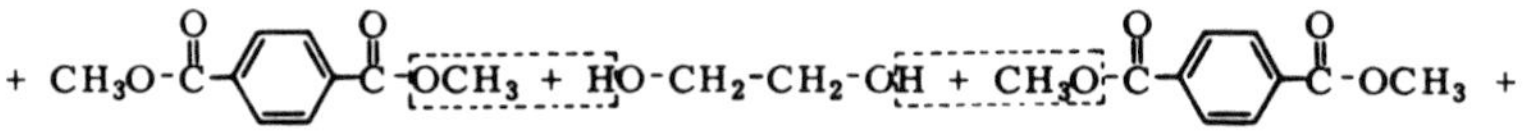

The reaction product is a *polyester*. Give the formula for the product formed from two moles of each compound.    →   $\boxed{12}$

---

$\boxed{23}$     No; – ethane is a *saturated* compound:

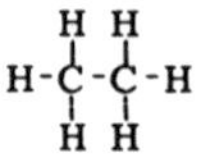

Ethylene (ethene) is the simplest unsaturated molecule. It has two carbon atoms and four hydrogen atoms.

Draw the formula for ethylene.    →    ⟦2⟧

---

**24**    Polyethylene was the first polymer described in this program. It was also mentioned that other polymers can be prepared from propylene, vinyl chloride, and styrene.

Give the names of the four polymers considered so far.    →    ⟦17⟧

---

**25**    Aminocaproic acid has the formula

$$H_2N–CH_2–CH_2–CH_2–CH_2–CH_2–COOH$$

Actually, *caprolactam* is used in place of aminocaproic acid in the preparation of polyamides. Caprolactam is the cyclic compound formed by *intramolecular elimination of water* between the two end groups in aminocaproic acid.

Draw the formula for caprolactam. What is the size of the ring?

● Five-membered ring        →    ⟦35⟧

● Six-membered ring         →    ⟦41⟧

● Seven-membered ring       →    ⟦14⟧

● Eight-membered ring       →    ⟦44⟧

● I don't know              →    ⟦49⟧

---

**26**    Ethane is a saturated compound:

```
    H H
    | |
H - C - C - H
    | |
    H H
```

Ethylene (ethene) is the simplest unsaturated compound. It has two carbon atoms and four hydrogen atoms.

Give the formula for ethylene.    →    ⟦2⟧

---

**22**

**27**    The polymerization of ethylene can be formulated as follows:

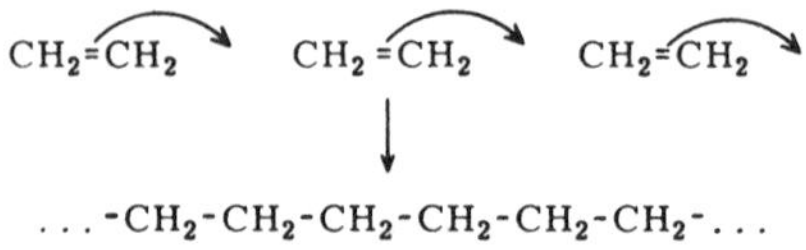

$$CH_2{=}CH_2 \quad CH_2{=}CH_2 \quad CH_2{=}CH_2$$

$$\downarrow$$

$$\ldots\,{-}CH_2{-}CH_2{-}CH_2{-}CH_2{-}CH_2{-}CH_2{-}\ldots$$

polyethylene

How many monomer units are required in order to obtain the polymer chain shown above?    →    |16|

---

**28**    Note the following formulas and names:

| $CH_3{-}$ | $CH_3{-}Cl$ | $CH_2{=}CH{-}$ |
|---|---|---|
| methyl group | methyl chloride | vinyl group |

Now give the formula for vinyl chloride.    →    |5|

---

**29**    *Catalysts*

The materials obtained in the two processes have different physical and chemical properties: the low pressure polyethylene is mainly composed of linear molecules,

$$\ldots\ldots CH_2{-}CH_2{-}CH_2{-}CH_2{-}CH_2 \ldots\ldots$$

whereas the high pressure material is cross-linked:

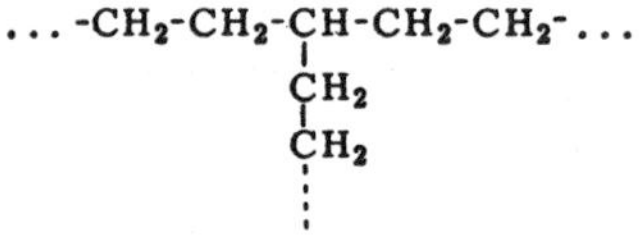

$$\ldots\,{-}CH_2{-}CH_2{-}CH{-}CH_2{-}CH_2{-}\ldots$$
$$\qquad\qquad\qquad |$$
$$\qquad\qquad\quad CH_2$$
$$\qquad\qquad\quad |$$
$$\qquad\qquad\quad CH_2$$
$$\qquad\qquad\quad \vdots$$

**22**

*Polypropylene* is a plastic similar to polyethylene. From what starting material is it made? Give the name and the formula and check your answer at  |18| , or, if you need help, go instead to  |8| .

---

**30**    One mole of phthalic acid reacts with two moles of the alcohol R–OH, giving the plasticizer and water:

$$\begin{array}{l}\text{[benzene ring]}\underset{\overset{\displaystyle C-OH}{\displaystyle \parallel O}}{\overset{\displaystyle O \atop \displaystyle C-OH}{}} + \text{2 R-OH} \longrightarrow \qquad + \text{2 H}_2\text{O}\end{array}$$

Give the formula for the ester formed.    →    $\boxed{6}$

---

$\boxed{31}$    Vinyl chloride has the formula

$CH_2 = CH - Cl$.

Vinyl cyanide is a related molecule in which Cl has been replaced by the cyano group, $-CN$.

Give the formula for vinyl cyanide (=acrylonitrile).    →    $\boxed{19}$

---

$\boxed{32}$    No; in order to obtain a polyamide, $NH_2$- and COOH-groups must be present.

The name *amino*-caproic *acid* clearly indicates that only these two groups are present in the molecule. Complete the formula:

$-CH_2-CH_2-CH_2-CH_2-CH_2-$                    →    $\boxed{25}$

---

$\boxed{33}$    A carboxylic amide is derived from a carboxylic acid and an amine:

$$R-C\overset{O}{\underset{OH}{\Big\backslash}} + H_2N-R \longrightarrow R-C\overset{O}{\underset{NH-R}{\Big\backslash}} + H_2O$$

primary amine

or

$$R-C\overset{O}{\underset{OH}{\Big\backslash}} + HN\overset{R}{\underset{R}{\Big\backslash}} \longrightarrow R-C\overset{O}{\underset{\underset{R}{N}}{\Big\backslash}}R + H_2O$$

secondary amine

Does a tertiary amine react in the same way?

$$R-N\overset{R}{\underset{R}{\Big\backslash}}$$

tertiary
amine

**22**

● Yes          →    $\boxed{11}$

● No           →    $\boxed{9}$

---

**34** *Therephthalic* acid is an aromatic dicarboxylic acid in which the two acid functions are *para* to each other. The dimethyl ester of this diacid is required. *Ethylene glycol* is formally derived from ethylene, $CH_2=CH_2$, by the addition of an OH group to each end of the double bond.

Write the formulas for the two compounds. → 22

**35** Your answer is wrong.

Read the explanation at 49 .

**36** No; an *ester* is the product formed from a carboxylic acid and an alcohol:

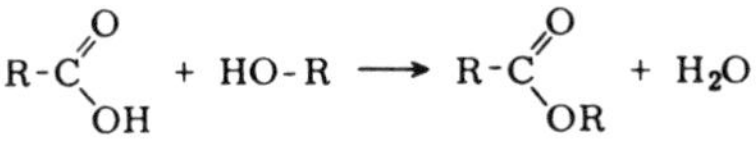

$$R-C\overset{O}{\underset{OH}{\big\langle}} + HO-R \longrightarrow R-C\overset{O}{\underset{OR}{\big\langle}} + H_2O$$

Repeat 19 .

**37** Correct; the compound is a polyester:

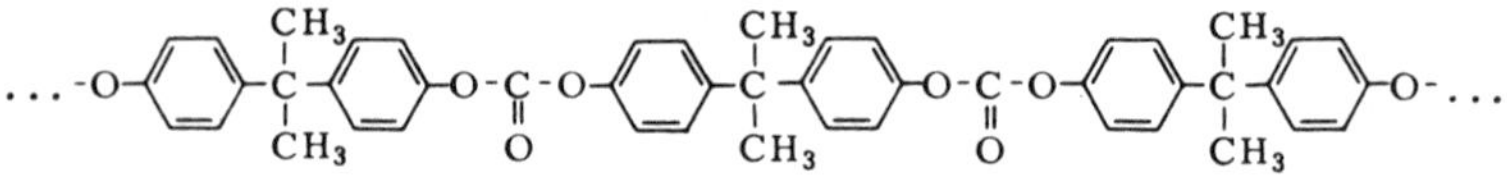

... -O-⟨benzene⟩-C(CH₃)(CH₃)-⟨benzene⟩-O-C(=O)-O-⟨benzene⟩-C(CH₃)(CH₃)-⟨benzene⟩-O-C(=O)-O-⟨benzene⟩-C(CH₃)(CH₃)-⟨benzene⟩-O- ...

It is prepared from a phenolic starting material known as "bisphenol A". This has the formula

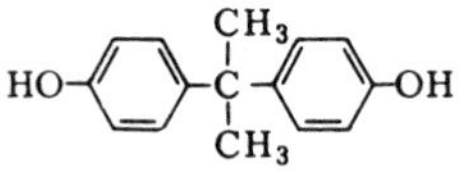

HO-⟨benzene⟩-C(CH₃)(CH₃)-⟨benzene⟩-OH

"Bisphenol A" has to be treated with a dibasic acid in order to obtain the polymeric ester.

Which acid is needed?

● Adipic → 53

● Carbonic → 48

● Terephthalic → 57

● I need more information → 74

**22**

**38**       No; in order to obtain a polyamide, $NH_2-$ and COOH-groups must be present.

The name *amino*-caproic *acid* clearly indicates that only these two groups are present in the molecule. Complete the formula:

$$-CH_2-CH_2-CH_2-CH_2-CH_2-$$

→ ☐25

---

**39**       *Ten* moles of water are eliminated.

A cross-linked polymer can be schematically formulated as follows:

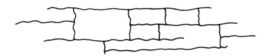

Cross-linked polymers are *infusible* and *insoluble*.

Cross-linking can also be achieved in the polymerization of olefins. Although simple olefins lead to linear polymers,

$$\begin{array}{c} H \ H \\ | \ \ | \\ C=C \\ | \ \ | \\ H \ H \end{array} \quad \rightarrow \quad \ldots-\begin{array}{c} H \ H \\ | \ \ | \\ C-C \\ | \ \ | \\ H \ H \end{array}-\ldots$$

olefins possessing *two* double bonds can cross-link since there are four reactive sites:

$$\begin{array}{c} H \ H \ H \ H \\ | \ \ | \ \ | \ \ | \\ C=C-C=C \\ | \ \ \ \ \ \ \ \ \ | \\ H \ \ \ \ \ \ \ H \end{array} \quad \rightarrow \quad \ldots-\begin{array}{c} H \ H \ H \ H \\ | \ \ | \ \ | \ \ | \\ C-C-C-C \\ | \ \ \vdots \ \ \vdots \ \ | \\ H \ \ \ \ \ \ \ H \end{array}-\ldots$$

**22**

Are cross-linked polymers thermoplastics?

- ● Yes          →    ☐63

- ● No            →    ☐55

- ● I don't know    →    ☐76

| 40 |
|---|

A carboxylic amide is derived from a carboxylic acid and an amine:

$$R-C\overset{O}{\underset{OH}{\diagup}} + H_2N-R \longrightarrow R-C\overset{O}{\underset{NH-R}{\diagup}} + H_2O$$

primary
amine

or

$$R-C\overset{O}{\underset{OH}{\diagup}} + H-N\overset{R}{\underset{R}{\diagup}} \longrightarrow R-C\overset{O}{\underset{N\diagdown R}{\diagup}}\overset{R}{} + H_2O$$

secondary
amine

Does a tertiary amine react in the same way?

$$R-N\overset{R}{\underset{R}{\diagup}}$$

tertiary
amine

● Yes          →     ⬚11⬚

● No           →     ⬚9⬚

---

| 41 |
|---|

Your answer is wrong.

See the explanation at   ⬚49⬚   .

---

| 42 |
|---|

A poly*amide* is a *nitrogen* compound.

There is no nitrogen in the compound at   ⬚12⬚   , however!

Repeat the problem.

---

**22**

| 43 |
|---|

Since fumaric acid is an unsaturated compound, it can indeed polymerize. However, the problem was concerned with the reaction between fumaric acid and ethylene glycol. What kind of reaction is this?      →   ⬚54⬚

---

| 44 |
|---|

Your answer is wrong.

See the explanation at   ⬚49⬚   .

---

**45**     Your answer is wrong.

See the explanation at   56   .

---

**46**     Note the following formulas and names:

$$R-O-R \qquad\qquad R-\overset{\overset{O}{\|}}{C}-O-R$$

ether            ester

Repeat   22   .

---

**47**     Your answer is wrong.

We have not yet discussed polyaddition reactions.

Repeat   65   .

---

**48**     Correct, carbonic acid,

$$HO-\overset{\overset{}{\underset{\underset{O}{\|}}{C}}}-OH$$

is one of the components needed to synthesize the polymeric ester. Because they are carbonic acid esters, such polymers are called *polycarbonates*. The free carbonic acid is, however unstable, and only derivatives are used in the manufacture of the polycarbonates.

Give the formulas of the five components used in the preparation of polycondensates (i.e., terephthalic acid, adipic acid, hexamethylenediamine, ethylene glycol, and "bisphenol A").    →   59

---

**22**

---

**49**     Aminocaproic acid has the formula

$$H_2N-CH_2-CH_2-CH_2-CH_2-CH_2-COOH.$$

If we write the carboxyl group in a more detailed form and "bend" the molecule, we get

$$H_2C\overset{\overset{\displaystyle CH_2-CH_2-C\overset{\diagup O}{\diagdown OH}}{\diagup}}{\underset{\underset{\displaystyle CH_2-CH_2-NH_2}{\diagdown}}{}}$$

Now form an amide by eliminating water between the carboxyl and amino groups.   →   $\boxed{14}$

---

$\boxed{50}$   The two reactions used are

(a) polycondensation and
(b) polymerization.

We shall now describe the third process used in the preparation of polymers:

(c) *polyaddition.*

In the *polyaddition* reaction one of the starting materials which must possess labile hydrogen atoms (amino or hydroxyl groups) is *added* to a reactive double bond in the other starting material. Therefore, the hydrogen atoms do not have the same position in the product as in the precursor molecule.

In contrast, no displacement of atoms takes place in the *polymerization* of unsaturated compounds.

In a *polycondensation* reaction, the monomer units are joined by the elimination of water or other small molecules; i.e. the polycondensate has a chemical composition different from that of the starting materials.

The most important unsaturated starting materials in *polyaddition* reactions are the *isocyanates.*

The isocyanates were described in detail in Program 21. They are obtained by treating a primary amine with phosgene:

$$R\text{-}NH_2 + \begin{matrix} Cl \\ \diagup \\ C=O \\ \diagdown \\ Cl \end{matrix} \longrightarrow \qquad + \ 2\ HCl$$

Complete the equation.   →   $\boxed{64}$

**22**

---

$\boxed{51}$   Note the following formulas and names:

$$R\text{--}O\text{--}R \qquad\qquad R\text{-}\overset{\overset{\displaystyle O}{\|}}{C}\text{-}OR$$
ether                        ester

Repeat   $\boxed{22}$   .

---

**52**

$$O=C=N-CH_2-CH_2-CH_2-CH_2-CH_2-CH_2-N=C=O$$

1,6-diisocyanatohexane

Diols (or triols) are often used as reaction partners in polyaddition reactions of isocyanates. The addition step is:

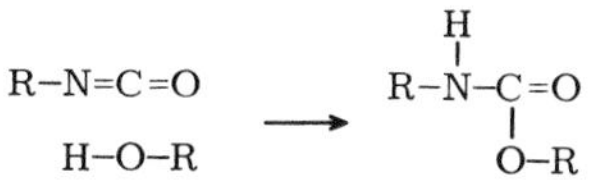

$$\begin{array}{c} R-N=C=O \\ H-O-R \end{array} \longrightarrow \begin{array}{c} H \\ | \\ R-N-C=O \\ | \\ O-R \end{array}$$

i.e., the alcohol is *added* to the C=N double bond in such a way that the hydrogen atom goes to nitrogen and the alkoxy group to the carbon atom of the isocyanate.

Thus, the polyaddition can be formulated schematically as shown below:

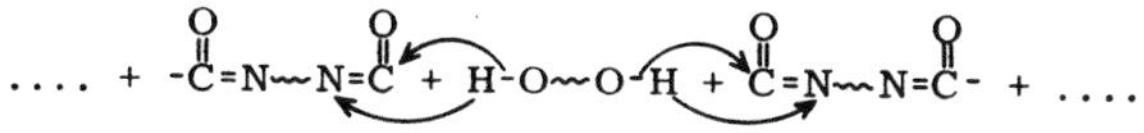

$$\cdots \cdots \; + \; \overset{O}{\overset{\|}{-C}}=N\!\sim\!\!N=\overset{O}{\overset{\|}{C}} \; + \; H\text{-}O\!\sim\!O\text{-}H \; + \; \overset{O}{\overset{\|}{C}}=N\!\sim\!\!N=\overset{O}{\overset{\|}{C}}- \; + \; \cdots \cdots$$

The wavy lines between the nitrogen and oxygen atoms symbolize the hydrocarbon chains.

Complete this polyaddition reaction, giving the formula for the product.   →   83

**53**      Your answer is wrong.

See the explanation at   74   .

**54**      Correct; the reaction between fumaric acid and ethylene glycol is a poly*condensation* reaction.

What is the structure of the product?

● A linear polymer          →      69

● A cross-linked polymer     →      78

● I don't know              →      75

**22**

---

**55**    Correct; cross-linked polymers are infusible and therefore *not* thermo-plastics.

To illustrate the cross-linking reaction we shall describe the formation of *unsaturated polyester resins*. The first step is a polycondensation between *fumaric acid* and *ethylene glycol*.

Give the formulas for these two starting materials and check your result at 65 , or, if you need help, go instead to 73 .

---

**56**    Linear polymers are obtained from *monoolefins* only, i.e. unsaturated compounds containing *one* double bond:

$$\begin{array}{cc} H & H \\ | & | \\ C = C \\ | & | \\ H & H \end{array} \longrightarrow \ldots \ldots \begin{array}{cc} H & H \\ | & | \\ -C-C- \\ | & | \\ H & H \end{array} \ldots \ldots$$

When *two* double bonds are available, there are *four* possible reactive sites, and cross-linking can take place:

$$\begin{array}{cccc} H & H & H & H \\ | & | & | & | \\ C = C - C = C \\ | & & & | \\ H & & & H \end{array} \rightarrow \ldots \ldots \begin{array}{cccc} H & H & H & H \\ | & | & | & | \\ -C-C-C-C- \\ | & | & | & | \\ H & & & H \end{array} \ldots \ldots$$

The polyester in question has *several* double bonds. Which kind of copolymer, therefore, is formed on reaction with styrene?

● A linear copolymer          →    45

● A cross-linked copolymer    →    79

---

**57**    Your answer is wrong.

See the explanation at 74 .

---

**58**    Your answer is wrong.

See the explanation at 66 .

[59]

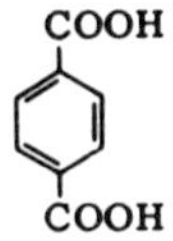

terephthalic acid                adipic acid

$HOOC-CH_2-CH_2-CH_2-CH_2-COOH$

$H_2N-CH_2-CH_2-CH_2-CH_2-CH_2-CH_2-NH_2$

hexamethylenediamine

$HO-CH_2-CH_2-OH$

ethylene glycol                "bisphenol A"

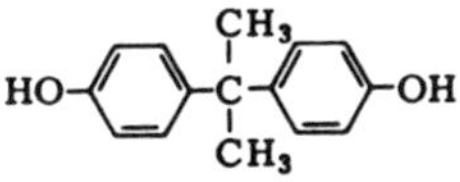

These five substances belong to the following classes of compounds:

| | |
|---|---|
| dicarboxylic acids | HOOC 〜〜〜 COOH |
| diamines | H₂N 〜〜〜〜 NH₂ |
| diols | HO 〜〜〜〜 OH |

*Linear* polymers are formed from precursors such as these.

If *more* than two functional groups are present in the precursor molecules, *cross-linking* will take place. The following scheme shows the ways in which polymer formation between a dicarboxylic acid and a triol can occur:

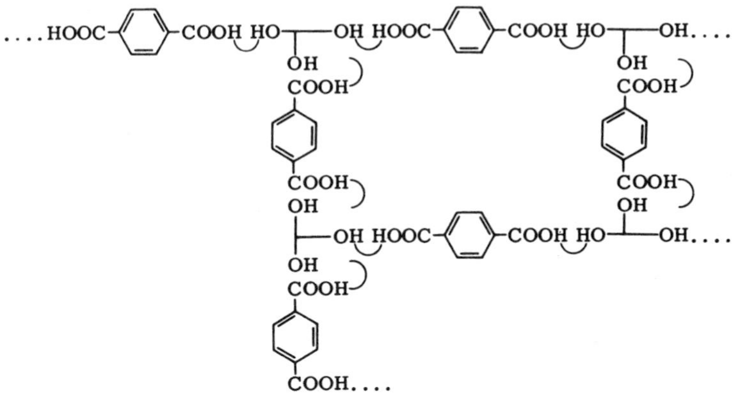

22

How many moles of water will be eliminated in the particular polycondensation shown?   →   [39]

Your answer is correct: cross-linked polymers are infusible and insoluble. Hence, once solidified, they cannot be moulded. The cross-linking reaction is therefore carried out in such a way that the solid product has the desired form. The strength of polyester resins can be improved by glass fibers in much the same way that concrete is strengthened by irons rods. In this procedure a mat of glass fibers is soaked in a solution of the unsaturated polyester and styrene. The mixture is pressed into the desired form. The cross-linking between the unsaturated polyester and styrene is initiated by heating.

Give the names of the reactions used in the preparation of (a) the unsaturated polyester, and (b) the cross-linking of the resin with styrene. Check your answer at [50] or, if you need help, go instead to [55] .

toluene

2,4-Diisocyanatotoluene is derived from toluene by substitution of –NCO groups in positions 2 and 4. Draw the formula.    →    [71]

 Your answer is wrong.

See the explanation at [66] .

**63** Thermoplastics are polymers that soften (i.e. become plastic) on heating.

Cross-linked polymers are infusible and *not* thermoplastic.    →    [55]

**64** Isocyanates are compounds of the type

$$R–N=C=O.$$

In order to achieve a *poly*addition reaction, at least two isocyanato groups must be present in the starting molecule.

1,6-Diisocyanatohexane is one such compound. It is obtained by reaction of hexamethylenediamine with two moles of phosgene.

Develop the formula for the diisocyanate and check it at [52] or, if in doubt, go instead to [77] .

---

**65**

    HOOC–CH=CH–COOH      HO–CH$_2$–CH$_2$–OH

    fumaric acid                  ethylene glycol

What kind of reaction can take place between these two compounds?

- A polymerization    →    43
- A polycondensation    →    54
- A polyaddition    →    47

---

**66**

    *Linear* macromolecules are *fusible* and *soluble*.

*Cross-linked* polymers are *infusible* and *insoluble*. These polymers cannot, therefore, be moulded by heating or dissolution.    →    60

---

**67**

    (a) *linear*
    (b) *cross-linked*

The cross-linking can, of course, be more or less pronounced. There is a wide spectrum of possibilities between the linear and completely cross-linked polymers.

A polymer can also be *branched* without being cross-linked, e.g.:

These polymers are thermoplastics.

The possible structures for cross-linked polymers are shown below:

**B$_1$**                **B$_2$**

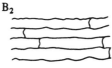

**22**

These can be described as weakly or strongly cross-linked polymers, respectively. Complete the following:

    B$_1$: .................... cross-linked
    B$_2$: .................... cross-linked.    →    80

---

**68**      Correct; a thermoplastic polymer is obtained from 1,6-diisocyanato-hexane and a diol.

The isocyanate function is of particular importance because it can react not only with alcohols, but also with amines or carboxylic acids:

$$-N=C\overset{\overset{O}{\|}}{\underset{H}{\phantom{C}}}\quad \overset{H}{\underset{H}{N}}-R \longrightarrow$$

$$-N=C\overset{\overset{O}{\|}}{\phantom{C}} + \ H\text{-}O\text{-}\overset{\overset{O}{\|}}{C}\text{-}R \longrightarrow$$

Complete the two equations.    →   81

---

**69**      Correct; a linear material is formed in the polycondensation between fumaric acid and ethylene glycol:

$$\ldots\text{-}\overset{\overset{O}{\|}}{C}\text{-CH=CH-}\overset{\overset{O}{\|}}{C}\text{-O-CH}_2\text{-CH}_2\text{-O-}\overset{\overset{O}{\|}}{C}\text{-CH=CH-}\overset{\overset{O}{\|}}{C}\text{-O-CH}_2\text{-CH}_2\text{-O-}\overset{\overset{O}{\|}}{C}\text{-CH=CH-}\overset{\overset{O}{\|}}{C}\text{-}\ldots$$

This is an *unsaturated* polyester containing several double bonds. Being unsaturated, it can undergo a further copolymerization with styrene.

What kind of copolymer will be formed?

● A linear one          →   45

● A cross-linked one     →   79

● I don't know         →   56

**22**

---

**70**      Your answer is wrong.

A *linear* macromolecule is obtained when 1,6-diisocyanatohexane reacts with a diol.

In general, linear macromolecules are fusible and, therefore, thermoplastic.    →   68

---

**71**

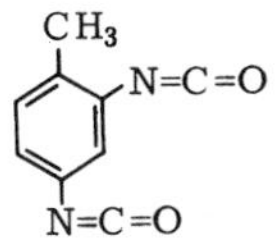

$$CH_3$$

2,4-diisocyanatotoluene

It should be noted that isocyanates are very reactive toward water:

$$R-N=C=O + H_2O \longrightarrow R-NH-C\underset{OH}{\overset{O}{\diagup}}$$

The aminocarboxylic acid formed is *unstable* and decomposes with the elimination of carbon dioxide:

$$R-NH-C\underset{OH}{\overset{O}{\diagup}} \longrightarrow$$

Complete the equation.   →   82

---

**72**     A *linear* macromolecule is obtained from 1,6-diisocyanatohexane and a diol.

In general, linear macromolecules are fusible and, therefore,

thermoplastic.   →   68

---

**73**     Fumaric acid is an olefin

$$-CH=CH-$$

carrying two carboxyl (COOH) groups. Complete the formula!

Ethylene glycol is formally derived from ethylene, $CH_2=CH_2$, by the addition of an OH group to each end of the double bond. Derive the formula!   →   65

**22**

---

**74**     The material is a *polyester*. If this polyester is hydrolyzed according to the following scheme:

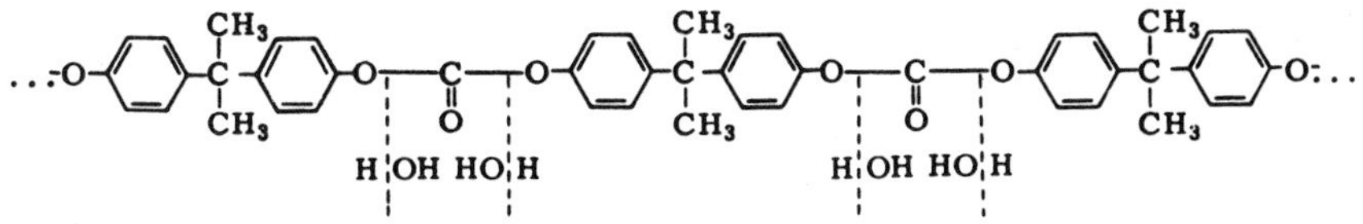

it leads to the formation of the acid:

$$HO-\underset{\underset{O}{\|}}{C}-OH \qquad or \qquad H_2CO_3$$

What is the name of this acid?

Repeat [37] .

---

**75**  In order to appreciate the difference between linear and cross-linked polymers, continue reading at  [59] .

---

**76**  *Thermoplastics* soften on heating.

*Cross-linked* polymers are infusible and do *not* soften on heating. They are *not* thermoplastic.

Continue at  [55] .

---

**77**  1,6-Hexamethylenediamine has the formula

$$H_2N-CH_2-CH_2-CH_2-CH_2-CH_2-CH_2-NH_2$$

Now give the formula for the corresponding diisocyanate.  →  [52]

---

**78**  Your answer is wrong.

In order to appreciate the difference between linear and cross-linked polymers, continue reading at  [59] .

---

**79**  Correct; a *cross-linked* polymer is formed.

Can such a material be moulded?

● Yes, by melting                                →  [58]

● Yes, by dissolution                          →  [62]

● No, neither by melting nor by dissolution    →  [60]

● I don't know                                  →  [66]

**22**

**80**

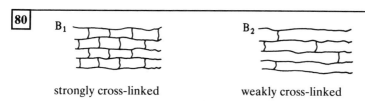

B$_1$                                          B$_2$

strongly cross-linked              weakly cross-linked

Weakly cross-linked polymers have properties similar to those of natural *rubber.*

We have only given some examples of the many polymeric materials known. The ability to manufacture

linear,
branched, and
cross-linked

polymers of varying complexity has led to a range of synthetic materials with widespread applications in modern life.

This is the end of Program 22.

---

**81**

$$-N{=}\overset{\overset{O}{\|}}{C} \quad \overset{H}{N}{-}R \longrightarrow -NH{-}\overset{\overset{O}{\|}}{C}{-}NH{-}R$$

$$-N{=}\overset{\overset{O}{\|}}{C} + H{-}O{-}\overset{\overset{O}{\|}}{C}{-}R \longrightarrow NH{-}\overset{\overset{O}{\|}}{C}{-}O{-}\overset{\overset{O}{\|}}{C}{-}R$$

2,4-Diisocyanatotoluene is a particularly important isocyanate. The prefix "isocyanato" for the $-N{=}C{=}O$ group is used in the same way as "chloro", "bromo", etc.

Give the formula for this compound and check your result at $\boxed{71}$ or, if you need help, go instead to $\boxed{61}$ .

**22**

---

**82**

$$R{-}NH{-}\overset{\overset{O}{\diagup}}{\underset{\cdot\ OH}{C}} \longrightarrow R{-}NH_2 + CO_2$$

This reaction is used in the formation of *polymer foams:* the diisocyanate is allowed to polymerize with a diol (or triol) in the presence of water. Some of the isocyanato groups will react with water, giving $CO_2$. The liberation of $CO_2$

causes the polymer to expand as a foam. Polyurethane foams which are commercially available in spray-cans work according to this principle.

Now consider the two types of polymers with the following schematic structures:

A

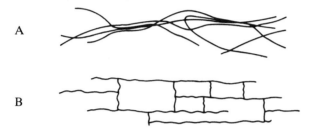

B

What are the general names of such polymers?

A: ...................
B: ...................

→  67

---

**83**  The polyaddition between a diisocyanate and a diol takes the course:

$$\ldots\ldots + \overset{O}{\overset{\|}{C}}{=}N{\sim}N{=}\overset{O}{\overset{\|}{C}} + H{-}O{\sim}O{-}H + \overset{O}{\overset{\|}{C}}{=}N{\sim}N{=}\overset{O}{\overset{\|}{C}} + \ldots\ldots$$

$$\downarrow$$

$$\ldots\ldots {-}O{-}\overset{O}{\overset{\|}{C}}{-}NH{\sim}NH{-}\overset{O}{\overset{\|}{C}}{-}O{\sim}O{-}\overset{O}{\overset{\|}{C}}{-}NH{\sim\sim}NH{-}\overset{O}{\overset{\|}{C}}{-}O{-}\ldots\ldots$$

Is the material obtained a thermoplastic polymer?

● Yes          →  68

**22**  ● No          →  70

● I don't know          →  72

# Program 23

## Natural Products and Heterocyclic Compounds

---

| 1 |      Most of the organic compounds described so far in this book are found in nature and are produced synthetically as well. In the early history of chemistry the chemical compounds were discovered first in nature, and then synthesized on laboratory scale. Today there are many more organic compounds brought into existence by increasingly sophisticated synthetic methods than substances found in nature.

Anyway, we want to stress that there is no difference in an orgainc substance found in nature or made on laboratory or industrial scale. Both vitamin C in lemons and vitamin C from an industrial process have the same structural formula and therefore the same properties: both compounds are identical.

Most of the organic compounds occurring in nature contain the same functional groups already presented, although their overall structure is usually quite complicated. We can therefore select only a few examples from this fascinating area of organic chemistry.

There are three classes of naturally occurring organic compounds which are of the utmost importance in our every day living: *proteins*, *fats*, and *carbohydrates*.

*Amino acids* are the basic units from which proteins are formed. Aminoacetic acid is the simplest amino acid; it is derived from acetic acid by replacing an aliphatic hydrogen atom by an $NH_2$ group.

Give the formula for aminoacetic acid.   →   | 14 |

---

| 2 |      Correct; glucose has the empirical formula $(CH_2O)_6$ or $C_6H_{12}O_6$.

The structure of glucose is based on a six-carbon chain in which the first member is an aldehyde group:

The other five carbons carry one hydroxyl group each. Complete the formula for glucose, including the hydrogen atoms and the hydroxyl groups, and consider then how many primary and secondary OH-groups there are in this molecule.

**23**

- Five secondary                     →     | 22 |

- Five secondary and one primary     →     | 26 |

- Four secondary and one primary     →     | 18 |

- Five primary                       →     | 31 |

---

---

**3**

$$\overset{3}{C}H_3 - \overset{2}{C}H - \overset{1}{C}OOH$$
$$| $$
$$NH_2$$

2–aminopropionic
acid

$$\overset{3}{C}H_2 - \overset{2}{C}H_2 - \overset{1}{C}OOH$$
$$| $$
$$NH_2$$

3–aminopropionic
acid

The naturally occurring amino acids are all 2-aminocarboxylic acids. These are also called α-aminocarboxylic acids or, for short, α-amino acids.

Proteins are formed by the polycondensation of α-amino acids. Let us now consider this reaction in detail:

Two amino acid molecules can form an amide by eliminating a molecule of water between the carboxyl group of one molecule and the amino group of the other, e.g.:

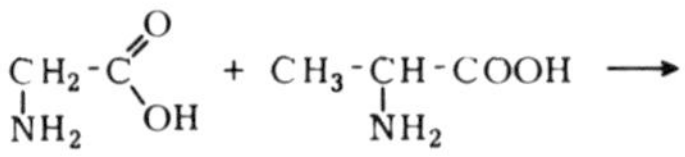

$$
\begin{array}{cc}
CH_2 - C\overset{\displaystyle O}{\underset{\displaystyle OH}{\diagup}} & + \; CH_3 - CH - COOH \longrightarrow \\
| & \qquad\qquad | \\
NH_2 & \qquad\quad NH_2
\end{array}
$$

aminoacetic        aminopropionic
acid               acid

Complete the equation.     →     15

---

**4**

Glycerol:   $CH_2OH$
$\quad\quad\quad\;\; |$
$\quad\quad\quad CHOH$
$\quad\quad\quad\;\; |$
$\quad\quad\quad CH_2OH$

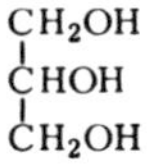

Fats are the products of the esterification of long-chain carboxylic acids with glycerol. Abbreviating the carboxylic acid as R–COOH, you can complete the reaction equation:

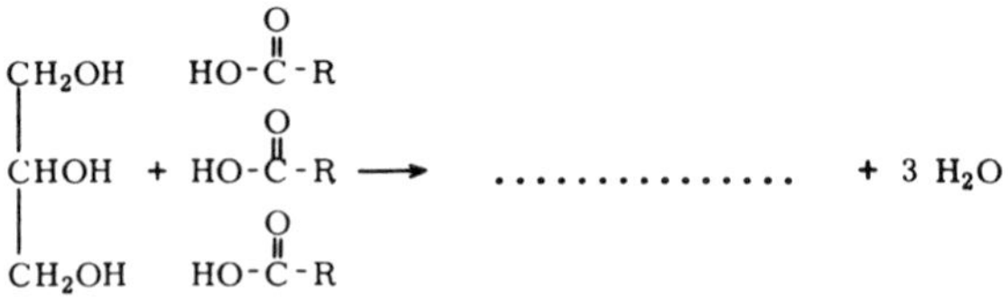

$$
\begin{array}{ll}
CH_2OH & HO-\overset{\displaystyle O}{\overset{\|}{C}}-R \\
| & \qquad O \\
CHOH & + \; HO-\overset{\|}{C}-R \longrightarrow \quad \ldots\ldots\ldots\ldots\ldots \quad + \; 3\;H_2O \\
| & \qquad O \\
CH_2OH & HO-\overset{\|}{C}-R
\end{array}
$$

→   19

**23**

---

**5**     Your answer is wrong.

See the explanation at   9   .

---

| **6** |

It is a *cross-linked* polymer.

Indeed, the properties of cellulose are such as can be expected for a cross-linked polymer. For example, it is infusible. The cross-linking is due to the presence of numerous hydroxyl groups which hold different polymer strands together through hydrogen bonds:

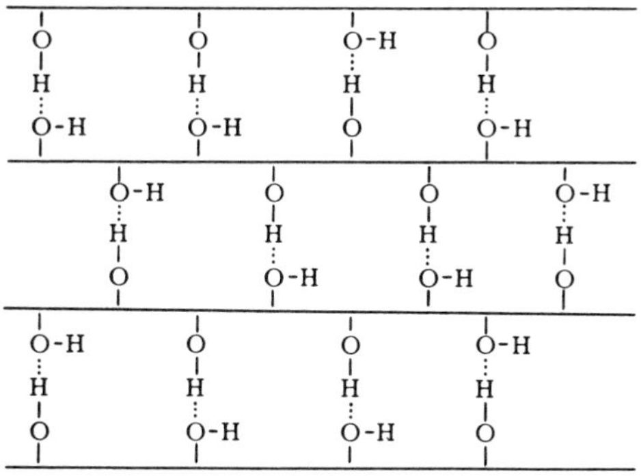

Which chemical reaction is required in order to transform cellulose into a *thermoplastic* polymer?

Mention one or two such reactions and check your answers at [16] , or, if you need help, go instead to [24] .

| **7** |

Your answer is wrong.

See the explanation at [9] .

| **8** |

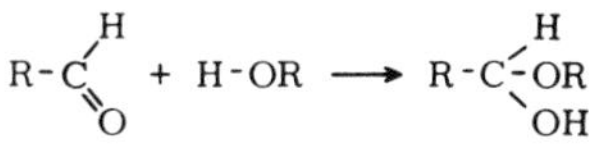

$$R-C\overset{H}{\underset{O}{\diagup}} + H\text{-}OR \longrightarrow R-C\overset{H}{\underset{OH}{\diagdown}}OR$$

The glucose molecule possesses *one* aldehyde group and several hydroxyl groups. Therefore, it can form an intramolecular hemiacetal:

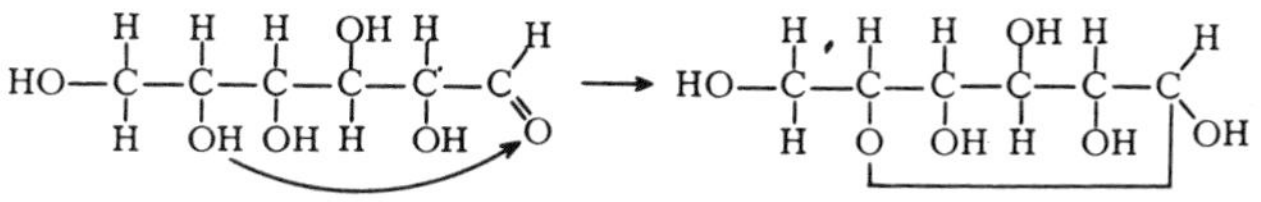

**23**

The cyclic molecule is often designated as follows:

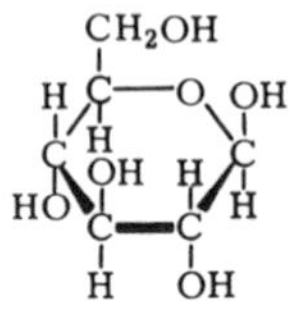

This is the normal form of glucose. Glucose is one of the most common carbohydrates. A more complex carbohydrate is obtained by condensing two molecules of glucose. Complete the equation using empirical formulas only:

$$2 \; C_6H_{12}O_6 \rightarrow \quad \ldots\ldots\ldots\ldots \quad + \; H_2O \qquad\qquad \rightarrow \quad \boxed{20}$$

---

**9**    Glucose is composed of the elements carbon, hydrogen, and oxygen in the ratio

$$C : H : O = 1 : 2 : 1$$

That is, it is a multiple of

$CH_2O.$

In order to determine the correct formula, it is necessary to know the molecular mass. This turns out to be 180. Thus, glucose is composed of six $CH_2O$ units:

$(CH_2O)_6.$

From this information derive the empirical formula for glucose.   $\rightarrow$  $\boxed{2}$

---

**10**    The *amino acids* contain *carboxyl* groups.

When amino acids are transformed into proteins, the carboxyl group is changed into another functional group. Which one?

Repeat  $\boxed{27}$ .

**23**

---

**11**    Glycerol is a trivalent alcohol derived from propane,

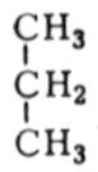

$$\begin{array}{l} CH_3 \\ | \\ CH_2 \\ | \\ CH_3 \end{array}$$

Derive the formula for glycerol by replacing a hydrogen atom at each carbon in propane by an OH group.   $\rightarrow$  $\boxed{4}$

|12|          Correct; the amino acids are joined through the *amide groups* in the proteins.

There are over 30 naturally occurring amino acids. Each protein is composed of several *different* amino acids. The *order* in which the different amino acids are joined is characteristic of the individual protein. Obviously, there is an enormous number of ways of joining the 30 or so amino acids in different sequences. Because of this it is possible for every living being to have *its own* specific proteins. The proteins taken up in the food are cleaved, in the body, to the fundamental amino acids from which the new, specific, proteins are synthesized. This is in fact what makes organ transplantation so difficult: the body may reject the foreign organ because it contains foreign proteins.

Amino acids are the key substances for the genetic code, the means by which the manufacture of proteins in the living cell is controlled.

The proteins are chemically related to a class of *synthetic* polymers described in the preceding program.

If you know the name of these synthetic polymers, check it at |23| ; if not, go to |29| .

---

|13|          *Amino acids* obviously contain *amino groups*. Since these partake in the protein forming reaction, however, a new functional group is formed. What is it called?

Repeat |27| .

---

|14|          Aminoacetic acid:

$$CH_2\text{-}COOH$$
$$|$$
$$NH_2$$

There are *two* amino acids derived from propionic acid,

$$CH_3\text{--}CH_2\text{--}COOH$$

Give the formulas for the two aminopropionic acids.

Check your result at |3| , or go to |25| if help is required.

**23**

---

|15|
$$\begin{array}{cc} O & CH_3 \\ \| & | \end{array}$$
$$CH_2\text{-}C\text{-}NH\text{-}CH\text{-}COOH$$
$$|$$
$$NH_2$$

There is still a carboxyl group in this compound which can react with the amino group of a third amino acid molecule. The reaction can, therefore, continue till a *macromolecule* is formed.

What kind of reaction is this?

● A polymerization          →          30

● A polycondensation        →          27

● A polyaddition            →          34

● I don't know              →          38

---

**16**     The formation of either esters or ethers reduces the number of OH groups in cellulose and therefore removes its ability to cross-link through hydrogen bonding. Consequently, cellulose ester and cellulose ethers are valuable thermoplastic or soluble polymers.

Cellulose is hydrolyzed into its component molecules by acid treatment. The components have the formula $C_6H_{12}O_6$. What is the name of this compound?
→     35

---

**17**     You confused the *amide* and the *ester* groups. The two functional groups are:

$$-C\overset{\nearrow O}{\underset{\searrow OR}{}}$$

carboxylic acid
*ester*

$$-C\overset{\nearrow O}{\underset{\searrow NH-R}{}}$$

carboxylic acid
*amide*

Repeat     27   .

**23**

**18**     Correct; glucose has *four secondary* and *one primary* hydroxyl group:

$$HO-\overset{\overset{H}{|}}{\underset{\underset{H}{|}}{C}}-\overset{\overset{H}{|}}{\underset{\underset{OH}{|}}{C}}-\overset{\overset{H}{|}}{\underset{\underset{OH}{|}}{C}}-\overset{\overset{H}{|}}{\underset{\underset{OH}{|}}{C}}-\overset{\overset{H}{|}}{\underset{\underset{OH}{|}}{C}}-C\overset{\nearrow H}{\underset{\searrow O}{}}$$

This is still an incomplete structure, however. Like all carbohydrates, glucose possesses *asymmetric* carbon atoms.

How many asymmetric carbon atoms are there in glucose?

● Four                                              →   $\boxed{28}$

● Five                                              →   $\boxed{37}$

● Six                                               →   $\boxed{41}$

● I first need to know the definition of an
  asymmetric carbon atom                            →   $\boxed{33}$

---

$\boxed{19}$

$$\begin{array}{l} CH_2-O-\overset{\overset{\displaystyle O}{\|}}{C}-R \\ | \quad\quad \overset{\displaystyle O}{\|} \\ CH-O-\overset{}{C}-R \\ | \quad\quad \overset{\displaystyle O}{\|} \\ CH_2-O-\overset{}{C}-R \end{array}$$

The fatty acid residue, R, is generally a 16 or 18 carbon atom chain.

The glycerol esters formed from *saturated* fatty acids are usually solid or waxy; those from *unsaturated* fatty acids are liquids or oils.

One can easily transform an unsaturated fatty acid into a saturated one. Which chemical reaction is required?

● Hydrogenation        →   $\boxed{32}$

● Oxidation            →   $\boxed{36}$

● Condensation         →   $\boxed{40}$

● I don't know         →   $\boxed{42}$

---

$\boxed{20}$     $2\ C_6H_{12}O_6 \longrightarrow C_{12}H_{22}O_{11} + H_2O$

          glucose

**23**

Ordinary sugar has a similar composition.

When many glucose molecules combine in this manner, a *macromolecule* is obtained. These macromolecules form a larger network in which individual polymer strands are held together by hydrogen bonds between OH groups. This material is *cellulose,* a widespread natural product.

A special name for polymers which form networks was given in the preceding program. What is the name?    →    6

---

21

$$CH_2-O-\overset{\overset{O}{\|}}{C}-R$$
$$CH-O-\overset{\overset{O}{\|}}{C}-R \; + \; 3\;NaOH \longrightarrow \; \overset{CH_2OH}{\underset{CH_2OH}{CHOH}} \; + \; 3\;R-COONa$$
$$CH_2-O-\overset{\overset{O}{\|}}{C}-R$$

The sodium salts of the fatty acids are *soaps*.

Soap manufacture – like alcohol fermentation – is one of the oldest chemical reactions, known to man for thousands of years.

Having described the proteins and fats, now consider the *carbohydrates*. *Glucose,* also called grape sugar, is the most important of the carbohydrates. Let us develop its formula step by step.

A chemical analysis shows that glucose contains the elements carbon, hydrogen, and oxygen. A quantitative analysis gives the ratio of these elements:

$$C : H : O = 1 : 2 : 1$$

This gives the empirical formula as:    $(CH_2O)_x$

A measurement of the molecular mass of the compound gives the figure 180. From this calculate the value of x.

23

● 6                          →    2

● 30                         →    7

● 180                        →    5

● I cannot answer the question    →    9

**22**     The five carbon atoms in glucose to which the OH groups are attached are not identical!

A *primary* hydroxyl group must be bonded to a *primary* carbon atom, i.e. a carbon atom having only *one* other carbon neighbor. A *secondary* hydroxyl group is bonded to a *secondary* carbon atom having *two* carbon neighbors.

Repeat   2  .

**23**     Both *proteins* and *polyamides* are held together by *amide* groups.

The organisms achieve the synthesis of proteins under very mild conditions, i.e. body temperature. The chemist, in contrast, needs temperatures above 200 °C in order to prepare synthetic polymers.

We now turn to the next group of foodstuffs, the *fats*.

*The fats are glycerol esters of the fatty acids.*

Give the formula for glycerol and check your result at   4  . If you need help, go instead to   11  .

**24**     The number of hydrogen bonds in cellulose can be reduced by *reducing the number of hydroxyl groups*. Which chemical reactions can be used in order to replace the hydrogen atoms of OH groups by other residues?   →   16

**25**     *Two amino acids* can be derived from propionic acid, $CH_3-CH_2-COOH$.

By replacing a hydrogen atom the amino group can be attached to either the $CH_3$ group or the $CH_2$ group.

Give the formulas for the two amino acids formed.   →   3

**23**

**26**     There are only *five* hydroxyl groups in glucose.
Repeat   2  .

---

**27**     Correct; a protein is formed from amino acid molecules in a *polycondensation* reaction. A section of a protein is shown below:

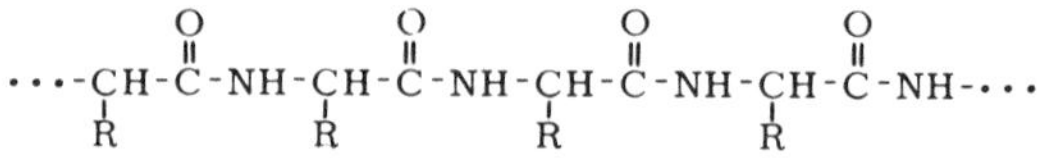

$$\cdots-\text{CH}-\overset{\overset{\textstyle O}{\|}}{\text{C}}-\text{NH}-\text{CH}-\overset{\overset{\textstyle O}{\|}}{\text{C}}-\text{NH}-\text{CH}-\overset{\overset{\textstyle O}{\|}}{\text{C}}-\text{NH}-\text{CH}-\overset{\overset{\textstyle O}{\|}}{\text{C}}-\text{NH}-\cdots$$
$$\quad\;\; \underset{R}{|}\qquad\qquad \underset{R}{|}\qquad\qquad \underset{R}{|}\qquad\qquad \underset{R}{|}$$

The formula has been simplified using the symbol R for all organic residues.

Give the name of the groups in the protein which are formed by uniting the amino acids.

- Carboxyl groups     →     10

- Amine groups     →     13

- Amide groups     →     12

- Ester groups     →     17

---

**28**     Correct; glucose has four asymmetric carbon atoms. These are indicated by stars in the formula:

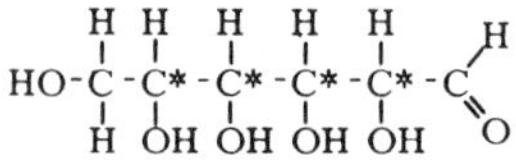

$$\text{HO}-\overset{\overset{\textstyle H}{|}}{\underset{\underset{\textstyle H}{|}}{\text{C}}}-\overset{\overset{\textstyle H}{|}}{\underset{\underset{\textstyle OH}{|}}{\text{C*}}}-\overset{\overset{\textstyle H}{|}}{\underset{\underset{\textstyle OH}{|}}{\text{C*}}}-\overset{\overset{\textstyle H}{|}}{\underset{\underset{\textstyle OH}{|}}{\text{C*}}}-\overset{\overset{\textstyle H}{|}}{\underset{\underset{\textstyle OH}{|}}{\text{C*}}}-\text{C}\!\!\overset{\displaystyle H}{\underset{\displaystyle O}{\diagup\!\!\!\diagdown}}$$

The groups attached to an asymmetric carbon atom possess a specific steric orientation and cannot be interchanged by simple rotation. In glucose *one* of the hydroxyl groups is oriented to the side opposite from the remaining three:

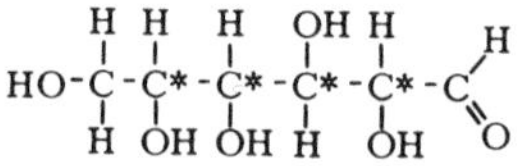

$$\text{HO}-\overset{\overset{\textstyle H}{|}}{\underset{\underset{\textstyle H}{|}}{\text{C}}}-\overset{\overset{\textstyle H}{|}}{\underset{\underset{\textstyle OH}{|}}{\text{C*}}}-\overset{\overset{\textstyle H}{|}}{\underset{\underset{\textstyle OH}{|}}{\text{C*}}}-\overset{\overset{\textstyle OH}{|}}{\underset{\underset{\textstyle H}{|}}{\text{C*}}}-\overset{\overset{\textstyle H}{|}}{\underset{\underset{\textstyle OH}{|}}{\text{C*}}}-\text{C}\!\!\overset{\displaystyle H}{\underset{\displaystyle O}{\diagup\!\!\!\diagdown}}$$

(The orientation of the fifth OH group is unimportant since it is not bonded to an asymmetric carbon atom.)

The fact that glucose is an aldehyde has important consequences. Remember the reaction between simple aldehydes and compounds of the type H–X:

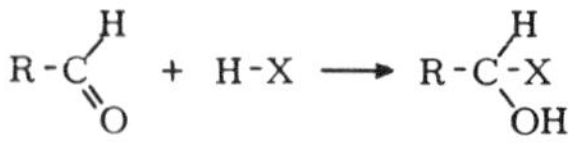

$$\text{R}-\text{C}\!\!\overset{\displaystyle H}{\underset{\displaystyle O}{\diagup\!\!\!\diagdown}} \; + \; \text{H-X} \;\longrightarrow\; \text{R}-\text{C}\!\!\overset{\displaystyle H}{\underset{\displaystyle OH}{\diagdown\!\!\!\diagup}}\text{-X}$$

**23**

The reaction with an alcohol is an example; the product is called a *hemiacetal:*

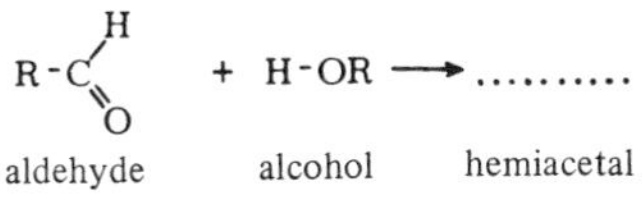

$$R-C\overset{H}{\underset{O}{\diagup}} \quad + \quad H-OR \longrightarrow \dots\dots\dots$$

aldehyde          alcohol        hemiacetal

Complete this equation.    →    $\boxed{8}$

---

$\boxed{29}$     *Polyamides* were described in the program on synthetic polymers. Obviously, polyamides contain amide groups. This is also true for proteins.

Continue at    $\boxed{23}$  .

---

$\boxed{30}$     Your answer is wrong. Read the recapitulation at    $\boxed{38}$  .

---

$\boxed{31}$   ˙ The five carbon atoms in glucose to which the OH groups are attached are not identical!

A *primary* hydroxyl group must be bonded to a *primary* carbon atom, i.e. a carbon atom having only *one* other carbon neighbor. A *secondary* hydroxyl group is bonded to a *secondary* carbon atom having *two* carbon neighbors.

Repeat    $\boxed{2}$  .

---

$\boxed{32}$     Correct; the unsaturated fatty acids can be converted to saturated ones by *hydrogenation*. This is also true for the fats themselves. In other words, the liquid unsaturated fats are hydrogenated to the more useful solid fats. Margarine is obtained in such a process.

Being esters, fats can also be *saponified*, i.e. hydrolyzed by bases. The hydrolysis products are glycerol and a salt of the fatty acid:

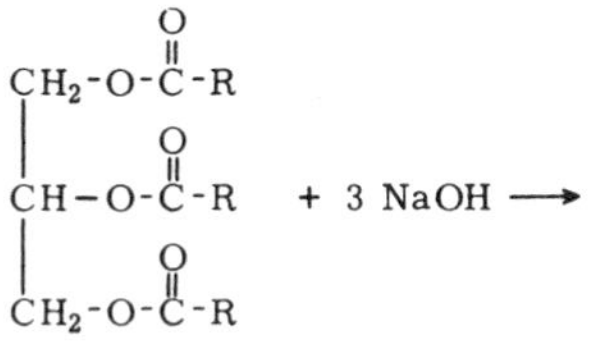

**23**

$$\begin{array}{l} CH_2-O-\overset{\displaystyle O}{\overset{\|}{C}}-R \\[4pt] CH-O-\overset{\displaystyle O}{\overset{\|}{C}}-R \quad + \ 3\,NaOH \longrightarrow \\[4pt] CH_2-O-\overset{\displaystyle O}{\overset{\|}{C}}-R \end{array}$$

Complete the equation.    →    $\boxed{21}$

**33**    A carbon atom is asymmetric when it carries *four different* substituents. Take the central carbon in lactic acid as an example:

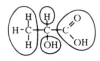

Examine each carbon atom in glucose and determine *how many* are asymmetric. Continue at   |18|  .

**34**    Your answer is wrong. Read the recapitulation at   |38|  .

**35**    *Glucose* is obtained by the cleavage of cellulose.

To terminate the presentation of the basic foodstuffs required by man, repeat the names of the three classes of compounds.   →   |48|

**36**    You are mistaken. See the explanation at   |42|  .

**37**    Your answer is wrong. Study the definition of an asymmetric carbon atom at   |33|  .

**38**    There are three basic processes used in the formation of macro-molecules:

(a) *polymerization* of unsaturated precursors. No atoms change place in this reaction.

(b) *polycondensation* in which the precursor molecules are joined through the elimination of water or other small molecules;

(c) *polyaddition* in which one of the precursor molecules is added to a double bond in the other. The polymer does not have the same arrangement of atoms as in the precursor molecules.

When you have learnt these facts, continue at   |27|  .

|39|

$$HC = CH$$
$$HC_{\diagdown O \diagup} CH$$     or shorter:

furan

As seen from the structure, furan is an *ether*. The corresponding thioether, called *thiophene,* is obtained from furan vapor and $H_2S$ at 400 °C in the presence of a catalyst. In other words, one heterocyclic compound is transformed into another:

 + $H_2S$ $\xrightarrow[\text{cat.}]{400\,°C}$ ..... + $H_2O$

furan                    thiophene

Give the formula for thiophene.     →     |53|

|40|     You are mistaken. See the explanation at |42| .

|41|     Your answer is wrong. Study the definition of an asymmetric carbon atom at |33| .

|42|     The transformation of an *unsaturated* fat or fatty acid into a saturated one is a reaction of the type:

$$>C=C< + H_2 \longrightarrow >\overset{\overset{H}{|}}{C}-\overset{\overset{H}{|}}{C}<$$

What is the name of this reaction?     →     |32|

|43|     Correct; in the oxidation of nicotine the non-aromatic, hydrogenated pyrrole ring is preferentially attacked. Thus, the pyrrole ring behaves as an alkyl side-chain.

**23**

$$\begin{array}{c} H_2C - CH_2 \\ \overset{H}{\underset{}{C}} \diagdown N \diagup CH_2 \\ | \\ CH_3 \end{array}$$

The methyl group in toluene can be oxidized to a carboxyl group:

$CH_3$ ⬡ $\xrightarrow{\text{oxidation}}$ $COOH$ ⬡

Similarly, the hydrogenated pyrrole ring in nicotine is transformed into a carboxyl group by mild oxidizing agents. Give the formula for the carboxylic acid formed.  →  70

---

**44**

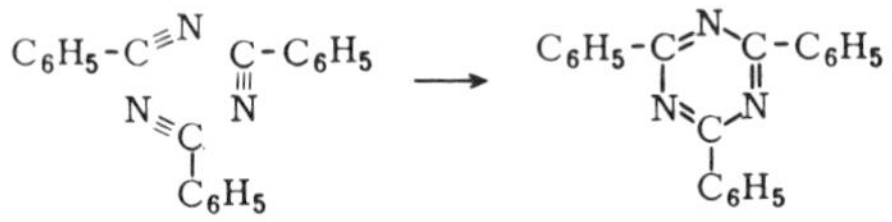

$$C_6H_5-C{\equiv}N \quad C-C_6H_5 \quad \longrightarrow \quad C_6H_5-C{\overset{N}{\cdots}}C-C_6H_5$$

Heterocyclic compounds containing two or more rings (identical or different) are also known. Quinoline is a simple example:

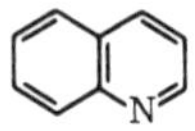

From which two ring systems is quinoline derived?

- Benzene and furan        →   55

- Benzene and thiophene    →   57

- Benzene and pyrrole      →   59

- Benzene and pyridine     →   66

---

**45**     Your answer is wrong. Admittedly, the question was not easy. See the explanation at   49   .

---

**46**

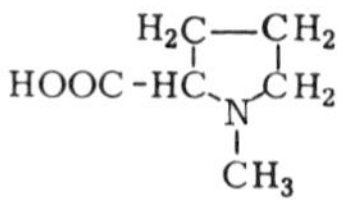

$$\begin{array}{c} H_2C{-}CH_2 \\ HOOC{-}HC{\underset{N}{\diagdown}}CH_2 \\ CH_3 \end{array}$$

**23**  The structure of this degradation product of nicotine was already known, thereby allowing its ready identification.

The elucidation of the structure of nicotine is summarized below:

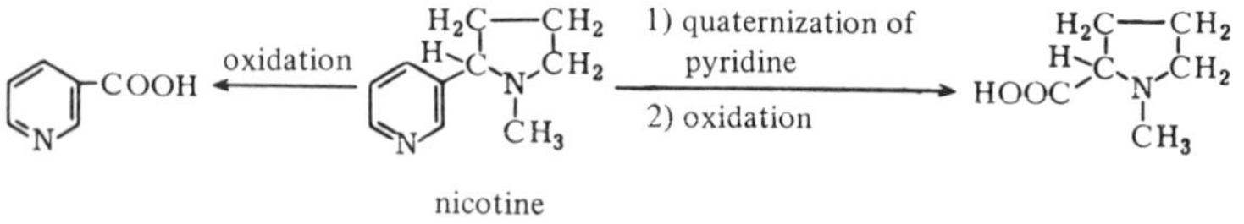

nicotine

The identification of these two degradation products allows the *deduction of the full structure of nicotine*. The final proof of the structure was obtained by *synthesis* of nicotine from known starting materials.

Structure determination through degradation and synthesis was very important in classical organic chemistry. Today, these procedures have been largely replaced by modern spectroscopic methods. This chapter has permitted us a very limited look at the chemistry of natural products. Hopefully, the elucidation of the structure of nicotine has helped illustrate the fascinating complexity of organic chemistry.

This is the end of Program 23.

---

**47**     Your answer is wrong. Admittedly, the question was not easy.

See the explanation at    49  .

---

**48**     *Proteins, fats,* and *carbohydrates* are the three basic components of food.

We now need to describe a class of organic compounds not so far considered. The two major classes descirbed previously were

1) aliphatic compounds,
2) .................... compounds.

The third class is
3) heterocyclic compounds.

Note the names of these three classes.    →    61

---

**49**     In nicotine,

the *pyridine* ring is *aromatic*. Consequently, it is quite stable.

The second ring in nicotine, the hydrogenated pyrrole ring, is *not* aromatic and has, therefore, no aromatic stability. It behaves much like a simple alkyl side-

**23**

chain and is susceptible to oxidation in the same way that toluene can be oxidized to benzoic acid:

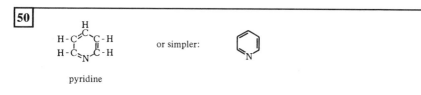

What is the product formed when the hydrogenated pyrrole side-chain in nicotine is *oxidized to a COOH group?*    →    50

---

50

$$H-C\overset{\overset{\displaystyle H}{\displaystyle C}}{\underset{\underset{\displaystyle N}{\displaystyle C}}{}}\begin{matrix}C-H\\ \\C-H\end{matrix}$$       or simpler:

pyridine

The molecule is analogous to benzene:

benzene

which it resembles in many of its reactions. Like benzene, pyridine is an *aromatic* compound.

The distinguishing feature in pyridine is, however, the *nitrogen atom.* What type of nitrogen atom is it?

- Like in a primary amine                →    60

- Like in a secondary amine              →    64

- Like in a tertiary amine               →    62

- Like in a quaternary ammonium salt     →    69

- I don't know                           →    74

---

| 51 |
|----|

Pyrrole:    HC———CH    or simpler:

Pyrrole is only very weakly basic. Thus, not all secondary amines are bases; the nature of a compound can depend to a large extent on the interactions between neighboring groups or atoms within the molecule. This is the case in pyrrole: the nitrogen lone pair interacts with the four electrons in the two C=C double bonds, thereby forming a six-electron system similar to that in benzene:

For this reason, the amine lone pair is not available to the same extent as in normal secondary amines, and pyrrole does not form a salt.

*Furan* is the five-membered heterocycle derived from pyrrole by replacement of NH by O.

Give the formula.    →    |39|

---

| 52 |
|----|

     Pyrrole is a *five-membered* heterocyclic compound containing one *nitrogen* atom:

Haemoglobin (the red colorant in blood) and
chlorophyll (the green pigment in plants) are complex derivatives of pyrrole.

Heterocyclic compounds are present in large numbers in the *alkaloids*. The alkaloids are isolated from plants and many are used as drugs.

Cocaine, quinine, caffeine, and nicotine are examples. These and many other such compounds of physiological and chemical interest were isolated and investigated many years ago. The structures were established by classical organic chemistry methods: degradation to known compounds, and total synthesis.

**23**

Let us consider nicotine:

$$\begin{array}{c} H_2C———CH_2 \\ H\text{-}C \diagdown \diagup CH_2 \\ N \\ | \\ CH_3 \end{array}$$

Two ring systems are present in nicotine. What are they called?

- Pyridine and pyrrole          →   71

- Hydrogenated pyrrole and pyridine     →   68

- Hydrogenated pyridine and
  pyrrole                          →   73

- I need more information          →   58

---

**53**

                 HC——CH
Thiophene:    HC⟍S⟋CH      or simpler:

Thiophene is an aromatic compound. This is seen, for example, in the reaction with chlorine. What type of reaction will occur?

- An addition reaction      →   63

- A substitution reaction     →   65

- I cannot tell            →   75

---

**54**

$$H_2C{\overset{\overset{\displaystyle H_2}{C}}{\diagup}}{}^{CH_2}\ ...$$

$$H_2C\diagdown_{N}\diagup CH_2 \quad H_2C—CH_2 \quad H_2C—CH_2 \quad H_2C—CH_2$$

Are these heterocyclic compounds?

- Yes     →   67

- No      →   72

---

**23**

**55**     Your answer is wrong. Go to   65   where the heterocyclic compounds are summarized.

---

**56**

$$H_2C—CH_2 \atop H_2C\ \ \ CH_2 \atop HO\ \ \ OH \quad \xrightarrow{-H_2O} \quad H_2C—CH_2 \atop H_2C\diagdown_O\diagup CH_2 \quad \xrightarrow{-2\ H_2} \quad HC——CH \atop HC\diagdown_O\diagup CH$$

This reaction constitutes one of the pieces of evidence for the structure of furan.

Most heterocyclic compounds are five- or six-membered rings. They may contain one, two, three, or more heteroatoms.

As an example consider the six-membered heterocycle with three nitrogen atoms formed by the trimerization of benzonitrile:

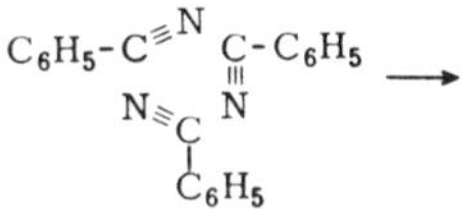

$$C_6H_5-C{\equiv}N \quad C-C_6H_5 \longrightarrow$$
$$N{\leqslant}_C N$$
$$\overset{|}{C_6H_5}$$

The compound formed is called triphenyl*triazine*.

Complete the equation.   →   44

---

**57**     Your answer is wrong. Go to   65   where the heterocyclic compounds are summarized.

---

**58**     The compounds mentioned were:

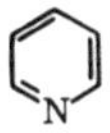

pyridine

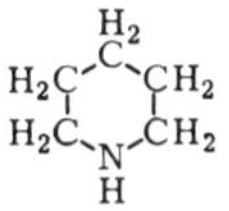

hydrogenated pyridine

pyrrole

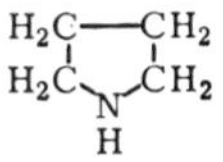

hydrogenated pyrrole

Nicotine has the formula

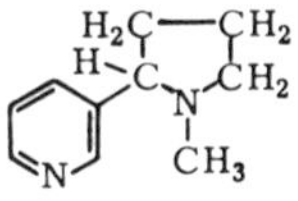

Which rings are present in nicotine?

● Pyridine and pyrrole     →   71

● Hydrogenated pyrrole and pyridine     →   68

● Hydrogenated pyridine and pyrrole     →   73

**23**

---

**59**     Your answer is wrong. Go to   65   where the heterocyclic compounds are summarized.

| 60 | Your answer is wrong.

See the explanation at  74 .

---

| 61 | 1) Aliphatic compounds.
  2) *Aromatic* compounds.
  3) Heterocyclic compounds.

The heterocyclic compounds (or *heterocycles*) are cyclic molecules incorporating an element other than carbon in the ring. The most common *heteroatoms* are nitrogen, oxygen, and sulfur. There can be one or more heteroatoms in a ring.

*Pyridine* is a *six*-membered ring composed of five carbon and one nitrogen atoms. The compound has three double bonds as in benzene and can be derived from benzene by replacing a CH group by N.

Give the formula for pyridine.  →  50

---

| 62 | Yes, pyridine resembles a *tertiary amine*. It is a water-soluble liquid with an unpleasant smell. It forms salts with strong acids.

*Pyrrole* is another nitrogen heterocycle. The molecule is a five-membered ring with one nitrogen atom. It has two C=C double bonds, and all five ring members carry a hydrogen atom.

Derive the formula for pyrrole  →  51

---

| 63 | Your answer is wrong.

See the explanation at  75 .

---

| 64 | Your answer is wrong.

See the explanation at  74 .

---

| 65 | Yes, the reaction is

$$\text{[ring]}_S + Cl_2 \longrightarrow \text{[ring]}_S\text{-Cl} + HCl$$

Thiophene undergoes *substitution,* thus reacting like an aromatic compound.

The heterocycles considered above can all be hydrogenated:

Pyridine:

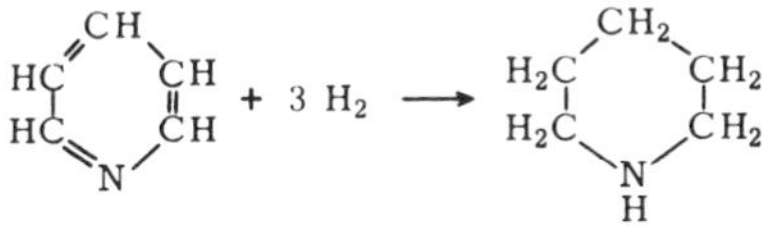

$+ \ 3 \ H_2 \longrightarrow$

Pyrrole:

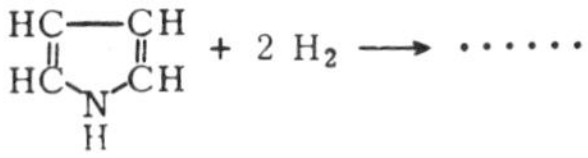

$+ \ 2 \ H_2 \longrightarrow \ \cdots\cdots$

Furan:

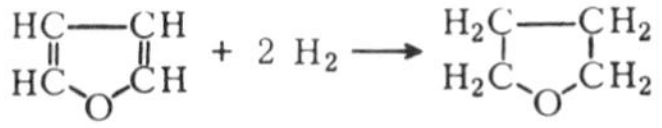

$+ \ 2 \ H_2 \longrightarrow$

Thiophene:

$+ \ 2 \ H_2 \longrightarrow \ \cdots\cdots$

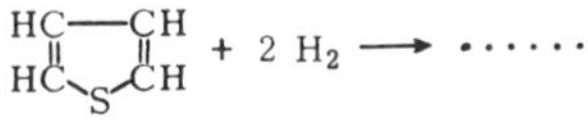

Draw the structures of the four reaction products.     →     54

---

**66**     Yes, quinoline is composed of a benzene- and a pyridine ring:

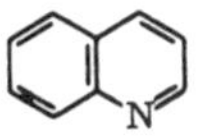

Numerous natural products are heterocyclic compounds. For example, hemoglobin and chlorophyll (the red pigment in blood, and the colorant in green leaves, respectively) are rather complicated *pyrrole* derivatives.

What size of ring is *pyrrole,* and which heteroatom does it contain?     →     52

---

**67**     Correct.

The hydrogenation products of pyridine, pyrrole, furan, thiophene, etc., are all heterocyclic compounds. The hydrogenated heterocyclic compounds behave like *aliphatic* compounds. Conversely, the fact that a compound is heterocyclic does not make it aromatic.

Heterocycles are often synthesized from aliphatic starting materials. The elimination of water from 1,4-butanediol is an example. The initial product is tetrahydrofuran (= hydrogenated furan), and this can be dehydrogenated to furan itself:

**23**

$\xrightarrow{\ -H_2O\ } \ \cdots\cdots\cdots \ \xrightarrow{\ -2\ H_2\ }$

Complete the equation.     →     56

---

| 68 |
|---|

Yes, nicotine is composed of a pyridine ring and a hydrogenated pyrrole ring.

The pyridine ring is aromatic; the hydrogenated pyrrole is not.

From this, does it follow that one of the two rings is particularly susceptible to oxidation?

● Yes, the pyridine ring                    →    | 45 |

● Yes, the pyrrole ring                     →    | 43 |

● No, the two rings are equally reactive    →    | 47 |

● I don't know                              →    | 49 |

| 69 |
|---|

Your answer is wrong.

See the explanation at    | 74 | .

| 70 |
|---|

Since the pyridinecarboxylic acid formed is a known substance, this reaction gives us the structure of one half of the nicotine molecule. We now need to determine the structure of the second half.

**23**

To do so, a degradation method must be found which *leaves the hydrogenated pyrrole ring intact.* To this end nicotine is first methylated to give a quaternary salt of the pyridine nucleus:

This renders the *pyridine* ring susceptible to oxidative degradation to a COOH group.

What, therefore, is the carboxylic acid formed in this oxidation?     →     46

---

**71**     Your answer is wrong.

See the explanation at     58 .

---

**72**     The hydrogenated heterocycles are, of course, heterocyclic compounds in their own right!

The definition of a heterocycle says only that the substance must contain a heteroatom in a ring. It must not necessarily be aromatic. The most common heteroatoms are N, O, and S.

Continue at     67 .

---

**73**     Your answer is wrong.

See the explanation at     58 .

---

**74**     Examine the following formulas:

primary amine:         $R-NH_2$

secondary amine:
$$\begin{array}{c} R \\ \diagdown \\ \phantom{R}NH \\ \diagup \\ R \end{array}$$

tertiary amine:
$$\begin{array}{c} R \\ \diagdown \\ \phantom{R}N-R \\ \diagup \\ R \end{array}$$

quaternary ammonium salt:
$$\left[ \begin{array}{c} R \\ | \\ R-N-R \\ | \\ R \end{array} \right]^{\oplus} X^{\ominus}$$

**23**

Pyridine has a trivalent nitrogen atom which carries no hydrogen atom. What type of amine is it?

- A primary amine    →    [60]

- A secondary amine    →    [64]

- A tertiary amine    →    [62]

---

[75]    The C=C double bonds in *aromatic* compounds do not behave in the same way as normal *olefins*.

You will remember that chlorine *adds* to an *olefinic* double bond:

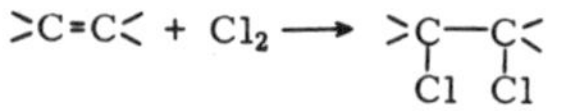

$$>C=C< + Cl_2 \longrightarrow >\underset{Cl}{C}-\underset{Cl}{C}<$$

In contrast, *aromatic* compounds react by *substitution:*

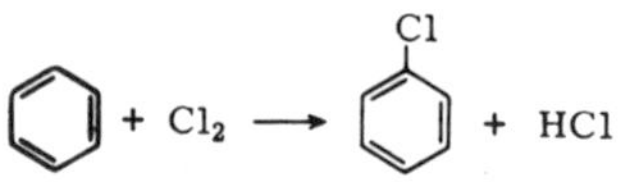

From this can you predict how the aromatic substance thiophene will react with chlorine?

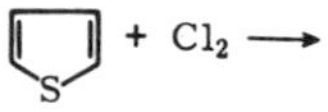

- By substitution    →    [65]

- By addition    →    [63]

---

**23**

# Index

The structure of this book is such that each program should be studied in a continuous manner. The index is intended only to allow the user a quick repetition of particular subjects. The **boldface** numbers refer to the programs, the normal print to the sections.